Towards Sustainable and Trustworthy 6G

Challenges, Enablers, and Architectural Design

Ömer Bulakçi, Xi Li, Marco Gramaglia,
Anastasius Gavras, Mikko Uusitalo,
Patrik Rugeland and Mauro Boldi
(Editors)

now
the essence of knowledge

Published, sold and distributed by:
now Publishers Inc.
PO Box 1024
Hanover, MA 02339
United States
Tel. +1-781-985-4510
www.nowpublishers.com
sales@nowpublishers.com

Outside North America:
now Publishers Inc.
PO Box 179
2600 AD Delft
The Netherlands
Tel. +31-6-51115274

ISBN: 978-1-63828-238-9
E-ISBN: 978-1-63828-239-6
DOI: 10.1561/9781638282396

Suggested citation: Ömer Bulakçı, Xi Li, Marco Gramaglia, Anastasius Gavras, Mikko Uusitalo, Patrik Rugeland and Mauro Boldi. (2023). *Towards Sustainable and Trustworthy 6G: Challenges, Enablers, and Architectural Design*. Boston–Delft: Now Publishers

Table of Contents

Dedications

To my parents, siblings, niblings, my wife Anna, and our kids Yunus and Ela for their continuous love and support, and to the victims of Türkiye-Syria earthquakes.

Ömer Bulakçı

To my lovely daughter Lina, my family, and friends for their love and support, and to my team from the 6G networks group, who provided great contributions to this book, and to all the people who are striving their effort and dedications in research to make our dreams into reality.

Xi Li

To my family; my friends and colleagues; and all the scholars, academics, and curious minds who seek to deepen their understanding of this field.

Marco Gramaglia

To my family, friends, colleagues, and all experts with whom I have been collaborating in the last years, for I was conscious that I knew practically nothing.

Anastasius Gavras

To my family, especially my wife Nina, as well as all the people aiming to make the world a better place for us and those after us.

Mikko Uusitalo

To my family, especially my wife Jing, and to our sons Alexander and Victor who will grow up and experience the benefits of 6G to its fullest.

Patrik Rugeland

To my family, and to the memory of all those killed and involved in wars.

Mauro Boldi

Acknowledgements

Since the start of the 5th generation public–private partnership (5G PPP) projects in mid-2015, there have been three major phases contributing to the design and further development of the 5G system. In the same period, 5G has moved from vision to actual deployments, and the further evolution of 5G, known as 5G Advanced, is already being specified. As 5G has become a commercial reality, attention in research and development has been shifting towards the 6th generation (6G).

This book is based on the outcome of Phase 3 projects within the 5G PPP framework primarily coming from the Architecture Working Group and the flagship Hexa-X project, and complemented by contributions from various additional experts. We would like to thank all the contributors for the substantial effort and engagement invested into this book. In particular, we would like to thank the main chapter editors for consolidating the diverse contents originated from different projects into a coherent structure and story, namely, Mårten Ericson and Dr. Bahare Masood Khorsandi for Chapter 2, Dr. Mir Ghoraishi for Chapter 3, Dr. John Cosmas for Chapter 4, Dr. Marco Gramaglia, Dr. Xi Li, and Dr. Gines Garcia-Aviles for Chapter 5, Dr. Agapi Mesodiakaki and Dr. Md Arifur Rahman for Chapter 6, Dr. Dimitris Tsolkas for Chapter 7, Dr. Alexandros Kostopoulos for Chapter 8, and Mauro Boldi, Dr. Mikko Uusitalo, and Dr. Patrik Rugeland for Chapter 9. Considering that many contributors have also used their free time to finalize the book in parallel to technology development and project work, we would also like to thank the families of the contributors for their continuous patience and support.

Naturally, we would like to thank the European Commission for funding the projects that have led to this book and, in particular, Dr. Peter Stuckmann for his personal support of the book. We would also like to thank the Smart Networks and

Services (SNS) joint undertaking (JU) and the 6G infrastructure association (IA) for their support, and in particular Dr. Colin Willcock for his personal support.

Beyond the researchers who have been directly involved in the projects, there are of course many more people involved in our home organisations. We would thus like to thank all our colleagues in the mobile communications industry, research institutes, and universities for inspiring discussions, the contribution of ideas, and the help on various tasks.

Dr. Bulakçı and Dr. Uusitalo would like to thank Peter Merz, Dr. Peter Vetter, Dr. Harish Viswanathan, and Horst Angerer from Nokia for their support in the preparation of this book. Dr. Bulakçı would also like to thank Dr. Simone Redana from Nokia for the support in the preparation of the book and for the great contributions to the Architecture WG, where he founded the working group in 2015 and acted as the chairman until 2021.

We would also like to thank David Kennedy from Eurescom for facilitating the publication of this book as an open access publication.

Last but not least, we would like to thank *now publishers* for their pleasant collaboration and continuous support throughout the writing and production process of this book.

Ömer Bulakçı, Xi Li, Marco Gramaglia, Anastasius Gavras,
Mikko Uusitalo, Patrik Rugeland, and Mauro Boldi
On behalf of the book contributors

Foreword by 6G Infrastructure Association

This book represents an important step towards future 6th generation (6G) networks. Created by the 5th generation public–private partnership (5G PPP) Architecture Working Group (WG) together with the European 6G flagship project Hexa-X, it contains the latest results from the European Research Community, compiling the outcomes from many projects into a coherent book that provides the reader with essential information about the main trends for the development of the new generation of 6G networks.

Modern telecommunication networks play a critical role in all aspects of everyday life. During the 2010s, the world witnessed a dramatic improvement in telecommunication services with the arrival of 4th generation (4G) networks. Since the early 2010s, scientists and organizations worldwide have laboured to design and deploy 5G networks. In addition to enhanced mobile broadband, the target was to provide an advanced set of new services that would lay the foundations for the digital transformation of various vertical markets (e.g., smart industry, energy, automotive, transportation and logistics, health, media, gaming, etc.). The 5G story is still near its beginning, and it will remain the key mobile network technology for many years to come. However, due to the complexity and long lead time, research on 6G has already started.

The book you hold (or digitally explore) contains the latest results from the European Research Community on 6G networks. This work includes the research outcomes for beyond 5G and 6G architectural design from a significant number of research projects in the context of European Union (EU) activities. The first group of projects has been working for the long-term vision of 6G networks and the realization of pervasive mobile virtual services. The second group has been building the foundation of beyond 5G/6G networks. Among this group of projects, the Hexa-X project has acted as the European flagship, leading the European 6G research activities.

The editorial team has successfully compiled the outcome from all these projects into a coherent book that provides the reader with essential information about 6G networks. The editors and authors have carefully selected all the technical areas where 6G networks will be different compared to previous generations and presented them in a comprehensive way. More specifically, the book analyses the key strategic goals and requirements to develop 6G networks and discusses what effect these will have on the overall architecture. Notably, the role of sustainability in 6G networks is elevated in a dedicated chapter. Key topics like the further evolution of the access network, the expected importance of accurate positioning solutions, the native support of Artificial Intelligence/Machine Learning (AI/ML) in the network, and the expansion of programmability in telecommunication functions are presented in detail. Moreover, as security and privacy are expected to play a fundamental role in 6G networks, the book authors explain how this can be achieved while also focusing on important aspects like the trustworthiness of these solutions.

Each chapter provides information on current solutions and future research trends. Overall, I expect that this book, written by several top professionals of the European information and communications technology (ICT) sector, will be a reference point for the future research activities on 6G networks and thus extremely useful for professionals and academics.

Dr. Colin Willcock, Chairman of the Board,
6G Infrastructure Association

Foreword by European Commission

Recent years have shown us the importance of resilient and high-speed communications infrastructure. Trust and acceptance in connectivity infrastructure have grown, as global societies have discovered their benefits. Indeed, it offers possibilities not only for remote working but also for citizens' daily lives. Also, businesses have understood the critical importance of high-speed networks and technologies in maintaining operations and processes.

These developments illustrate both the potential that 5th Generation (5G) networks have to provide in terms of the connectivity basis for the digital and green recovery in the short to mid-term, and the need to build technology capacities for the following generation – 6th Generation (6G) – in the long term.

5G technology and standards will evolve in several phases over the next few years as deployment advances. Operators worldwide have launched commercial 5G networks with a focus on cities. This early deployment builds on 4th Generation (4G) networks and primarily aims to enhance mobile broadband services for consumers and businesses. Huge investments need to be unlocked for a more comprehensive deployment covering all urban areas and major transport paths by 2025.

5G networks have already started employing "standalone" 5G core networks, enabling gigabit speeds and industrial applications, such as connected and automated mobility (CAM) and Industry 4.0. These will be a first step towards digitising and greening our entire economy. The growth potential in economic activity enabled by 5G and later 6G networks and services has been estimated to be in the order of €3 trillion by 2030 (McKinsey Global Institute, 2/2020). For such critical services, we need to ensure that 5G networks will be sufficiently secure.

Research & Innovation (R&I) initiatives focusing on 6G technologies have been kicked off in leading regions worldwide. The first products and infrastructure are expected at the end of this decade. 6G systems will offer a new step change in performance, moving us from gigabit towards terabit capacities and sub-millisecond response times. This will enable new critical applications, such as real-time automation or extended reality ("Internet of Senses"), by collecting and providing the sensor data for nothing less than a digital twin of the physical world.

Moreover, new smart network technologies and architectures will be needed to drastically enhance the energy efficiency of connectivity platforms despite major traffic growth, and keep electromagnetic fields under safe limits. They will form the technology base for a human-centric next-generation internet, and they will address sustainable development goals, such as accessibility and affordability of technology.

All parts of the world are starting to be heavily engaged in 6G developments. There will be opportunities and challenges concerning new business models and players through software networks with architectures, such as Open-RAN, for more open and interoperable interfaces in radio access networks (RAN). This is part of the convergence with new technologies in the areas of cloud and edge computing, Artificial Intelligence (AI), and components and devices beyond smartphones.

Success in 6G will first depend on the extent to which regions succeed in building a solid 5G infrastructure, on which 6G technology experiments and, later, 6G deployments can be built. In this context, building 5G ecosystems will be of key importance. Furthermore, we must bear in mind that industry R&I investments tend to relocate to where markets are more advanced.

Secondly, 6G will require taking a broader value chain approach, ranging from connectivity to components and devices beyond smartphones. This includes devices that make up the Internet of Things (IoT) and connected objects like cars or robots. They also exist on the service side, with edge computing integrated into connectivity platforms and cloud computing enabling advanced service provisioning, e.g., for big data and AI.

One important success factor in creating and seizing such opportunities is that Europe is a standard-setter in 6G and related technology fields. Both future users and suppliers need to shape key technology standards in the field of radio communications but also in next-generation network architecture. This will ensure the delivery of advanced service features while meeting energy-efficiency requirements, for example, through the effective use of software technologies and open interfaces.

Spectrum resources are another key factor that will determine success in 6G. Bands currently allocated for mobile communications will be reused for 6G; new frequency bands will be identified and harmonized. Industry and governments need to identify the opportunities related to spectrum that can be suitable for 6G and be made available with the potential to be harmonized at a global level.

6G technology has the potential to take a further step towards a multi-purpose service platform replacing legacy radio services for dedicated applications. This could help progress in defragmenting the radio spectrum and drastically enhance spectrum efficiency that will in turn free up new bands for 6G or other purposes.

Such outcomes in global standardization and spectrum harmonization need to be prepared by proactive and effective international cooperation at government and industry levels. This includes regular dialogues with leading regions and possible focused joint initiatives in R&I, standardization, or regulation.

The issues at stake call for a strategic R&I roadmap to be set out and followed by a critical mass of European actors. So, we have created the Smart Networks and Services Joint Undertaking (SNS JU) to implement research activities on 6G technology under Horizon Europe. Commission funding of €900 million is to be matched by the same amount through co-funding by the industry.

Other world regions are moving; there is no time to waste. In Europe, a first set of 6G projects[1] was launched in 2021, and we recently scaled up the 6G research portfolio[2] to activities worth around €300 million in total.

The Hexa-X project is part of this portfolio and a good illustration of its potential. The flagship is developing the first 6G system concept, imagining the technology of the future with near-instant and unrestricted wireless connectivity to enable embedding ourselves in entirely virtual or digital worlds. One possible vision is an x-enabler fabric of empowered connected intelligence, networks of networks, and sustainability aspects to address the major challenges of our society, with trustworthiness ingrained as a fundamental design principle.

Furthermore, the 5G public–private partnership (5G PPP) Initiative has established working groups (WGs) that provide collaboration platforms for the European projects to attain a joint view on the key technology areas. In this regard, the overall goal of the Architecture WG is to consolidate the main technology enablers and leading-edge design trends in the context of the architecture. As a result, it provides a consolidated view of the architectural efforts developed in the European projects and other research efforts, including standardization. This effort serves not only to review the current state of the art, but also to identify promising trends towards the next generation of mobile and wireless communication networks, namely, 6G. For instance, since October 2020, 45 Phase III 5G PPP projects have contributed to the evolving architecture discussions over the various editions of the white paper[3] prepared by the WG.

1. https://smart-networks.europa.eu/5g-innovations-and-beyond-5g-calls/

2. https://smart-networks.europa.eu/europe-scales-up-6g-research-investments-and-selects-35-new-projects-worth-e250-million/

3. https://5g-ppp.eu/white-papers/

This current book, as a joint effort between the Hexa-X project and the Architecture WG, is the culmination of the European architecture work as a whole. It highlights the latest requirements on the future architecture along with the architectural design principles to respond to technical, economical, and societal needs. Moreover, it elevates the perspective from the long-term evolution of the 5G technologies towards the introduction of the 6G system. It thus provides a reference point for future 6G architecture work to continue in the SNS JU.

We count on the Hexa-X flagship as well as collaborative facilities under the 5G PPP and SNS JU, such as the Architecture WG, to continue creating the critical mass in Europe towards this vision.

I am looking forward to the creativity and ambition of the global research and innovation community to shape the new generation of communication technology throughout this decade.

Peter Stuckmann
Head of Unit, Future Connectivity Systems, European Commission

Glossary

Symbols

3G - *3rd generation mobile network*. 43

3GPP - *3rd Generation Partnership Project*. 2, 4, 7, 29, 41, 42, 52, 59, 65, 68, 80, 81, 84, 85, 123, 125, 161, 171–173, 180, 181, 224, 273, 277, 278, 280, 287, 291, 299, 301–304, 306, 307, 364, 365

4D - *4 dimensional*. 27

4G - *4th Generation mobile network*. 2, 24, 43, 283, 288

5G - *5th Generation mobile network*. 2–7, 11, 13–15, 17–20, 24, 33, 41–44, 46, 52, 65, 66, 79–84, 93, 111, 123, 125, 127–130, 134, 149, 150, 164, 168, 172, 189, 193, 200, 205, 219, 225, 239, 242, 276–278, 280–283, 286–289, 291–294, 303, 307, 308, 315, 316, 318–320, 325, 328, 335, 363–367

5G PPP - *5th Generation Public Private Partnership*. 4, 5, 7

5G-NR - *5G-New Radio*. 254

6D - *6 dimensional*. 21

6G - *6th Generation mobile network*. 3–9, 11–13, 15, 16, 18–25, 27, 29–36, 41–45, 47, 52, 56, 65, 70, 80, 81, 87, 101, 107, 110, 111, 123–126, 142, 144, 145, 149, 157, 161–163, 170–172, 178, 179, 187–198, 200, 215, 216, 218, 219, 224, 225, 228, 235–240, 242, 254, 261, 264, 266, 267, 274–278, 286–289, 292–294, 299, 320, 325–327, 332, 333, 335, 338, 341, 347–351, 353, 361–367, 369

6G IA - *6G Industry Association*. 369

6gNB - *6th Generation NodeB*. 126–128

A

ACLR - *Adjacent Channel Leakage Ratio*. 208

ACT - *Actuator*. 179

ACT-S - *Actuating Functions Sublayer*. 181

ADC - *Analogue to Digital Converter*. 52, 92, 94, 95, 97, 101, 102

AE - *Analytics Engines*. 179, 192, 216–218, 224, 225, 227

AE-S - *Analytic Engines Sublayer*. 181

AF - *Application Functions*. 164

AGC - *Automatic Gain Control*. 92, 94, 97

AGV - *Automated guided vehicle*. 124, 125, 153, 264, 265, 337

AI - *Artificial Intelligence*. 5, 9, 15–18, 20, 22, 25, 27, 33, 35, 42, 125, 127, 128, 162–166, 170, 171, 174, 178, 179, 188, 190, 192, 200–203, 205, 208, 209, 215–217, 220, 225, 228, 248, 267, 275, 315, 318, 326, 338, 351–353, 365

AIaaS - *AI as a Service*. 35, 165

AMF - *Access and Mobility Function*. 283

ANN - *Artificial Neural Networks*. 201

AoA - *Angle of Arrival*. 104, 123–127, 138–140, 142, 143, 151, 211, 212

AP - *Access Point*. 41, 42, 44–54, 58–61, 64, 65, 67, 68, 75, 76, 126, 127, 139, 215

APD - *Avalanche photodiode*. 132

API - *Application Programming Interface*. 9, 23, 32, 124, 132, 165, 175–177, 180, 273–276, 279, 280, 286, 287, 289, 299–308, 310, 318–320, 341

APU - *Antenna Processing Unit*. 48, 49, 56

AR - *Augmented Reality*. 17, 18, 124, 125

AR/VR - *Augmented Reality/Virtual Reality*. 87, 219

ARP - *Adress Resolution Protocol*. 86

ASLT - *Asynchronous Sampling Localisation Technique*. 143

AT3S - *ATSSS*. 80, 81, 85

ATSSS - *Access Traffic Steering Switching and Splitting*. 80

AWGN - *Additive White Gaussian Noise*. 89, 102

B

B5G - *Beyond 5G*. 171, 178, 187–189, 239, 292, 325, 341

BER - *Bit Error Rate / Bit Error Ratio*. 89, 96, 134, 135, 206–208

BIOS - *Basic Input/Output System*. 342

BLER - *Block Error Rate*. 202

BRF - *Bayesian Recursive Filtering*. 138, 141

BS - *Base Station*. 43, 44, 46, 49, 62–65, 67, 68, 71, 73, 75–77, 79, 84, 88, 94, 127, 136, 148–150, 152–154, 203, 210, 211, 220, 221, 224, 256, 257

BSS - *Business Support System*. 181, 185, 224, 315

BW - *Bandwidth*. 107

C

C&C - *Command and Control*. 331

CAD - *Computer Aided Design*. 127, 146

CAPEX - *Capital Expenditure*. 237, 254

CDF - *Complementary Distribution Function*. 59, 107

CDN - *Content Delivery Network*. 332

CED - *Cumulative Energy Demand*. 243

CF - *Cell Free*. 44, 52–54, 197

ChE - *Channel Estimation*. 208

CI/CD - *Continuous Integration and continuous Delivery/Continuous Deployment*. 32, 33, 187, 309

CIR - *Channel Impulse Response*. 127, 138–140, 147

CKM - *Channel Knowledge Map*. 148, 149

CMOS - *Complementary Metal-Oxide-Semiconductor*. 98

cMTC - *critical Machine Type Communication*. 17

CN - *Core Network*. 2, 4, 5, 14, 15, 23, 25, 35, 64, 65, 79, 82, 111, 277

CNF - *Cloud-native Network Function*. 35, 181, 183, 242, 279

CNN - *Convolutional Neural Network*. 205, 207, 208, 225, 226

CoCoCoCo - *Connect-Compute-Control Co-design*. 261

CoT - *Chain of Trust*. 345

CP - *Control Plane*. 29

CPE - *Customer Premises Equipment*. 82

CPM - *Constant Phase Modulation*. 102

CPU - *Central Processing Unit*. 44–46, 48, 49, 51–53, 57, 181, 192–195, 216, 217, 256–260, 296, 344, 346

CQI - *Channel Quality Indicator*. 199, 216

CSI - *Channel State Information*. 41, 47, 52, 72, 134, 148, 149, 213, 245, 348

CSMF - *Consumer Service Management Function*. 35, 180, 181

CSP - *Cloud Service Provider*. 244, 274, 288, 346, 362

CU - *Centralised Unit*. 46, 52, 64, 65, 216, 217

CZF - *Centralized Zero-Forcing*. 58, 59

D

D-DRL - *Decentralized Deep Reinforcement Learning*. 192

D-MIMO - *Distributed Multiple-Input Multiple-Output*. 7, 24, 42–44, 46–54, 56, 57, 64, 68, 70, 110

D2D - *Device-to-device*. 71, 87

DA - *Decision Agents*. 219–221

DAC - *Digital to Analogue Converter*. 52, 62, 63, 92, 94, 97, 201

DDoS - *Distributed Denial of Service*. 186, 330

DDQN - *Double Deep Q-Learning*. 222

DE - *Decision Engines*. 179, 192, 218, 224, 227

DE-S - *Decision Engines Sublayer*. 181

DePF - *DNN-assisted Particle Filter*. 139

DFP - *Dynamic Function Placement*. 165–167, 169

DLIRL - *Deep Learning Integrated Reinforcement Learning*. 128

DMO - *Domain Management Orchestrator*. 181, 182

DMRS - *Demodulation Reference Signal*. 41, 205

DNN - *Deep Neural Network*. 127, 128, 139, 140, 173, 196, 213, 215, 221, 227, 288

DNS - *Domain Name System*. 242, 331

DoA - *Direction of Arrival*. 127

DoD - *Direction of Departure*. 127

DPD - *Digital Pre-Distortion*. 201–203

DRAM - *Dynamic Random Access Memory*. 345

DRL - *Deep Reinforcement Learning*. 192, 199, 219, 221, 222

DRL-NN - *Deep Reinforcement Learning-Neural Network*. 127

DRX - *Discontinuous Reception*. 248

DS-OMP - *Double-structured Orthogonal Matching Pursuit*. 211

DSF - *Domain Shared Functions*. 181, 182

DSP - *Digital Signal Processing*. 48, 95, 101

DT - *Digital Twin*. 16, 17

DTX - *Discontinuous Transmission*. 248

DU - *Distributed Unit*. 46, 52, 56, 64, 65, 174, 216

DZF - *Distributed Zero Forcing*. 58

E

E2E - *End-to-End*. 7, 11, 17, 20–22, 27, 32, 33, 35, 149, 162, 171, 180, 182, 187–189, 192, 205, 206, 216–218, 224, 225, 238, 240, 245, 246, 253, 254, 256–261, 285, 286, 292, 293, 349, 369

EC - *European Commission*. 4, 310

ECDF - *Empirical Cumulative Density Function*. 217

EDR - *Endpoint Detection and Response*. 330

EIRP - *Equivalent Isotropic Radiation Power*. 50, 91

ELPC - *Extremely Low-Power Communications*. 18, 251, 252

EM - *Electromagnetic*. 71

eMBB - *Enhanced Mobile Broadband*. 17, 82, 134, 173, 294

EMC - *Electromagnetic Compatibility*. 362

EMF - *Electric and Magnetic Field*. 14, 246, 247, 258–261, 366

eMMC - *Embedded MultiMediaCard*. 341

ENI - *Enhanced Network Management Interface*. 179, 216

ESPRIT - *Estimation of Signal Parameters via Rational Invariance Techniques*. 140

ETSI - *European Telecommunications Standards Institute*. 171, 172, 176, 179, 219, 224, 237, 239–242, 279, 280, 286, 289, 291, 318, 352, 364

EU - *European Union*. 361, 362, 366, 367

eURLLC - *extremely Ultra-Reliable and Low-Latency Communications*. 18, 251, 252

EVM - *Error Vector Magnitude*. 100, 202

F

F-DRL - *Federated Deep Reinforcement Learning*. 192

FaaS - *Function-as-a-Service*. 189

FDD - *Frequency Division Duplex*. 44

FeMBB - *Further enhanced Mobile BroadBand*. 18, 251, 252

FER - *Frame Error Rate*. 96

FHPPP - *Finite Homogeneous Poisson Point Process*. 66

FL - *Federated Learning*. 163, 216–218, 220–222

FlexRIC - *Flexible RAN Intelligent Controller*. 293

FMCW - *Frequency Modulated Continuous Wave*. 124

FoM - *Figure Of Merit*. 95

FW - *Firmware*. 30

G

GA - *Genetic Algorithm*. 69

GaAS - *Gallium-Arsenide*. 98

GCN - *Graph Convolutional Networks*. 226

GDF - *Gaussian density functions*. 141

GDP - *Gross Domestic Product*. 13

GDPR - *General Data Protection Regulation*. 332

GeSI - *Global Enabling Sustainability Initiative*. 236

GHG - *Green House Gases*. 237

GHz - *Gigahertz*. 109

gNB - *gNodeB (5G base station)*. 46, 84, 126–128, 254, 258, 274

GPS - *Global Positioning System*. 150, 154, 156

gRPC - *Google Remote Procedure Call*. 186

GTP - *Geometrical Theory of Propagation*. 127

H

HERO - *Heuristic for Energy-efficient VNF placement, traffic Routing and user assOciation*. 254–257

HetNet - *Heterogeneous Network*. 24, 43, 80

HPBW - *Half Power Bandwidth*. 107, 146

HW - *Hardware*. 30

I

I/O - *Input/Output*. 341, 344

IAB - *Integrated Access and Backhaul*. 7, 42, 43, 63–70, 110

IADZF - *Interference Aware Distributed Zero-Forcing*. 58, 59

IBN - *Intent-Based Network*. 275, 312, 313, 315, 316

ICNIRP - *International Commission on Non-Ionizing Radiation Protection*. 14, 366

ICT - *Information and Communications Technology*. 4, 13, 236, 237, 243, 363, 364, 367

ICT 20 - *Information and Communication Technologies 20*. 4

ID - *Identifier*. 143

IDM - *Infrastructure Domain Manager*. 182

IDMO - *Inter-Domain Manager and Orchestrator*. 180–182

IDSM - *Inter-Domain Slice Manager*. 181

IEC - *International Electrotechnical Commission*. 352, 362

IF - *Intermediate Frequency*. 99

IFFT - *Inverse Fast Fourier Transform*. 205

IL - *Insertion Loss*. 94

IMT - *International Mobile Telecommunications*. 68, 363

InP - *Indium-Phosphide*. 98, 332

IOMF - *Infrastructure Orchestrated Management Functions*. 182

IoT - *Internet of things*. 15, 71, 111, 125, 192, 219, 242, 243, 307, 326, 335, 336, 338, 349

IP - *Internet Protocol*. 81, 177, 281

IQ - *In phase / Quadrature*. 92, 96

IR - *Infrared*. 131

ISAC - *Integrated Sensing And Communication*. 24

ISG - *Industry Specification Group*. 291, 292, 364

ISM - *In-Slice Management*. 181

ISO - *International Organization for Standardization*. 352, 362

ITU - *International Telecommunication Union*. 109, 110, 236, 237, 243, 363–365

ITU-R - *International Telecommunication Union Radiocommunication Sector*. 18, 363

J

JCAS - *Joint Communication and Sensing*. 24, 126, 149, 150, 155, 157

JT-CoMP - *Joint Transmission Coordinated Multi-Point*. 44

JU - *Joint Undertaking*. 7, 366, 367

K

kHz - *Kilohertz*. 153

KPI - *Key Performance Indicator*. 20, 21, 42, 102, 167, 169, 172, 173, 177, 181, 185, 194, 197–199, 212, 224–227, 237, 246, 260, 267, 315, 346

KVI - *Key Value Indicator*. 21, 235, 236, 246, 260

KVM - *Kernel-based Virtual Machine*. 341

L

L-BRF - *Linearized BRF*. 138

LADN - *Local Area Data Network*. 239

LCA - *Life Cycle Assessment*. 243

LCM - *Life Cycle Management*. 178–180, 182

LD - *Location Database*. 142

LDHMC - *Long-Distance and High-Mobility Communications*. 18, 251, 252

LDPC - *Low-Density Parity-Check*. 201

LED - *Light Emitting Diode*. 131, 132

LIDAR - *Laser/Light Imaging, Detection and Ranging*. 29, 124, 126, 142, 143, 153

LL - *Low-Layer*. 81

LLR - *Log Likelihood Ratio*. 202

LMMSE - *Linear Minimum Mean Square Error*. 207

LNA - *Low Noise Amplifier*. 92–94, 98, 99

LNaaS - *logical network-as-a-service*. 280, 288

LO - *Local Oscillator*. 94, 96, 99

LoS - *Line of Sight*. 47, 69, 73, 77, 90, 103, 104, 126, 139, 140, 156, 157

LP - *Low Passs*. 92

LS - *Location Server*. 143

LSCPA - *Large Scale Cooperative Predictor Antenna*. 149

LTE - *Long Term Evolution*. 44

LTI - *Linear-Time-Invariant*. 262

LTV - *Linear-Time-Variant*. 262

M

M&O - *Management and Orchestration*. 7, 12, 27, 30, 32, 33, 35, 36

MA-DRL - *Multi-Agent Deep Reinforcement Learning*. 192, 195, 197–199

MA-RL - *Multi-Agent RL*. 196

MaaS - *Management as a Service*. 182

MAC - *Medium Access Control*. 46, 86, 181

MAE - *Mean Absolute Error*. 190

MANO - *Management and Orchestration*. 161, 163, 172, 176–183, 189, 191, 192, 215, 224, 242, 279, 348

MANO-MS - *MANO Monitoring System*. 182, 183

MAP - *Multi Antenna Processing*. 54

MAPE - *Monitor-Analyse-Plan-Execute*. 179, 181

MAPE-K - *Monitor-Analyze-Plan-Execute over a shared Knowledge*. 174, 175

MARL - *Multi-Agent Reinforcement Learning*. 248

MBS - *Macro Base Station*. 64, 69

MCS - *Modulation and Coding Scheme*. 88, 89, 91, 202

MDP - *Markov Decision Process*. 196, 197

ME - *Mobile Edge*. 239

MEAO - *Mobile Edge Application Orchestrator*. 240, 242

MEC - *Multi-access Edge Computing*. 71, 125, 142, 189, 239–242, 245, 246, 253, 258, 287

MEO - *MEC Orchestrator*. 239, 242

MEPM - *MEC Platform Manager*. 242

MHz - *Megahertz*. 59

MILP - *Mixed-Integer Linear Programming*. 173

MIMO - *Multiple Input Multiple Output*. 41–44, 64, 70, 88, 96, 101, 134, 149, 200, 203, 204, 212, 213, 248

MISO - *Multiple Input Single Output*. 244

MJLS - *Markov Jump Linear System*. 262

ML - *Machine Learning*. 6, 9, 16, 22, 27, 33, 35, 52, 56, 66, 69, 162–164, 170, 177, 183, 185–187, 195, 203, 205, 207, 208, 213, 215, 220, 315, 316, 318, 326–328, 330, 338, 347, 351–353

mMIMO - *Massive MIMO*. 41, 43, 44, 53, 54, 70, 157

MMSE - *Minimum Mean Squared*. 208, 209

mMTC - *Massive Machine-Type Communications*. 17, 134, 294

mmWave - *millimeter Wave*. 24, 42, 51, 64, 65, 87, 93, 111, 123, 125, 127, 130, 142, 143, 145, 148, 153, 211, 254

MNO - *Mobile Network Operator*. 3, 32, 33, 332, 348, 350

MPC - *Multipath Component*. 137, 138, 151

MPTCP - *Multi-Path Transmission Control Protocol*. 80–82, 84–86

MRF - *Media Resource Function*. 35

MS - *Monitoring System*. 179, 182, 216, 217, 221, 224, 225

MS-S - *Monitoring System Sublayer*. 181

MSE - *Mean Square Error*. 190

MSLE - *Mean Aquared Logarithmic Error*. 190

MT - *Mobile Termination*. 65

MTTF - *Mean Time to Failure*. 263, 264

MU - *Mobile Unit*. 138, 139

MUSIC - *Multiple Signal Classification*. 139, 140

MVNO - *Mobile Virtual Network Operator*. 332, 333, 348, 350

N

N-MAPE-K - *Network MAPE-K*. 175

N3IWF - *Non-3GPP Interworking Function*. 82, 277

NAS - *Non-Access Stratum*. 283, 286

NBI - *Northbound API*. 180

NCC - *Network Centric Clustering*. 54

Near-RT RIC - *Near-Real Time RAN Intelligent Controller*. 54, 56

NEF - *Network Exposure Function*. 177, 287, 307, 308

NF - *Network Function*. 35, 162, 164–170, 274, 278, 279, 285, 287, 295–298

NFV - *Network Function Virtualization*. 35, 173, 176, 186, 219, 224, 239, 241, 242, 280, 318, 348, 364

NFV MANO - *Network Functions Virtualization Management and Orchestration*. 176

NFVI - *Network Functions Virtualization Infrastructure*. 181, 182, 239

NFVO - *Network Functions Virtualization Orchestrator*. 35, 181, 239, 251

NG - *Next Generation*. 283

NI - *Network Intelligence*. 6, 161, 162, 171–176, 182–187, 190, 191, 228

NIF - *Network Intelligence Function*. 162, 172–177, 187–189

NIF-C - *NIF Component*. 175–177

NIO - *Network Intelligence Orchestrator*. 172, 176

NIP - *Network Intelligence Plane*. 172, 190

NIS - *Network Intelligence Service*. 172–174, 176, 177

NIST - *National Institute of Standards and Technology*. 352

NLoS - *Non Line of Sight*. 51, 69, 103, 104, 139, 140, 157, 211

NMSE - *Normalized Mean Square Error*. 209, 212

NN - *Neural Network*. 176, 201–203, 208, 209, 211, 212, 222, 226, 227

NNRT - *Non-Near Real-Time*. 169

NP - *Non-deterministic Polinomial*. 193

NPN - *Non-Public Network*. 15, 276, 277

NR - *New Radio*. 2, 65, 82, 302

NRF - *Network Registry Function*. 177

NRT - *Near-Real-Time*. 169, 220

NS - *Network Service*. 23, 25, 27, 33, 319

NSA - *Non-Standalone*. 2, 239

NSaaS - *Network Slice as a Service*. 239

NSB - *Network Slicing Broker*. 334

NSD - *Network Service Descriptor*. 181

NSI - *Network Slice Instance*. 225, 318

NSM - *Network Service Mesh*. 166

NSMF - *Network Slice Management Function*. 180, 224

NSSMF - *Network Sub-Slice Management Function*. 181

NST - *Network Slice Template*. 181

NTN - *Non-Terrestrial Network*. 24, 88

NWDAF - *Network Data Analytics Function*. 172, 177

O

O-DU - *Open Distributed Unit*. 52–54, 56, 110, 129

O-RAN - *Open RAN*. 9, 171–173, 177, 220, 320, 364, 365

O-RU - *Open Radio Unit*. 52–54, 56, 110, 129

OAM - *Operation Administration and Maintenance*. 166

OCI - *Open Container Initiative*. 340

OFDM - *Orthogonal Frequency-Division Multiplexing*. 102, 124, 126, 129, 134, 135, 153, 155, 205, 207, 208, 248

OPEX - *Operational Expenditure*. 174, 237, 254

OS - *Operating System*. 339

OSM - *Open-Source Management and Orchestration*. 309, 315, 316, 318, 319

OSS - *Operations Support System*. 181, 185, 224, 315

OTFS - *Orthogonal Time Frequency Space*. 124, 126, 134, 135

OTT - *Over-The-Top*. 239, 318

OVS - *Open vSwitch*. 293, 294

OWC - *Optical Wireless Communications*. 123, 125, 131, 142, 145

P

PA - *Power Amplifier*. 97

PA - *Predictor Antenna*. 50, 59, 91–99, 149, 201, 202, 205–208

PAPR - *Peak to Average Power Ratio*. 101, 102

PCR - *Platform Configuration Register*. 344

PE - *Positioning Error*. 133

PEF - *Protected Execution Facility*. 345

PF - *Particle Filter*. 139

PGM - *Particle Gaussian Mixture*. 138–141

PHY - *Physical Layer*. 46, 104

PL - *Path Loss*. 244

PN - *Pseudo Random Noise*. 129

PNF - *Physical Network Function*. 35, 181

PPDP - *Privacy Preserving Data Publishing*. 327

PPP - *Public Private Partnership*. 33, 66

PRB - *Physical Resource Block*. 216, 217, 221, 252

ps - *Pico seconds*. 129

PSO - *Particle Swarm Optimization*. 348

Q

QAM - *Quadrature amplitude modulation*. 89, 207, 208

QoE - *Quality of Experience*. 32, 315, 316, 325, 332

QoS - *Quality of Service*. 15, 29, 32, 44, 58, 71, 166, 181, 189, 200, 248, 254, 261, 278, 287, 295, 307, 315, 316, 325

R

RA - *Receive Antenna*. 149

RAM - *Random Access Memory*. 344

RAN - *Radio Access Network*. 4, 5, 14, 15, 23, 25, 41–43, 53, 54, 64, 71, 80, 110, 143, 152, 174, 177, 181, 195, 197, 200, 215–217, 219–221, 224, 237, 238, 241, 245, 252, 267, 274, 277, 278, 284, 292, 293, 315, 316, 318, 364–366

RANO - *Radio Access Network Orchestrator*. 181

RE - *Resource Element*. 205

REM - *Radio-Environment Map*. 265, 266

RF - *Radio Frequency*. 43, 46, 51, 62, 63, 72, 79, 87–93, 95, 96, 99–102, 107, 111, 145–147, 200, 201, 238, 244, 364

RFC - *Request For Comments*. 81

RFIC - *Radio-Frequency Integrated Circuit*. 95

RIC - *RAN Intelligent Controller*. 53, 80, 220, 252, 284

RIS - *Reconfigurable Intelligent Surface*. 6, 7, 42, 43, 70–79, 111, 149, 200, 210–212

RISA - *RIS Actuator*. 72

RISC - *RIS Controller*. 72

RISO - *RIS Orchestrator*. 72

RL - *Reinforcement Learning*. 127, 128, 195–200, 221, 222, 249, 252, 319

RLC - *Radio Link Control*. 46

RMF - *Risk Management Framework*. 352

RNTI - *Radio Network Temporary Identifier*. 130

ROM - *Read Only Memory*. 344

RoT - *Root of Trust*. 344, 345

RRH - *Remote Radio Head*. 43

RRM - *Radio Resource Management*. 212, 213, 348

RSS - *Received Signal Strength*. 123–126, 131, 142, 155

RSSI - *Received Signal Strength Indicator*. 132

RT - *Ray Tracing*. 127, 128, 145–148, 169, 171, 185, 197, 220, 221

RTT - *Round Trip Time*. 84

RU - *Remote Unit*. 46, 52, 54, 56, 64, 88, 127, 128, 252

Rx - *Receiver*. 106

S

SA - *Standalone*. 2, 364

SaaS - *Sensing as a Service*. 24

SAI - *Securing Artificial Intelligence*. 352

SBA - *Service-Based Architecture*. 14, 239, 278, 287, 307

SBL - *Sparse Bayesian Learning*. 137

SBS - *Small Base Station*. 64, 66, 67, 69

SBTi - *Science Based Targets initiative*. 236

SC - *Small Cell*. 254

SC-FDE - *Single-Carrier Frequency Domain Equalization*. 102

SCP - *Service Communication Proxy*. 168, 280

SD - *Secure Digital card*. 341

SDG - *Sustainable Development Goal*. 13, 21, 235, 236, 367

SDK - *Software Development Kit*. 9, 293, 305, 320

SDN - *Software-Defined Networking*. 7, 25, 181, 183–186, 219, 286, 289, 291, 318, 339

SDO - *Standards Developing Organization*. 289, 362–365

SE - *Spectral Efficiency*. 59–61

SEV - *Secure Encrypted Virtualization*. 345

SFC - *Service Function Chain*. 253, 254, 256

SFL - *Slice Functional Layer*. 181

SiGe - *Silicon-Germanium*. 98

SINR - *Signal to Interference plus Noise Ratio*. 56, 59, 60, 127, 256, 257

SISO - *Single-Input Single-Output*. 43

SLA - *Service-Level Agreement*. 190, 215–218, 221, 224, 312

SLAM - *Simultaneous Localisation and Mapping*. 16, 123, 125, 150

SmartNICs - *Smart Network Interface Cards*. 285

SME - *Small and Medium-sized Enterprises*. 6

SML - *Slice Management Layer*. 181

SNR - *Signal to Noise Ratio*. 58, 65, 88–91, 94–96, 100–102, 128, 151, 202, 207

SNS - *Smart Networks and Services*. 7, 366, 367

SNS JU - *Smart Networks and Services Joint Undertaking*. 361, 367

SoC - *System-on-Chip*. 48

SOD - *Slice Orchestration Domain*. 181

SRTT - *Smooth Round Trip Time*. 81

stdev - *Standard deviation*. 129

StFL - *Statistical Federated Learning*. 218

SW - *Software*. 30, 300, 301

SWO - *Software Ontology*. 188

T

TaaS - *Trust as a Service*. 326, 349, 350

TCG - *Trusted Computing Group*. 338

TCO - *Total Cost of Ownership*. 237

TCP - *Transmission Control Protocol*. 81, 82, 84

TDD - *Time Division Duplex*. 45, 50, 59, 64, 65, 93

TDoA - *Time Difference of Arrival*. 125, 126, 138, 142

TDX - *Trust Domain Extensions*. 345

TEE - *Trusted Execution Environment*. 338, 344–346

TFS - *TeraFlowSDN*. 184, 185, 289, 291

THz - *Terahertz*. 124, 125

TNSaaS - *Transport Network Slice as a Service*. 288, 289

ToA - *Time of Arrival*. 123, 124, 126, 128, 129, 134, 138, 151

ToF - *Time of Flight*. 126, 127

TPM - *Trusted Platform Module*. 338, 343, 344

TRP - *Transmission/Reception Point*. 216

TSN - *Time-Sensitive Networking*. 276, 278

Tx - *Transmitter*. 107

U

UAV - *Unmanned Aerial Vehicle*. 29, 71, 78, 79, 111, 134, 264–266

UCC - *User Centric Clustering*. 54

UDM - *Unified Data Management*. 350

UE - *User Equipment*. 29, 43–46, 51–54, 56, 58–60, 63–69, 74, 79–81, 85, 110, 111, 123, 124, 126–130, 134, 135, 138–143, 148–152, 163, 193, 199, 203, 239, 240, 254–258, 281–289, 299–302, 308, 364

UEFI - *Unified Extended Firmware Interface*. 342

UMi - *Urban Micro*. 206

umMTC - *Ultra-massive Machine-Type Communications*. 18, 251, 252

UN - *United Nations*. 13, 21, 235, 236

UP - *User Plane*. 29, 286

UPF - *User Plane Function*. 80, 81, 239–241, 286

URA - *Uniform Rectangular Array*. 136

URLLC - *Ultra-Reliable Low Latency Communications*. 17, 83, 173, 240, 241, 286

UWB - *Ultra-Wideband*. 150, 153

V

V2X - *Vehicle-to-Everything*. 27, 219, 303

VAE - *Vertical Application Enablers*. 274, 287, 303

VC - *Verifiable Credential*. 336, 337

VCO - *Voltage Controlled Oscillator*. 99

VIM - *Virtual Infrastructure Manager*. 239, 241

VLAN - *Virtual Local Area Networks*. 181

VLP - *Visible Light Positioning*. 131, 132, 142

VM - *Virtual Machine*. 81, 82, 181, 239, 242, 253, 301, 339–341

VMM - *Virtual Machine Monitor*. 341

VNA - *Vector Network Analyser*. 109, 136

VNF - *Virtual Network Function*. 35, 142, 174, 181, 192, 216, 239, 241, 242, 253–257, 279, 315, 318, 319

VR - *Virtual Reality*. 17, 18

vRAN - *virtualised Radio Access Network*. 174, 238, 239

W

W3C - *World Wide Web Consortium*. 336

WAT - *Wireless Access Technology*. 80, 82

WDM - *Wavelength-Division Multiplexing*. 56

WG - *Working Group*. 5, 364

WPT - *Wireless Power Transfer*. 238, 242, 244

WRR - *Weighted Round Robin*. 83

X

XR - *Extended Reality*. 17–19, 143

Z

ZSM - *Zero-touch network and Service Management*. 35, 178, 179, 216, 219, 220, 364

ZXM - *Zero-Crossing Modulation*. 102

DOI: 10.1561/9781638282396.ch1

Chapter 1

Introduction

By Ömer Bulakçı, Mikko Uusitalo, Patrik Rugeland, Marco Gramaglia, Xi Li, Mauro Boldi, Anastasius Gavras, et al.[1]

1.1 Architecting the 6th Generation of Mobile and Wireless Communications System

Since early generations, mobile and wireless communications systems have played a crucial role in the establishment of essential connectivity. This has enabled easy access to information from anywhere and anytime and, thus, has fostered accelerated knowledge build-up and timely value-adding actions based on that beyond telecommunications ecosystems in all sectors of society. The COVID-19 pandemic has additionally proven the importance of the wireless communication infrastructure during these challenging times and has supported global stability as well as faster recovery via maintaining a stable technological environment and seamless societal connection despite quarantine regulations.

Considering such critical role and importance, the relevant technological advancements within and for the telecommunications ecosystem are captured by a non-backward-compatible set of features specified in a new generation of mobile and wireless communications systems. As advancing from a previous generation to the new one requires significant efforts, the need for a new generation shall be

1. The full list of chapter authors is provided in the Contributing Authors section of the book.

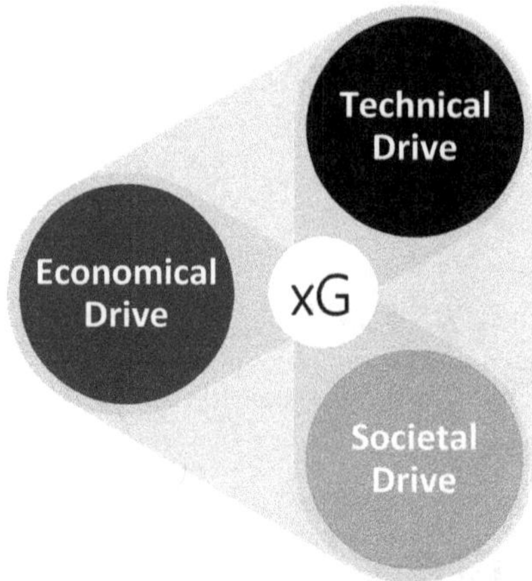

Technical Drive	• Increase in the mobile data traffic • Diverse Services with Broad Range of Requirements • Rapid growth of wirelessly-connected Devices
Societal Drive	• Demand for Seamless Connectivity Anywhere & Anytime • Demand for Affordable Access to Broadband • Demand for Sustainable Access
Economical Drive	• The need for cost-efficient cellular communications • The need for energy-efficient cellular communications • The need for new business areas

Figure 1.1. Drivers for developing new generation of mobile networks.

justified factoring in the technical, societal, and economical drives (see Figure 1.1). These drives are not mutually independent, i.e., one boost in any of the drives can accelerate the whole process. For instance, the need for attaining new business areas also implies the need for supporting a diverse set of induced requirements.

Accordingly, the development of mobile and wireless networks has followed around a 10-year cycle, where, up to the 5th generation (5G), previous generations have been designed for particular use case categories. With 5G, an inherent flexibility has been introduced, and the target customer space has been extended from mobile broadband users towards vertical industries with diverse requirements [1]. The framework of network slicing has been introduced to cope with such different requirements. The commercial deployments of the 5G system have been underway since 2019 with an original focus on non-standalone (NSA) architecture, while more and more standalone (SA) architecture has been employed globally. The NSA deployment implies that 5G new radio (NR) is anchored with the 4th generation (4G) system, and the 4G core network (CN) is in use. In the case of the SA deployments, where 5G NR connects to the 5G CN, the full potential of the 5G system can be unleased, e.g., by means of the utilization of the network slicing. Moreover, 5G-Advanced enhancements, e.g., extended reality (XR) optimizations, are already being specified in the 3rd Generation Partnership Project (3GPP) Release 18 [2].

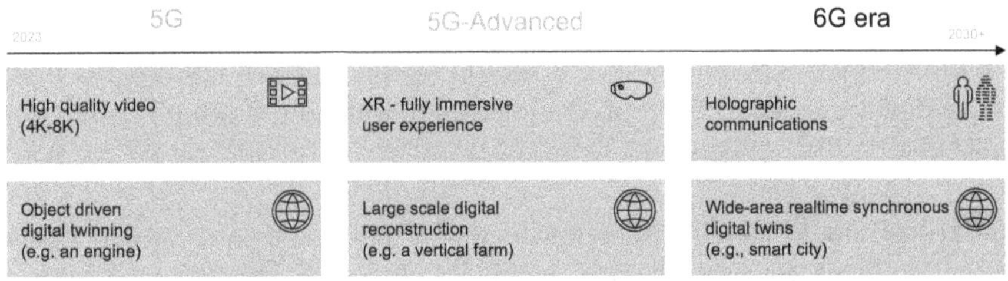

Figure 1.2. Example evolution of use case families from 5G towards 6G era.

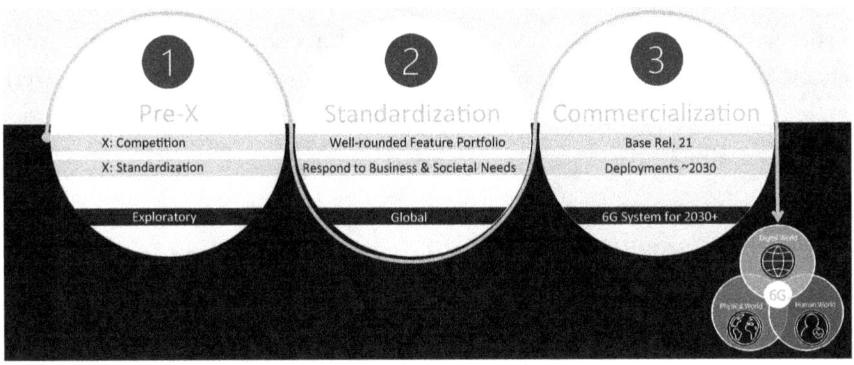

Figure 1.3. Three-phase approach to realize the complete value chain of 6G.

While 5G deployments and 5G-Advanced standardization are progressing well, as highlighted in Chapter 9, research into the new generation of mobile and wireless communications systems, i.e., the 6th Generation (6G), has already started. Like in the case for other generations, the creation of 6G has started with the identification as well as prediction of the use cases and the associated requirements (see Section 2.1). An illustration of the evolution of use case families is shown in Figure 1.2. As depicted, digital twinning of objects is already supported by a 5G system, and this support can be expanded via 5G-Advanced. Nevertheless, it is foreseen that a wide-area synchronous digital twinning can be possible only during the 6G era. Accordingly, it can be stated that 5G-Advanced provides the stepping stone towards 6G, where 6G will enable new use cases as well as existing use cases at scale. It is worth noting that the wireless networks are designed to be flexible to address the requirements of new use cases that may not be predicted today.

On this basis, a three-phase approach is being followed to realize the complete value chain of the 6G system, as depicted in Figure 1.3. These phases are outlined in the following.

Pre-X Phase: This is the exploratory research phase, where X can refer to competition or standardization. During this phase, the key stakeholders, e.g., academia, vendors, diverse industries, and mobile network operators (MNOs), come together to set the vision, design guidelines, and the foundation for the 6G system. The main motivation for such a phase is the build-up and expansion of the 6G ecosystem. This phase is essential for rapid and efficient standardization process that will follow.

Standardization Phase: In this phase, a well-rounded feature portfolio is specified, e.g., radio access network (RAN) and CN by 3GPP, that shall respond to the business and societal needs, as highlighted in Figure 1.1, which have been identified during the Pre-X Phase as well as the standardization phase. It is expected that, differently from 5G specification, many architecture options shall be avoided to enable the full potential of 6G already in the early deployments. Moreover, to make use of the economies of scale and the proven benefits of previous generations, a global 6G standard should be aimed for.

Commercialization Phase: Based on the established standardization, which is envisioned to be 3GPP Release 21, the first commercial 6G deployments can be seen around 2030. It should be noted that the 6G system will be designed at least for the full next decade (2030s) and even beyond.

With the seamless execution of all three phases, the promises of the 6G system on the integration of digital, physical, and human worlds can be realized, which can transform the world while maximizing human potential. The details on the standardization and regulatory processes are provided in Chapter 9.

1.2 Approach and Timing of the Book

This book is a result of the collaborative work performed at European level during the Pre-X phase of the 6G system's creation. In particular, the main content of the book has been provided by the 5G Public Private Partnership (5G PPP) Phase 3 research projects [3]. 5G PPP Phase 3 projects are categorized under different calls that pertain to specific requirements set by the European Commission (EC); see the corresponding book preface. Among the Phase 3 calls, the most relevant ones for the scope of the book have been the information and communication technologies 20 (ICT 20) and ICT 52 calls, where the former has comprised projects that work on the longer-term vision of 5G, i.e., 5G evolution [4], and the latter has comprised projects that work on the foundation of the beyond 5G/6G system [5]. As part of the ICT 52 call, the Hexa-X project is the system flagship project [6]. Yet, since various enhancements considered for the 5G system by other Phase 3 projects could be employed by 6G with its full potential, such enhancements are also captured herein.

Figure 1.4. White papers released by the 5G PPP Architecture WG.

Moreover, a key part of the 5G PPP framework is a set of cross-project working groups (WGs). The outcome of the work from these groups is presented in white papers [7]. As highlighted in the prefaces of this book, the Architecture WG brings the research projects together to build a consolidated view of the architectural efforts, including the overall architecture of mobile and wireless communications networks across different network domains, such as RAN, Transport, and CN, as well as cloud or edge infrastructure.

The outcome of the Architecture WG has been published in a series of white papers and presented during various technical workshops at international conferences and webinars. As shown in Figure 1.4, the latest white paper of the WG is "*The 6G Architecture Landscape – European Perspective*" [8], which is Version 6.0. The first version of the architecture white paper from the Architecture WG was back in July 2016. Since then, this effort has continuously captured the technology trends as developed by the different phases of 5G PPP projects: the first phase (Phase 1) laid the foundation of the network slicing-aware operations we are seeing these days; the second phase (Phase 2) provided the first proof of concepts; and the third phase (Phase 3) has targeted the first large-scale platforms. All these efforts were captured in the subsequent releases of the white paper (Version 2 in January 2018, Version 3 in February 2020, and Version 4 in November 2021). It is worth noting that Version 6.0 is intentionally the next version after Version 4.0 as an indication of the focus on 6G.

Capitalizing on the Version 6.0 of the white paper, this book is a joint effort by the Architecture WG and the Hexa-X project, extending significantly the concise content of the published white paper [8]. Within the framework of the Architecture WG and the joint work, the content of the book has been based on the following projects (in alphabetical order) [3]:

- **5G-COMPLETE**, which aims at revolutionizing beyond 5G architecture, by efficiently combining compute and storage resource functionality over a unified, ultra-high capacity converged digital/analogue Fiber-Wireless (FiWi) RAN.
- **5G-CLARITY**, which brings forward the design of a system beyond 5G private networks to address challenges in spectrum flexibility, delivery of critical services, and autonomic network management using heterogeneous wireless access that integrates 5G, Wi-Fi, and LiFi technologies managed through novel Artificial Intelligence (AI)-based autonomic networking.

- **5G ERA**, which aims at providing third-party application developers with an experimentation facility as a playground to test and qualify their applications.
- **5GASP**, which aims at shortening the idea-to-market process through the creation of a European testbed for Small and Medium-sized Enterprises (SMEs) that is fully automated and self-service, in order to foster rapid development and testing of new and innovative 5G network applications.
- **6G BRAINS**, which brings reinforcement learning into radio-light network for massive connections.
- **AI@EDGE**, which develops a connect-compute fabric – specifically leveraging the serverless paradigm – for creating and managing resilient, elastic, and secure end-to-end slices.
- **ARIADNE**, which proposes to exploit bandwidth-rich D-band, the capabilities of Reconfigurable Intelligent Surfaces (RIS), and powerful Machine Learning (ML) tools in order to realize a novel Communication Theory framework beyond Shannon and design suitable technology enablers for highly reliable and reconfigurable 6G connectivity.
- **DAEMON**, which develops and implements innovative and pragmatic approaches to Network Intelligence (NI) design that enable high performance, sustainability, and an extremely reliable zero-touch network system.
- **DEDICAT 6G**, which addresses techniques for achieving and maintaining efficient dynamic connectivity and intelligent placement of computation in the mobile network.
- **EVOLVED-5G**, which designs and develops an open facility for the long-term support of third-party applications that interact with the network core.
- **Hexa-X**, which is the European 6G flagship research project, defining the vision, and developing technological enablers for connecting the physical, digital, and human worlds.
- **MARSAL**, which targets the development of a complete framework for the management and orchestration of network resources in 5G and beyond, by utilizing a converged optical wireless network infrastructure in the access and fronthaul/midhaul segments.
- **MonB5G**, which targets zero-touch management and orchestration in support of network slicing at massive scales for 5G evolution and beyond.
- **REINDEER**, which develops a new smart connect-compute platform with a capacity that is scalable to quasi-infinite, and that offers perceived zero latency and interaction with an extremely high number of embedded devices.
- **RISE 6G**, which aims at investigating innovative solutions that capitalize on the latest advances in the emerging technology of RISs and offers dynamic and goal-oriented radio wave propagation control, enabling the concept of the wireless environment as a service.

- **TeraFlow**, which aims to deliver a new generation open-source cloud-native Software-Defined Networking (SDN) controller to provide smart connectivity services to beyond 5G networks.

1.3 Scope and Structure of the Book

This book highlights the related research work of the contributing 5G PPP Phase 3 projects and presents all the key elements and key architecture enablers and solutions of future 6G network design – a design that is deeply rooted in real needs and can profoundly benefit humanity in the mid-to-long term. Specifically, a high-level view of the 6G End-to-End (E2E) architecture as well as a functional view of the 6G reference architecture are introduced, taking into consideration the new stakeholders in the mobile network ecosystem and how the architectural work is taking into account their requirements in all the domains of the network. The key architecture enablers, which will form the backbone of future sustainable and trustworthy 6G network architecture, include all the related technological solutions for building intelligent, flexible, energy efficient, secure, programmable networks and enabling versatile radio technologies, localization, and sensing in the 6G networks.

As 5G PPP Phase 3 consists of the last calls of the Horizon 2020 programme, this book is aimed to lay the architectural foundation for the next European programme towards 6G, i.e., smart networks and services (SNS) joint undertaking (JU).

The rest of this book is structured into the following eight chapters.

Chapter 2 – Architecture Landscape draws the envisioned system view of the overall 6G architecture associated with a functional view. The presented architecture is built up considering the key design principles that are populated based on the envisioned use cases, requirements, and trends as highlighted in Section 2.1. The discussions on the management and orchestration (M&O) as well as security and privacy architecture complete the big picture. A brief overview of the architectural enablers is also captured, which sets the guidance for the following detailed chapters enumerated from three to eight.

Chapter 3 – Towards Versatile Access Networks presents the envisioned enhancements for the wireless networks, including the efficient utilization of 3GPP and non-3GPP access networks. Such enhancements include the distributed multiple-input multiple-output (D-MIMO) implementation, integrated access and backhaul (IAB) deployments, RIS, multi-access connectivity, and sub-THz access. It is argued that for the limitless connectivity requirement of 6G, D-MIMO can provide expected macro diversity, design flexibility, and interference management; IAB can offer cost-efficient densification without the need of fibre to connect the small cell sites; and RIS can provide means to fine-tune the wireless configuration environments. Multi-access multi-connectivity will remain an important feature for 6G to enhance the wireless link's throughput, reliability, or even latency. In addition, sub-THz bands can help fulfil the requirements of the high data rate applications in short distances or can offer wireless connectivity for the backhaul.

Chapter 4 – Towards Joint Communications and Sensing presents the incorporation of sensing capabilities into the mobile networks. These capabilities can be separated into three different categories. First, the 6G network can efficiently collect, store, and analyse sensing data from various different sensors, and provide localization, sensing, and mapping information to applications and users to enhance different 6G services. Second, the localization, sensing, and mapping information can be provided to the radio network itself to improve the communication, e.g., through location or environment-aware beamforming or pre-emptive handover if an impending obstacle is detected. Finally, the purpose of the radio interface itself is reimagined, where the radio signals are used for both communication and sensing, either simultaneously or only using common hardware, potentially providing access to a plethora of transmitters and receivers in a more densified network that can sense and map the environment in a radar-like fashion without the need for dedicated hardware deployment.

Chapter 5 – Towards Natively Intelligent Networks presents the recent efforts in the design of an architecture that is natively capable of incorporating all the elements required by network functions empowered by Artificial Intelligence (e.g., data gathering, representation, decision enforcement), as well as some examples of such Network Intelligence Functions and their application in different domains of the network, such as the radio access and the orchestration.

Chapter 6 – Towards Sustainable Networks presents targeted metrics as well as the main technological enablers towards ensuring high sustainability in next-generation networks. The chapter focuses on two main principles, "sustainable 6G," i.e., how to make the 6G networks sustainable, and "6G for sustainability," i.e., how 6G can be leveraged so as to ensure sustainability in other markets and value chains. In this context, a broad analysis of the key network sustainability enablers is provided spanning across different levels, i.e., from enablers that include architectural or hardware innovations, and enablers at management/orchestration level or at service/application level that target at network operation efficiency maximization to cross-layer sustainability enablers, which include innovations in more than two layers.

Chapter 7 – Towards Continuously Programmable Networks presents design principles and technology enablers towards realizing programmability frameworks, i.e., frameworks that abstract the underlay network infrastructure and capabilities so that the network is dynamically controlled and configured. Standardized solutions and research enablers for such abstraction are indicated, organized in three levels, namely, service/application provisioning level, network and resource management level, as well as network deployment and connectivity level. Indicative approaches include the deployment of common Application Programming Interface (API) managers, the exploitation of P4-programmable switches, the usage of open interfaces of Open-RAN (O-RAN), and the design of Software Development Kits (SDKs) for providing network slices as a service.

Chapter 8 – Secure, Privacy-Preserving, and Trustworthy Networks presents aspects related to network privacy and security for information sharing among different tenants and cloud-stored data, as well as end-users' network security. Moreover, it investigates the application of blockchain-based platforms for network slicing by using smart contracts, as well as for industrial Internet of Things networks. Finally, it focuses on trusted execution, trust-as-a-service, as well as trustworthy ML/AI.

Chapter 9 – 6G Outlook and Timeline presents the most recent picture about European perspectives on the current status of 6G in terms of standardization, research and regulation initiatives. This final chapter presents also concluding remarks of the book.

References

[1] P. Marsch, Ö. Bulakci, O. Queseth, M. Boldi, Eds. *5G System Design – Architectural and Functional Considerations and Long Term Research*, Wiley 2018.

[2] 3GPP, "Release 18 Overview," 2022. Accessed: April 6, 2023. [Online]. Available: https://www.3gpp.org/specifications-technologies/releases/release-18.

[3] 5G PPP, "5G PPP Phase 3 Research Projects," 2022. Accessed: April 6, 2023. [Online]. Available: https://5g-ppp.eu/5g-ppp-phase-3-projects/.

[4] 5G PPP Phase 3 ICT 20 Call," 2022. Accessed: April 6, 2023. [Online]. Available: https://5g-ppp.eu/5g-ppp-phase-3-4-projects/.

[5] 5G PPP, "5G PPP Phase 3 ICT 52 Call," 2022. Accessed: April 6, 2023. [Online]. Available: https://5g-ppp.eu/5g-ppp-phase-3-6-projects/.

[6] Hexa-X, "A flagship for B5G/6G vision and intelligent fabric of technology enablers connecting human, physical, and digital worlds," 2021. Accessed: April 6, 2023. [Online]. Available: https://cordis.europa.eu/project/id/101015956.

[7] 5G PPP white papers, 2023. Accessed: April 6, 2023. [Online]. Available: https://5g-ppp.eu/white-papers/.

[8] 5G PPP Architecture WG, "The 6G Architecture Landscape: European Perspective," White paper v6.0, 2023. Accessed: April 6, 2023. [Online]. Available: https://5g-ppp.eu/wp-content/uploads/2023/02/Whitepaper-final-version-rev1.pdf.

DOI: 10.1561/9781638282396.ch2

Chapter 2

Architecture Landscape

By Mårten Ericson, Bahare Masood Khorsandi, et al.[1]

2.1 Introduction

The network architecture evolution journey will carry on in the years ahead, driving a large scale adoption of 5th Generation (5G) and 5G-Advanced use cases with significantly decreased deployment and operational costs, and enabling new and innovative use-case-driven solutions towards 6th Generation (6G) with higher economic and societal values. The goal of this chapter, thus, is to present the envisioned societal impact, use cases and the End-to-End (E2E) 6G architecture. The E2E 6G architecture includes summarization of the various technical enablers as well as the system and functional views of the architecture.

The design of the 6G architecture is based on the analysis of the societal, economic, regulatory, and technological trends, which are discussed in Section 2.1.2. A summary of the use cases envisioned for 6G is also introduced in Section 2.1.3. Accordingly, a set of architectural principles has been drawn, upon which the presented architecture is built. Herein, the main highlights of the 6G system design are provided, while the details on the various network domains are given in the subsequent chapters.

1. The full list of chapter authors is provided in the Contributing Authors section of the book.

In Section 2.2, the overall architecture description discusses the new stakeholders in the mobile network ecosystem, and how the architectural work is taking into account their requirements in all the domains of the network. Specific design principles that need to be factored in for the new architecture are also described. Section 2.3 discusses the components of the security architecture, which are required and must be applied to have security as a design principle for the 6G architecture. A deep dive into the Management and Orchestration (M&O) architecture is then presented in Section 2.4. Section 2.5 outlines the summary of this chapter and presents the outlook.

2.1.1 The Societal Impact of 6G

Since the invention of mobile telephony half a century ago, wireless network technology has undoubtedly transformed the everyday lives of billions of people on the planet, and profoundly shaped the economy and the evolution of human society to date. The mobility of communication and of access to information has allowed completely new ways of interacting, working, and evolving our communities. For each generation of wireless technology, the applications and usages have become increasingly ingrained in our societies and have become an established backdrop to our modern lives. 6G will continue to impact our societies and will enable that communication is always possible and information is always available.

2.1.2 Trends and Evolution Towards 6G

Today, when the world is facing several unprecedented challenges in parallel and the prosperity of human society and the long-term survival of mankind are in peril, access to information and the possibility to communicate everywhere are a must. From climate change to global pandemics, social inequalities, misinformation, and distrust in democracy, addressing any challenge that impacts today's global economic, societal, and political agendas requires further and sustainable digitalization of the global economy and society. Infused by emerging and disruptive digital technologies on the horizon, wireless networks are and will be the keystone for enabling such a transformation. The network evolution during this and the next decade will enable a large scale adoption of use cases to sustainably combat our challenges and enable higher economic and societal values at a significantly decreased operational cost.

As the Internet revolution played out over the past decades, with mobile broadband altering our interactions, professions, and habits in unforeseen ways, the true social impact of 6G can only be ascertained in hindsight. Nevertheless, the kernel of its potential can be considered through the current societal and economic trends towards 2030 and beyond, which will be analysed in the following sections. In addition, regulatory and technological trends that are critical for the design and deployment of future networks will be discussed, ensuring the vision and the research work encompass all the essential elements and will lead to a future network design that is deeply rooted in reality and profoundly benefits humanity in the mid-to-long term.

Societal trends towards 2030 and beyond

In 2015, 17 interlinked Sustainable Development Goals (SDGs) were collectively identified by the General Assembly of the United Nations (UN) [1]. Since then, all sectors of society have been called for working towards and delivering on these goals with a timeframe of 2030 and beyond. The Information and Communication Technology (ICT) and wireless network industries have positively contributed to many of the goal areas so far (e.g., to combat poverty and CO_2 emissions), and the potential of further contributing and successfully progressing towards the goals is huge. In developing future networks towards 2030, there is a consensus among major stakeholders from industry, academia, and policy makers around the world: network technology shall support and further accelerate this change for a better and sustainable world, and the network industry will increase its share of contributions and responsibilities to society, enabling significantly increased efficiency in the use of resources and facilitating new and sustainable ways of living in the next decades [2–9].

Economic trends towards 2030 and beyond

Wireless network technology has long been regarded as an important engine for driving global economic growth. As projected in [10], network technology that encompasses 5G and beyond will potentially trigger $13.2 trillion in global sales across ICT industry sectors by 2035, representing 5% of global real Gross Domestic Product (GDP), while 6G value chain will be able to generate 22.3 million jobs globally by 2035. This estimation did not even include the impact of connectivity on non-ICT sectors. As recognized in [11], "the next era of industry will be one where the physical, digital and human worlds are coming together," facing great economic and societal challenges towards 2030. Future networks will be a key enabler for such a revolution with advanced technological capability and human-centric design. New use cases will offer new growth opportunities assuming existing business models, and they can also drive and inspire new business models in an evolving revenue ecosystem.

Regulatory trends towards 2030 and beyond

While the telecommunications sector has been liberalized and privatized in the 1990s, sector regulation continues to be important in conjunction with efficient spectrum access rules, aspects of electromagnetic field (EMF), and assurance of level playing field with platform and cloud operators beyond telco context. Towards 2030, this trend will continue. For example, spectrum management is at the heart of future networks and any wireless technology development, and governments and regulators will have new opportunities due to a wide variety of spectrum bands with highly distinct deployment characteristics and spectrum access models with different levels and needs of spectrum sharing. Another rising issue is EMF exposure. The deployment of 5G technology has started in different areas of the world, and in some regions (including Europe), concerns over EMF exposure fuel the opposition of the public to its rollout [12, 13]. The exposure to EMF is and will be regulated, based on guidelines from the International Commission on Non-Ionizing Radiation Protection (ICNIRP) [14]. Since the beginning of telephony, regulations have played an important role in shaping innovation and the operation of the telecommunications industry, for example, setting the industry to be monopolies in the 1960s, liberalizing the sector with privatization in the 1990s, and setting up new regulations for 5G local and private networks. Future networks will likely combine a range of radio access network (RAN) technologies from macro cells to small cells with very high-capacity short-range links. This calls for refining regulations to resolve inconsistent local approval processes and frequency band assignments to enable dense small cell deployments.

Technological trends towards 2030 and beyond

Previous generations of mobile networks have incessantly increased the performance and capacity to connect and communicate, facilitating global mobile communication and an intertwined global economy accessible at our fingertips. The efficiency gains provided to industries facilitate complex global logistics chains, and the interactive media have engendered a plethora of novel services and industries. This trend is not foreseen to abate in the foreseeable future but is rather expected to expand and encompass even further aspects of our societies.

Technology evolution towards cost reduction and improved efficiency

- **Reimagined network architecture:** As the applications grow more demanding and diverse, the complexity of the deployments and management of the system increases, and the possibility to flexibly and dynamically scale and control the network resources in an efficient way becomes more important. In 5G, the core network (CN) was reimagined into a service-based architecture (SBA), which leverages the virtualization of network functions to only instantiate the functions needed at each instance. Current trends in network architecture point to an extension of the SBA further out into the RAN allowing more flexible and autonomous operation of the network. A new network architecture paradigm for the 6G era is driven by a decomposition of the architecture into platform, functions, orchestration, and specialization aspects [15]. Future network platform will be associated with an open, scalable, elastic, and agnostic heterogeneous cloud, which is data-flow centric and will include hardware acceleration options. Functionally, the convergence of RAN and CNs will help reduce architectural complexity. At the same time, options of flexible offload, extreme slicing, and flexible instantiation of subnetworks will drive the increased level of specialization of the architecture. Of high relevance for the open provision of services and the monetization of resources will be the transformation of orchestration architecture; cognitive closed loops and automation are likely to become pervasive. All future deployment scenarios will rely on a superior transport network and network fabric that is flexible, scalable, and reliable to support demanding use cases and novel deployment options, such as a mixture of distributed RAN and centralized/cloud RAN enabled by AI-powered programmability [2]. The network architecture shall provide the capability to facilitate all the AI operations in the network.

- **Improved network capacity:** The ever-increasing demand for network capacity necessitates the provision of additional bandwidth. The potential to utilize the higher frequency bands, such as the sub-THz (100–300 GHz) range, is currently being explored. However, the radio propagation is significantly attenuated at these frequencies, and the reduced diffraction makes the connection more susceptible to blockage. Coupled with the reduced power efficiency and increased noise at higher frequencies, this compounds to several technological challenges that need to be overcome to provide sufficient coverage.
- **New devices and interfaces:** Future networks will be connected to multitudes of devices and interfaces beyond mobile phones or computers, enabling novel human–machine/machine–machine. New human–machine interfaces created by a collection of multiple local devices will be able to act in unison [3]. In addition, the ubiquity and longevity of Internet of Things (IoT) devices will be further enhanced through zero-cost and zero-energy devices where printable, energy harvesting devices can be deployed anywhere.
- **Network of networks:** In order to capture local and specialized network and sub-network needs, 6G network-of-networks will cover multiple scales of – physical and virtual – networks. The evolution of private and 5G non-public network (NPN), such as campus networks, will expand to support many machines and process with strict requirements on quality of service (QoS) and connectivity, employing edge processing for further automation. With

digital twins (DTs), massive data harvesting from local sensors builds up capillary sub-networks handled by gateways, while in parallel the wide-area network must handle mobility and coverage.

- **Trustworthy networks:** As more and more aspects of our lives, societies, and industries become reliant on mobile connectivity, it becomes imperative to ensure the performance, reliability, and security of the networks so that the services can be used as intended, when needed, without undue connection disturbances or access to private data. This will require the network architecture design to consider the security implications at every step, to avoid a patchwork of solutions after the fact.
- **Sustainable 6G and 6G for sustainability:** As one of the major challenges facing our societies today, the sustainability of our environment, industries, and the society at large must be ensured to be able to reach the sustainable development goals set by, e.g., the United Nations. For 6G, this entails addressing both the first-order effect of the network, referring to the direct environmental impacts of the manufacturing and operating of the networks in terms of energy consumption, CO_2 emission and usage of scarce resources, as well as the second-order effect, referring to how the networks enable improvements in sustainability with, e.g., improved efficiency in industries, or a transition from business travels towards virtual business meetings. However, there are also higher-order effects, also known as rebound effects, that must be considered, where the improved functionality of the mobile networks induces a novel behaviour of the users, which could, e.g., increase the total energy consumption. Moreover, societal sustainability should be addressed, with new services enabled by 6G meeting societal needs and demands [16].

Disruptive technologies that will shape future connectivity

- **Convergence of communications, localization, imaging, and sensing:** With the use of wider bandwidth signals coupled with high band spectrum (>100 GHz) as well as the incorporation of simultaneous localization and mapping (SLAM) with communications at lower frequencies, future networks will be designed integrating high-precision localization (with centimetre-level accuracy), sensing (both radar-like and non-radar-like), and imaging (at millimetre-level) capabilities. This requires the development of highly novel approaches and algorithms to co-optimize communications, sensing, and/or localization.
- **Network intelligence:** The evolution of artificial intelligence (AI)/machine learning (ML) has progressed in the past decades and may bring major disruptions to future networks. Their applications are currently designed for specific tasks, but as the development progresses, more general-purpose applications

emerge. As this development occurs in parallel with and in conjunction with the development of the mobile networks, it is foreseen that there may be several synergies between them. By leveraging on the mobile access, the AI agents can operate in a distributed fashion, gathering, analysing, and acting upon data available across different localities on a much larger scale. At the same time, the AI functionality can be utilized to optimize and enhance the network operations to improve the performance and reduce the operating costs by impacting the design of air interface, data processing, network architecture, and management towards computing for achieving superior performance [3, 7]. It will become essential for the E2E network automation dealing with the complexity of orchestration across multiple network domains and layers [15]. This may also bring forth fundamental changes in how the mobile networks operate, when there are AI agents both managing and operating the networks, as well as transmitting and receiving the information being communicated, and the fundamental tenets of the network architecture design may need to be revisited. For instance, the AI agents may be able to optimize the network behaviour in near real time, which would necessitate the ability to reprogram the network functions. This programmability would go beyond the configurability available today and would allow the modification and introduction of novel functionality into already deployed equipment.

- **Digital twin:** A DT is a digital replica of a living or non-living entity, physical object, or process. The virtual representation reflects all the relevant dynamics, characteristics, critical components, and important properties of an original physical object throughout its life cycle. The creation and update of DTs relies on timely and reliable multi-sense wireless sensing (telemetry), while the cyber–physical interaction relies on timely and reliable wireless control [17] over many interaction points where wireless devices are embedded.

2.1.3 Use Cases: Revolution or Evolution?

In previous generations of mobile networks, the use cases were straightforward: how to provide voice and, later, data communication with increasing bit rates to mobile devices. For the 5G, the use cases were broadened beyond enhanced mobile broadband (eMBB) to include massive machine type communication (mMTC), requiring low data rates with very low power consumption to enable connectivity to billions of simple devices, as well as critical machine type communication (cMTC), later termed ultra-reliable low-latency communication (URLLC), instead requiring extreme reliability and low latencies [18].

With 5G Advanced, extended reality (XR) has been introduced as a prominent use case, which encompasses both virtual reality (VR) and augmented reality (AR).

In XR, different on-body or head-mounted devices can be used to experience the digital world either fully immersed (VR) or overlaid onto the physical world (AR).

With the recent development and deployment of the 5G networks, the current needs in the developed markets appear to be satiable with current technology at least at peak performance, primarily necessitating a network densification to meet the capacity and latency demands.

In 6G, it is foreseen that the incumbent use cases will continue to be prevalent, with access to mobile broadband extended even further, by incorporating non-terrestrial networks and satellites to cost-efficiently reach remote and underserved areas.

Considering again the example of XR use cases, this XR use case is projected to increase in relevance as the devices improve in performance and form factor, and improvements in the network can transition this use case from stationary use near a hotspot towards mobile outdoor use. To ensure the performance in terms of data throughput and low latency, while maintaining the small form factor, it is expected that the XR use case will have to leverage on computational offloading, where significant amounts of the data processing are completed in the network instead of on the device. Similar to this XR example, other 5G use cases will evolve towards extended range of usage, thanks to the developed capabilities offered by 6G, reaching more people and devices and allowing for usage in more extreme conditions.

When it comes to revolutionary use cases, the introduction of novel functionality, such as joint communication and sensing, integrated AI functionality, or energy-neutral devices relying on ambient energy harvesting, could enable unforeseen usages and applications of the future 6G networks. The development of "6G for sustainability" use cases, in collaboration with different sectors and verticals, could also open the way to revolutionary usages. Relying on 6G as a tool to contribute to the reduction of the environmental footprint of other sectors could indeed lead to new roles for 6G and mobile networks, beyond the traditional market of previous generations. Although the International Telecommunication Union Radiocommunication Sector (ITU-R) is working on defining the usage scenarios for 6G, a few possible extensions to the previous usage scenarios can be envisioned, e.g., further enhanced mobile broadband (FeMBB), ultra-massive machine-type communications (umMTC), extremely reliable and low-latency communications (eURLLC), long-distance and high-mobility communications (LDHMC), and extremely low-power communications (ELPC) [19, 20].

6G use cases

The societal and economic trends are driving the identification of relevant use cases for 6G. The Hexa-X project provides a vision on the role of 6G in the evolution

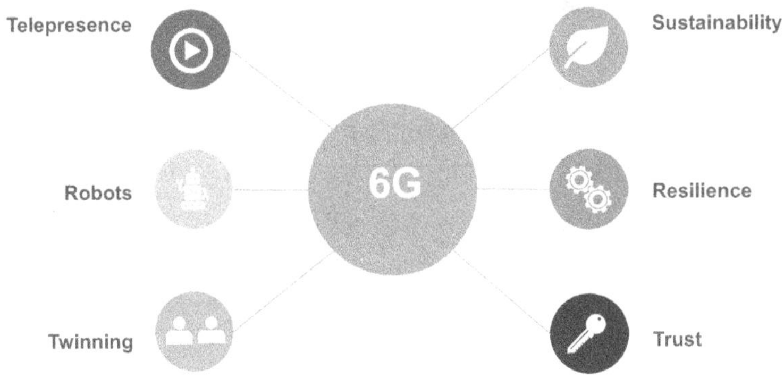

Figure 2.1. Generic use-case families for 6G.

of society [21–23], accounting for these trends and setting the baseline for the identification of use cases. Combining the sets of use cases identified by the various European projects provides an overview of the envisaged usage enabled by 6G. These different use cases can be clustered into broad and generic use case families, encompassing both evolutionary use cases and revolutionary ones, building on new functionalities. These generic use case families can be considered from the perspective of the type of end-user usage involved, as shown in Figure 2.1, such as:

- **Telepresence:** Immersive experience is a central theme for various use cases, with different degrees of immersion, from the evolution of XR experienced with 5G but with increased mobility, reliability, and scale to a fully immersive experience, fully merging physical and digital worlds, with various application areas, both professional and personal (travel, gaming, sports, etc.). This will leverage both the expansions and evolutions of existing technologies, providing connectivity as well as incorporating novel functionalities such as localization, sensing, and computational offloading.
- **Robots:** In parallel with the development of 6G, the evolution of robots and autonomous systems will continue, and robots are envisioned to become part of everyday life, both in professional and personal settings. They will collaborate and interact with each other, but also with humans. The generalization of robots will increase productivity but will also offer solutions to assist humans in their daily lives, meeting societal demands such as care of disabled persons. Although many aspects of this use case may be served by existing technologies, the increased demands for concurrent reliability, high bitrate, and low latency necessitate novel approaches.
- **Twinning:** The concept of DT will be extended in 6G, generalizing the use of the full digital representation of an environment to enhance control,

management, and maintenance of different flows and objects to various activity sectors. To capture, store, analyse, and distribute the digital representation of the environment, it will require a seamless network of unprecedented scale, incorporating sensing, computational offloading, and connectivity with low latency to numerous devices at the same time.

The generalization of these new services will also call for a new generation with increased capabilities to support large deployments.

Other use-case families can be identified according to the research challenges and values addressed:

- **Sustainability:** 6G can be a solution, for various verticals, to enable new use cases contributing to the reduction of their environmental impact (agriculture, industry, logistics, smart city, etc.). 6G can also help meeting societal demands by facilitating access to key institutions and enforcing human rights, such as access to healthcare, education, and reduction of inequalities.
- **Resilience:** Various 6G use cases are built upon resilient infrastructure, guaranteeing the delivery and quality of service despite the complexity of the network and possible situations and events. A resilient 6G network can be an asset to improve and develop key usages (e.g., in the automotive sector) or to develop new usages (e.g., facilitating public protection).
- **Trust:** A high level of trust in 6G networks is a prerequisite to various use cases, involving sensitive information or operations.

Other use-case families can be identified, related to capabilities offered by the network, either related to the management and operation of the network. New use cases can also be enabled, thanks to new capabilities introduced by 6G, such as sensing, positioning, AI processing, or compute capabilities.

Each use-case family encompasses a wide range of usages, from evolutionary ones, extending and enriching the 5G usages with new capabilities, to more disruptive ones, opening up new horizons where 6G could benefit and transform society. 6G use cases can also be evolutionary, relying on improvement of existing technologies, but also revolutionary when introducing new capabilities, such as sensing, AI, and compute capabilities as well as novel devices and interfaces.

6G requirements

Like the use cases, requirements for 6G can also be categorized into the evolution of key performance indicators (KPIs), e.g., higher throughput, lower latency, and revolution of novel KPIs (Figure 2.2). These novel KPIs explicitly focus on the E2E view required by novel 6G use cases, such as E2E dependability or service

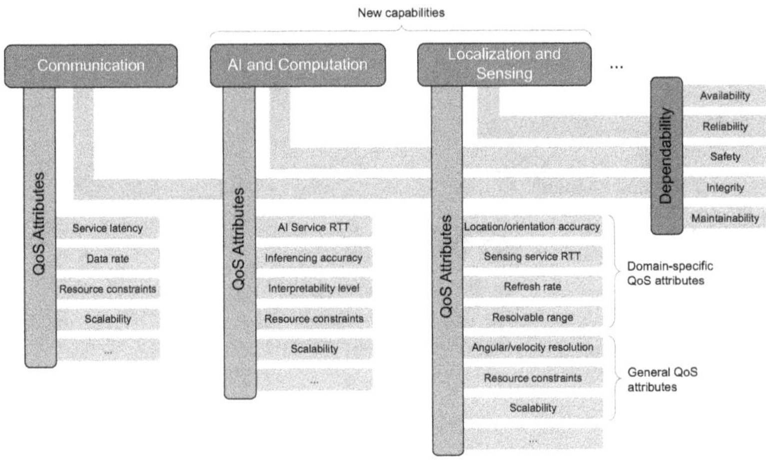

Figure 2.2. Classification of 6G KPIs.

availability. With the envisioned novel capabilities of 6G systems, such as ultra-precise 6D localization, sensing, and artificial intelligence functionality, additional KPIs for these capabilities need to be considered. KPIs for novel capabilities are discussed in [23].

In addition to KPIs, the social and economic trends towards 2030 motivate additional indicators for the fulfilment of key values, such as sustainability, inclusiveness, trustworthiness, and flexibility. To this end, the key value indicators (KVIs) and a methodology for value representation is introduced and described in [21, 22], and [23], with further alignment across different projects towards a unified methodology happening in the 6G-IA [24].

The main point in this methodology is that the use cases introduced at the beginning of this section can contribute to the key values, and a KVI is used to illustrate this. When feasible, it is proposed to use the target level of the UN SDGs as a preferred framework for identifying and detailing the value impact. In some cases, a KVI quantification may be challenging, and then a connection to a KPI can be made to grade or assess a value creation potential and contribution from a use case. For the key value of trustworthiness, a "level of trust" as a KVI is explored, and for flexibility, an association is made to proxies such as scalability requirements/KPIs.

A KVI analysis of a selected set of use cases is included in [23].

2.2 The Need for a New Architecture

This section presents the overall direction that the 6G architecture should move towards to fulfil the trends and technology evolutions. This is done by defining a set of architectural principles and followed by a high-level E2E architecture view.

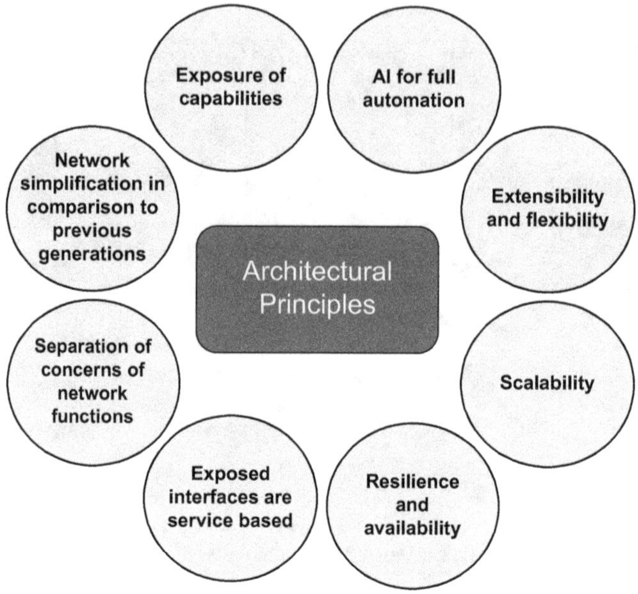

Figure 2.3. 6G architecture principles guiding the architecture design [1].

2.2.1 Architectural Principles

To serve as a guideline when developing the 6G architecture, eight different architectural principles are defined [24, 26]. The order or the numbering of the principles does not indicate the level of importance. Figure 2.3 shows a summary of the eight different principles.

Principle 1: Exposure of capabilities

The architecture solution shall expose new and existing network capabilities to E2E applications and management, such as predictive orchestration. The analytic information can, for example, be performance for predictions, such as latency and throughput, or it can also be localization and sensing information.

Principle 2: AI for full automation

The architecture should support full automation to manage and optimize the network without human interaction. The closed-loop network automation assumes the use of AI/ML.

Principle 3: Extensibility and flexibility

The ability of the network to adapt to various topologies without loss of performance while still enabling easy deployment. This can, for example, be the ability to adapt to new traffic demands, spectrum situations, private networks, and ad-hoc mesh networks.

Principle 4: Scalability

The network architecture needs to be scalable both in terms of supporting very small to very large-scale deployments, by scaling up and down network resources based on needs.

Principle 5: Resilience and availability

The architecture shall be resilient in terms of service and infrastructure provisioning using features, such as multi-connectivity and removing single points of failure.

Principle 6: Exposed interfaces are service based

Network interfaces should be designed to be cloud-native, utilizing state-of-the-art cloud platforms and IT tools in a coherent and consistent manner.

Principle 7: Separation of concerns of network functions

The network functions have a bounded context, and all dependencies among services are through their application programming interfaces (APIs) with a minimal dependency with other network functions, so that network functions can be developed, deployed, and replaced independently from each other.

Principle 8: Network simplification in comparison to previous generations

The network architecture should be streamlined to reduce complexity by utilizing cloud-native upper-layer RAN and CN functions with fewer (well-motivated) parameters to configure and fewer external interfaces.

2.2.2 End-to-end Architecture

Figure 2.4 depicts a high-level view of the envisioned 6G architecture and highlights the key technical enablers. The various building blocks are organized into three layers: **Infrastructure, Network Service (NS), and Application.**

The **infrastructure layer** comprises RAN (addressed more in Chapter 3), CN, and transport networks, which contain radio equipment, switches, routers, communication links, data centres, cloud infrastructure, and so on. The infrastructure layer provides the physical resources to host the NS and application layer elements.

The envisioned 6G infrastructure layer should also contain RAN improvements, such as extremely low latency, high reliability, high availability, high data rate, high capacity, affordable coverage, and high energy efficiency. Extremely high data rate links will be required in some very high-performance applications anticipated in 6G, e.g., immersive smart cities (a use case from telepresence use case family) and fully merged cyber-physical worlds (a use case from twinning use case family) (see Section 2.1). Most of those are related to highly advanced online imaging, including holographic communications as well as providing extreme data rates for high-capacity cells. In those cases, a throughput of 100 Gbit/s or even significantly higher can be required. This means bandwidths of several tens of GHz would be required to provide this. The architecture design, in particular the infrastructure layer, needs to ensure that such data rates can be brought to local small-scale base stations that will serve end users. More details can be found in Chapter 3.

Furthermore, due to the introduction of new use cases and their strict requirements, e.g., immersive smart city [22], the infrastructure layer envisioned for 6G should be able to accommodate new enablers, such as localization and sensing (addressed more in Chapter 4). Joint communication and sensing (JCAS), also known as integrated sensing and communication (ISAC), will be one of the main differentiators of the vision for 6G architecture with respect to 5G communication systems. Sensing not only includes positioning but also encompasses other novel functionalities that were not present in 5G, such as radar-type sensing and non-radar-type sensing using communication technologies, which in turn leads to new services, such as sensing as a service (SaaS), and landscape sensing [23].

The deployment of mobile networks has become increasingly complex and diverse with every new generation. The 6G network of networks should easily and flexibly adapt to new topologies to meet the requirements of both extreme performance and global service coverage well beyond what 5G is capable of. The 6G architecture incorporates different (sub)network solutions into a network of networks. The 6G network should also be able to support very small to very large-scale deployments, by scaling up and down network resources based on needs (see Chapter 5). 4G brought the so-called heterogeneous network (HetNet) solutions, i.e., how networks with both wide-area macro- and small-cell pico-base stations should cooperate. The extension of the radio spectrum into mmWave in 5G added yet another aspect to flexible deployment. 6G deployments will include nodes using even higher sub-THz spectrum (e.g., in the 100–300 GHz frequency range) with limited coverage as well as nodes at low frequencies with seamless coverage. Furthermore, the number of network solutions for capacity and coverage is also expected to increase in the 6G timeframe. These include solutions such as distributed multiple-input multiple-output (D-MIMO) networks, non-terrestrial networks (NTN), campus networks, mesh networks, and cloudification of the network elements. Thus, 6G will be a network of networks.

Even with the new 6G solutions mentioned, the increased use of mobile broadband and digital solutions may require a more densified network, in order to cope with the increased capacity needs. This could lead to an increase in overall emissions unless energy efficiency continues to be addressed.

The envisioned 6G architecture will employ a number of key sustainability enablers to complement the 6G sustainability targets; it is fundamental to jointly take into account all the sustainability aspects of networking, including hardware, planning, deployment, operations, and the entire equipment life cycle. These aspects can be effective in achieving sustainability in all layers and levels of the architecture, namely ***at deployment levels*** that include architectural and hardware innovations, ***at management and orchestration levels*** that target network operation efficiency maximization, ***at service/application layers***, such as application-aware networks, and ***at cross-layer sustainability enablers*** that include innovations in two or more layers. Detailed information on these enablers can be found in Chapter 6.

The **NS layer** is envisioned to be cloud- and micro-service-based with functions and microservices expanded from central cloud to the edge cloud (see Figure 2.4).

One of the key technology enablers of the NS layer is the introduction of the extreme edge cloud (see Figure 2.4)). Extreme edge cloud covers part of the network with high heterogeneity of devices with a wide variety of technologies, in terms of both hardware and software. These devices could be personal devices (smartphones, laptops, etc.) and a huge variety of IoT devices (wearables, sensor networks, connected cars, industrial devices, connected home appliances, etc.). The concepts of edge and extreme edge computing become more and more relevant for the 6G architecture and services. Microservice-based implementation can provide improvements towards a softwarized, intelligent, and efficient 6G architecture. Chapter 5 describes the enablers for an intelligent network. The ultimate target for the 6G architecture is to enable autonomous and adaptable networks, with no (or minimal) human intervention, leveraging cognitive, closed-loop control network functions that can be instantiated on an on-demand basis, even across network domain boundaries. In this sense, an intelligent 6G architecture should be able to define the underlying mechanisms to support embedded AI for 6G and ensure dynamic adaptability of the network architecture to new use cases while keeping the infrastructure and energy costs at acceptable and sustainable levels.

Another important aspect of a more flexible and intelligent network is programmability, addressed in detail in Chapter 7. Programmability can be a tool to introduce new features, especially to deployments that have a limited footprint due to limited hardware types and specific requirements. Over the last decade, programmability is significantly enhanced thanks to the software-defined networking (SDN) paradigm as well as the ongoing trend towards softwarization and cloudification. For 6G architecture, this trend is expected to continue in order to allow third-party developers to interact with the network in new ways, and 6G architecture is expected to ensure reusability and flexibility.

Furthermore, with a cloud-native approach, the RAN and CN architectures can be streamlined, e.g., reduce some complexity by removing multiple processing points for a certain message and removing duplication of functionalities among functions [30].

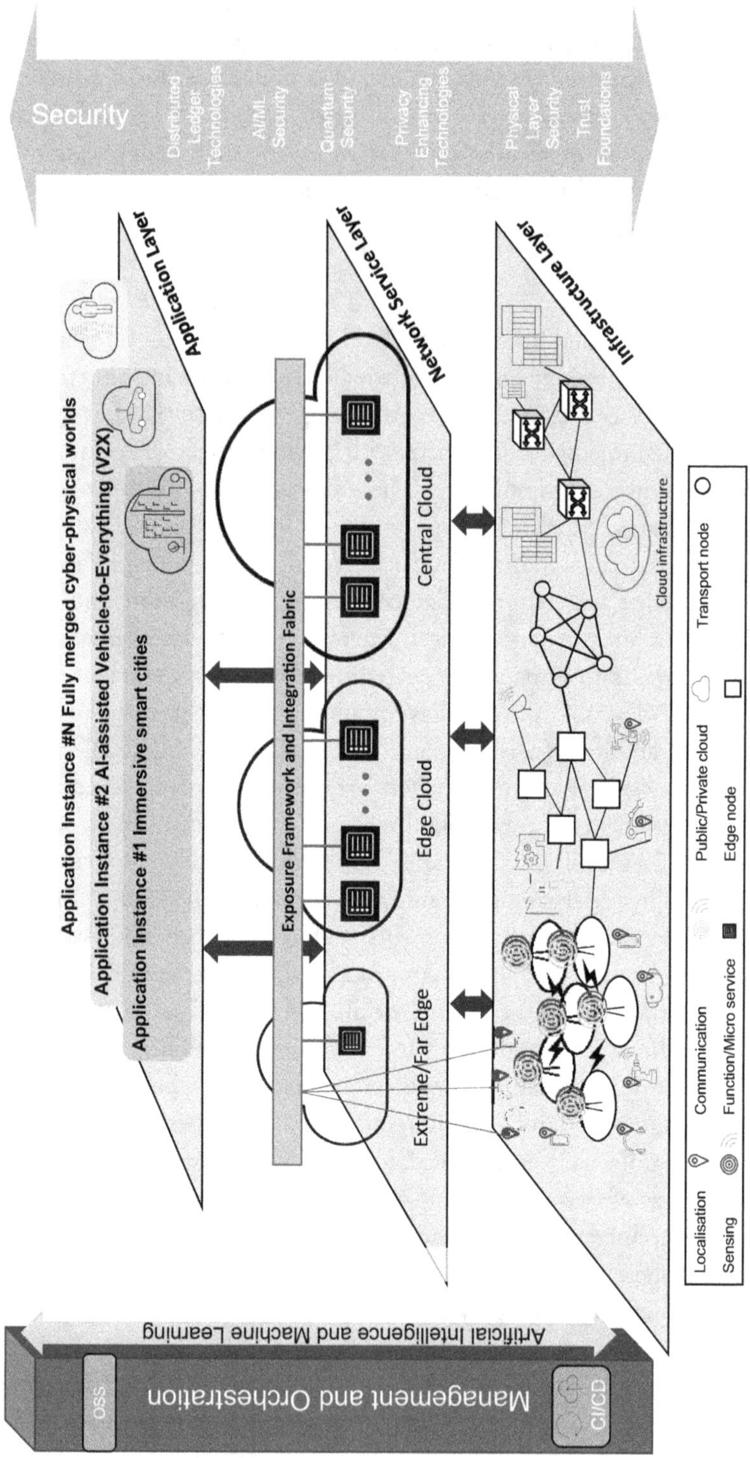

Figure 2.4 End-to-end system view of the 6G architecture [23].

Cloud-native technologies can enable the creation of cloudlets at the edge of the network, with application-to-application and function-to-function communications, which are capable to satisfy a large number of interconnected assets with flexible mesh topologies. Another important aspect of the NS layer is the *exposure framework and integration fabric*. It establishes a communication channel that enables seamless interoperation and networking across different domains.

The 6G architecture will be able to support the strict requirements of various use cases that have been envisioned for the next generation of mobile networks (see Section 2.1.3). In particular, use cases belong to massive twining and telepresence use-case family, e.g., immersive smart city [22] that can be a digital replica of a real traffic scenario of the city, the automated train operations, the control of the utilities (energy, water, gas, etc.), air quality and more are some of the aspects of the implementation of massive twining to city environments. An interactive 4D map can be used to plan utility management, such as public transport, garbage, piping, cabling, buildings, and heating, or to connect many parts of a factory that can be inspected and steered in detail. Similarly, AI-assisted vehicle to everything (V2X) is another example of use cases that can provide high level of safety and security for any transport system, especially road transport due to the prevalence of accidents. This motivates the need to further explore the potentiality of the AI algorithms for enhanced automotive services provided by future 6G networks, and it requires a solid architectural foundation.

Network M&O is gradually moving towards increasing the levels of automation and fully automated closed-loop control. This is supported by the parallel adoption of advancements in AI/ML technologies. The aim of this shift is to provide a framework to optimally support reliability, flexibility, resilience, and availability and addressing changes in the infrastructure, requirements, and failures. More details on the 6G M&O architecture envisioned for 6G can be found in Section 2.4.

Security and privacy mechanisms are integral parts of the overall architecture, affecting all network layers as well as the M&O domain. Figure 2.6 highlights the 6G security technology enablers across different layers [31].

Privacy-enhancing technologies are important on all layers where sensitive data are gathered or processed, and clearly also in the management domain. Similarly, AI/ML security is relevant for all functions making use of AI/ML, in the sense of specifically protecting this use, but also refers to AI/ML-driven security mechanisms, e.g., in the management domain [32]. Finally, distributed ledger technologies are relevant wherever it is required to establish "distributed trust," i.e., trust that is not anchored in a centrally trusted authority, as it may be the case in inter-domain management.

Figure 2.5 shows one possible functional view of the envisioned 6G architecture, which is depicted on the E2E system view of the architecture. It is hierarchically

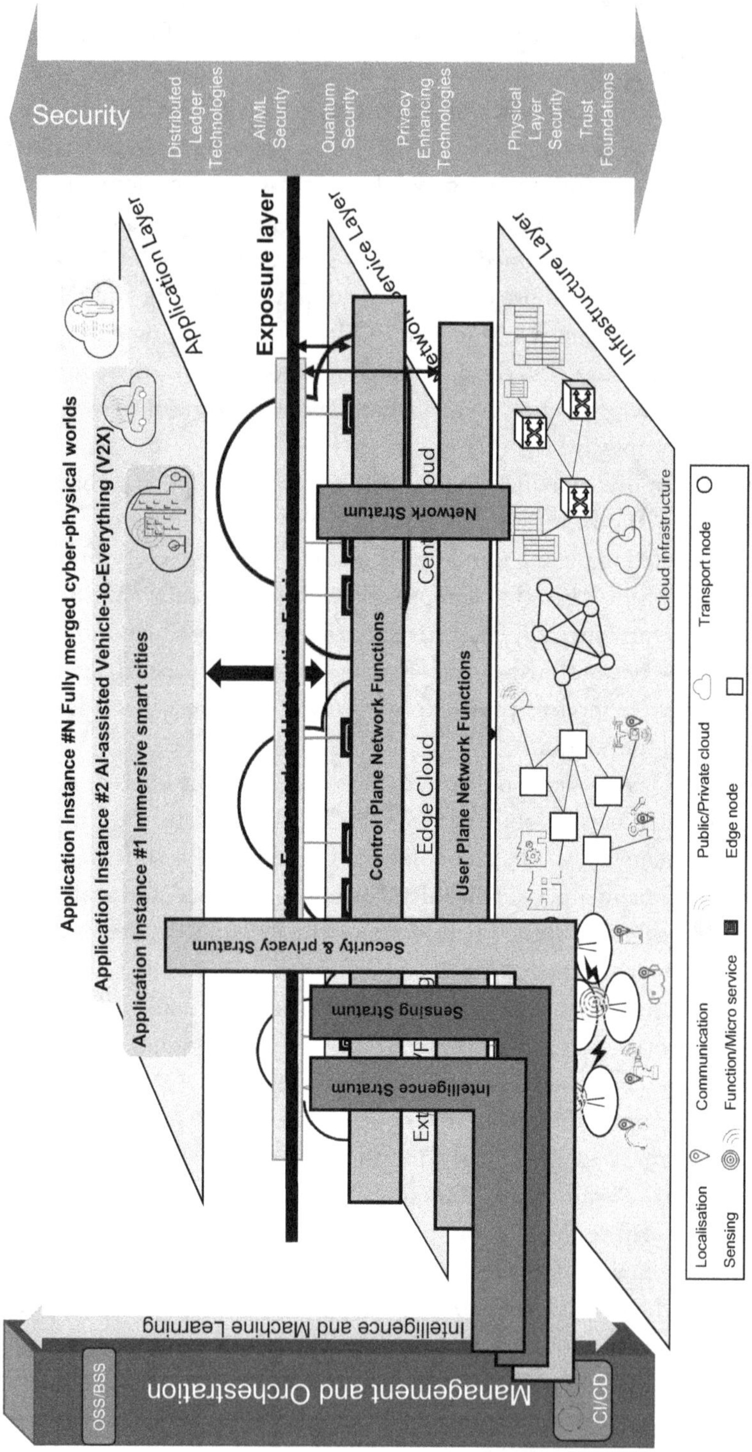

Figure 2.5 Functional view of the proposed 6G reference architecture with the focus on the stratums.

composed of the set of planes that traditionally build the mobile network architecture and has done so since the earliest releases of the 3rd generation partnership project (3GPP) standards. In this context, and by borrowing and extending the terminology from the 3GPP system, a stratum is defined as a set of coordinated functions that is running in different planes or domains of the network. For the proposed functional architecture, four stratums on *Network*, *Sensing*, *Security & Privacy*, and *Intelligence* have been introduced.

Network stratum is consisted of network functions in control plan (CP) and user plan (UP) that are responsible for delivering the expected QoS, efficiently allowing user equipment (UE) to exchange data with the network. CP and UP entail novel access technologies, which may also include the ones leveraging sub-terahertz bands and visible light communications; AI-native air interface, arranged in specific ways (e.g., cell free networks [33, 34]), and even including extreme edge functions like the ones that are managing and reconfiguring intelligent metasurfaces.

Traditionally, the non-access stratum included functions from the UE, UP, and CP. The **network intelligence stratum** encompasses and coordinates functions in all networks, ranging from the intelligent operation of network functions to their autonomous management and orchestration. The network intelligence stratum gathers data and analytics from the **infrastructure layer**.

The infrastructure can be extended to include environmental aspects (i.e., the environment where the infrastructure is deployed, and functions are executed) to allow a tight interaction between the network and the surrounding space. Properly steering beams at very high frequencies or using unmanned aerial vehicle (UAV) to extend the network's coverage requires a **sensing stratum** that can efficiently coordinate functions, harvesting data from fixed landmarks or dynamic laser/light imaging, detection, and ranging (LIDAR) scans, or even using the UP wireless technology as an additional source of sensing, possibly in an energy harvesting fashion.

The last stratum is the **security & privacy stratum**, which manages the cyber security and data privacy aspects of the network. This stratum coordinates functions in all the planes and domains of the network up to the vertical service provider one, which also benefits from the enhanced 6G security and cooperates with it to minimize the attack surface, while allowing the service customers to have full control over the data (including the network one).

Clearly, this richness in the available network functions, which have to be arranged and properly configured according to the network slices they belong to, poses new challenges to the **management plane** of the network, see Section 2.4.

This interaction is possible thanks to the **enhanced exposure interface** between the network and the vertical service providers on the **application layer**, through the use of network applications, which can leverage on data, functions, and procedures offered to support and enhance the user experience. Through the exposure interface, the traditional barrier between operators and service providers is removed, allowing a white-box customization of the vertical services.

2.3 Security & Privacy Architectural Components

Figure 2.6 shows the overall architecture, visualizing the applicable security and privacy components in all areas, and highlighting the specific 6G security technology enablers. While the focus lies on the technology enablers as new architectural components, a holistic 6G security architecture must also comprise today's well-proven security mechanisms, as far as they are still relevant in 6G. On this basis, Figure 2.6 summarizes these components, without the aspiration of exhaustiveness and depth of detail.

Figure 2.6 distinguishes among non-virtualized equipment (for radio access and optical transport), the cloud infrastructure, and the software running on it, including the virtualization layer, the logical network layer, and the management and orchestration functions, including security and risk management and inter-domain management. In each part, the figure shows the most relevant security and privacy building blocks or architectural components, with the new 6G security technology enablers highlighted in red, and the more traditional building blocks, like for example "*Secure SW*," in blue.

Many building blocks apply to multiple areas, e.g., "*Secure SW*" applies to the non-virtualized radio and optical equipment (as far as this equipment comprises software), to the virtualization layer, and to all the software running on it, including M&O functions. As another example, "*Trust foundations*" apply to all hardware, i.e., the radio and optical equipment as well as the cloud infrastructure. On the other hand, some building blocks appear in dedicated places only, like "*Distributed ledger technologies*" appearing at inter-domain management only, but this does not preclude the potential applicability of the building block in other areas. Also, when a building block appears in an area, this does not imply that the building block is always applicable. For example, certain non-virtualized radio access equipment may not have access to sensitive data, so no privacy-enhancing technologies may be required here. As another example, obviously not all transport equipment is required to support quantum key distribution.

The traditional security building blocks may be mostly self-explaining, but note the following:

- "*Secure SW*" refers to software with a low (close to zero) degree of vulnerability. "*Secure HW/FW*" has the same meaning for hardware or firmware. An

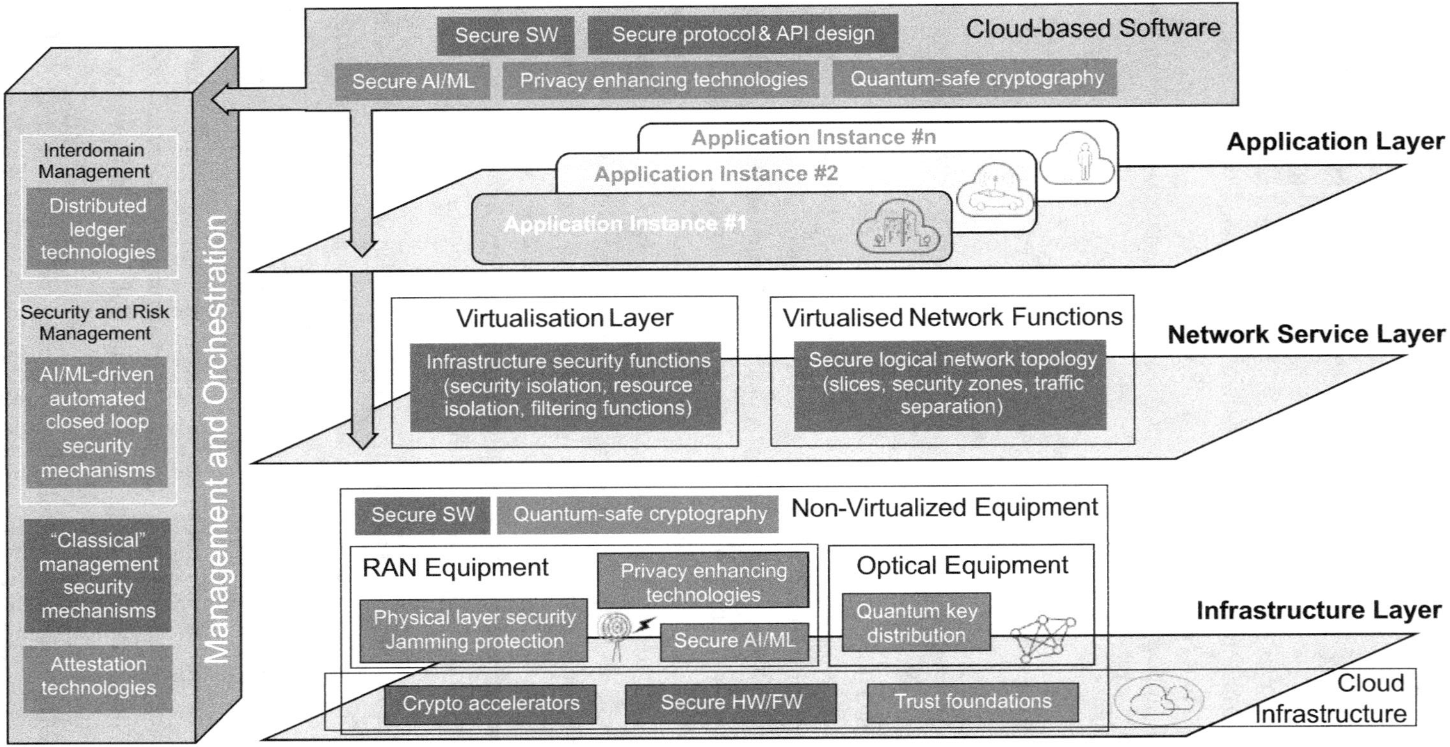

Figure 2.6 Overview of the essential 6G security architectural components [33]. The new 6G security technology enablers highlighted in red, and the more traditional building blocks are in blue.

example is the robustness of a processor against leaking information between different processes running on this processor in a (quasi-) parallel manner.

- "*Secure protocol and API design*" refers to robustness not only against external attackers (which is typically achieved by the use of cryptography), but also against erroneous or malicious behaviour of authorized peers.
- "*Classical management security mechanisms*" comprise well-established mechanisms, such as access control, role-based access, secure logging, isolation of management functions/traffic from all other traffic, etc.

Further details on the components of the security and privacy-preserving technologies are provided in Chapter 8.

2.4 Service Management and Orchestration

Service M&O deals with the deployment and operation of the NSs supplied through the mobile network operator (MNO) to their customers, preserving all of the contractual aspects associated to those services. It addresses the provision of services, QoS and quality of experience (QoE) fulfilment, or fault reporting, among others. In previous generations of the mobile communications systems, the customers of the MNOs have been mainly individuals consuming voice and messaging services. However, the market situation is much more complex now, including new data services and corporate customers, such as vertical industries, digital operators, hyper-scalers, or large-scale content providers, among others. It is anticipated that this trend, in terms of heterogeneity of stakeholders and provided services, could continue and even experience growth within the coming years.

To cope with this complexity, it is needed to enable the services M&O systems with the required capabilities to provide the necessary orchestration resources. Specifically, the following main capabilities have been identified for the future 6G M&O systems.

The adoption of the cloud-native principles also in the M&O system. This would be aligned with the E2E architectural concepts in Section 2.2.2, but from the M&O perspective, it would involve three main aspects: (i) the priority on using micro-services, i.e., light-weight self-contained, independent, and reusable components from different suppliers; (ii) the implementation of the *service mesh* concept, regarding the communication among the network components; and (iii) the enabling mechanisms for the NSs to be deployed/updated using "continuous" DevOps-like practises, e.g., implementing CI/CD workflows with a high automation degree.

Unified M&O across multiple domains that could be owned/administered by multiple stakeholders and featured with heterogeneous technology resources. This entails the definition of converging interfaces, the mechanisms to dynamically check and expose the different resources and capabilities from each domain, and the access control procedures for consuming the various primitives and services.

Increased degree of automation to strongly reduce manual interventions regarding the functionalities of service and network planning, design, provisioning, optimization, and operation/control, leveraging closed-loop and zero-touch responses. The M&O system needs to be able to identify, detect, or predict potential issues, triggering automatic reactions.

Adoption of data-driven and AI/ML techniques in the M&O system. AI/ML techniques could cover numerous optimization aspects and lifecycle actions concerning the services M&O, including resource allocation and slice sharing at provisioning time, service composition, scaling, migration, re-configuration, and re-optimization of NSs, among others.

Intent-based approaches for service planning and definition. In order to help with the extended complexity, the M&O system would implement automated mechanisms for translating service specifications and commands based on high-level intents, which might be expressed even in natural language (e.g., relying on AI/ML techniques).

To meet these main challenges, the M&O system is seen as a common functionality impacting all layers of the E2E architecture: from the infrastructure up to the applications (see Figure 2.7). In this regard, an initial high-level M&O architectural design for the future 6G networks has been produced. This architectural design takes the previous 5G architectural view from the 5G PPP Architecture Working Group as a baseline [34, 35] represents the structural view of this architecture, with the main building blocks grouped in different layers.

The NSs and slices at the service layer (top in Figure 2.7) are executed on the infrastructure layer (bottom) through the network functions at the network service layer (middle). All these elements (network functions, services, and slices) are designed and provided from the design layer (right).

A new layer, named the *design layer*, has been included to represent the M&O-related operations involving third-party software providers. This is intended to introduce the well-known DevOps-like practises (e.g., continuous integration and continuous delivery/continuous deployment (CI/CD)) in the telco-grade environment. Also, hyperscalers, private networks, and the extreme edge domain have been explicitly included as part of the infrastructure layer. New control loops have been included: (i) the "DevOps control loop," representing the automated continuous iterations (e.g., CI/CD) between the MNO scope (grey colour) and the external design layer (light blue colour); and (ii) the "infrastructure control loop," meant to automate the infrastructure discovery processes and the related monitoring methods targeting the extreme edge asset integration (which can be potentially asynchronous in terms of connection/disconnection of devices, so requiring special

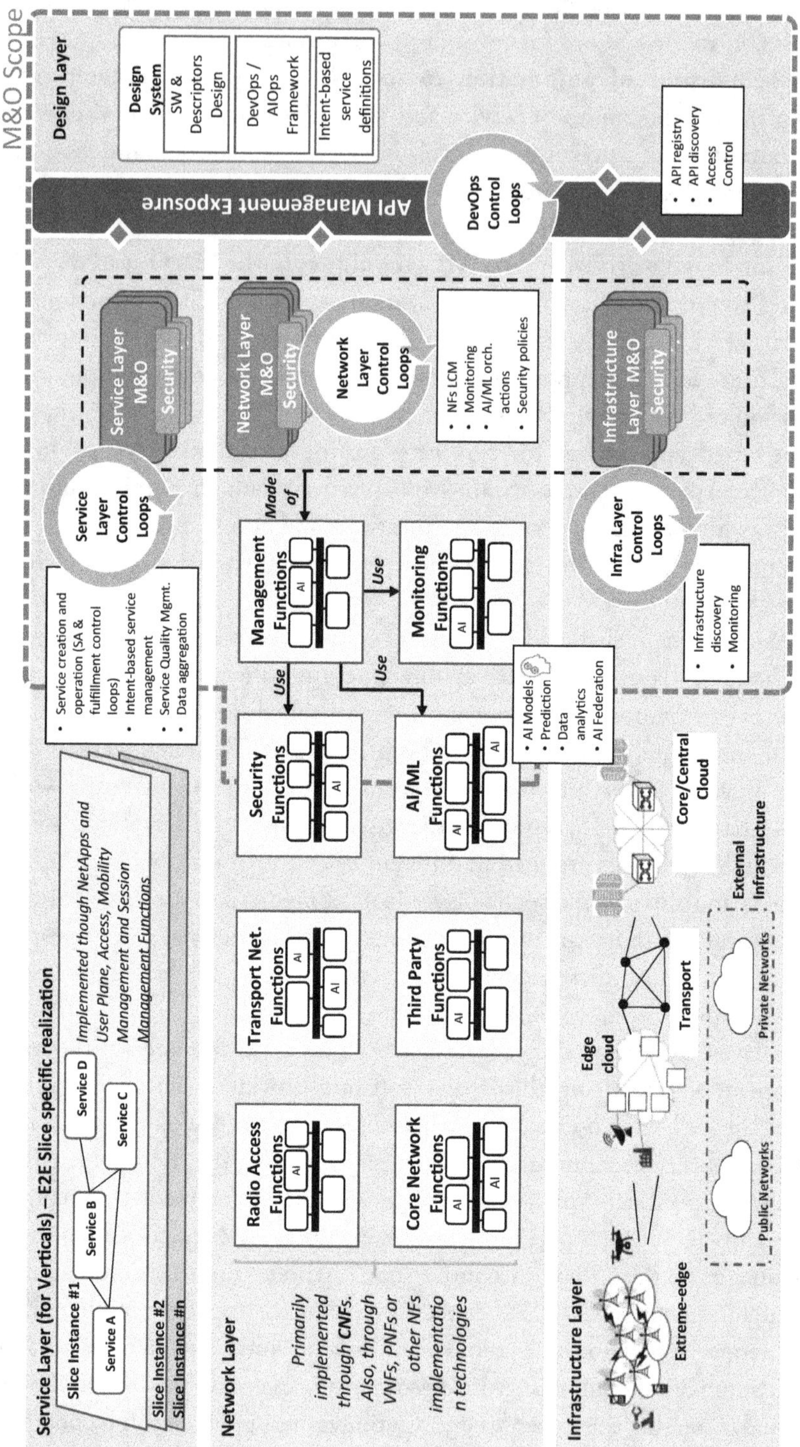

Figure 2.7 Proposed 6G management & orchestration system – structural view [34].

processes for their management). As in the baseline architecture in [35], NFs are associated in different groups at the network layer (e.g., radio access functions, CN functions, M&O functions, AI/ML functions, etc.). However, following the cloud-native practises, these functions would be primarily implemented through containerized NFs (CNFs), although also through virtualized NFs (VNFs), physical NFs (PNFs), or other NF implementation technologies (e.g., to ensure backward compatibility). It should be noticed here that, although some functions work only as managed resources (e.g., CN functions or third-party functions), others are specific M&O resources (e.g., the monitoring functions or the management functions themselves); however, other functions are *hybrid*: they can support M&O resources (e.g., certain security-related or AI/ML functions) or work as *pure* managed resources (e.g., certain AI as a Service (AIaaS) functions or security functions not in the M&O scope). Functions in the network layer are generic, i.e., instead of referring specific functions (e.g., communication service management function (CSMF), media resource function (MRF), NFV orchestrator (NFVO), etc.) as in [35], just generic blocks are provided. This is intentional, in order to consider the new functions that would be probably defined for the future 6G stack. A new set of AI/ML collaborative components have been distributed across the network covering both managing and managed scopes. M&O functions can be instantiated in the three different layers (service, infrastructure, and network layers), including specific security-related functions. Finally, and also aligned with the cloud-native approach, a new cross-layer API management exposure block has been included to communicate the different network elements in the different network layers. In short, it mimics the behaviour of the zero-touch service management (ZSM) *cross-domain integration fabric*, enabling the so-called *capabilities exposure* of the network of elements in the various architectural layers. It makes possible communicating the various M&O resources within and between administrative domains, although it could be applied more broadly to represent potential federated interactions.

2.5 Summary and Outlook

This chapter discusses the current architectural trends and technologies for the future 6G network. Motivated by the surge of new requirements stemming from societal trends and use cases, a set of architectural principles has been introduced, and new architectural and technical enablers needed for the 6G architecture have been identified. A high-level view of a possible E2E system of the 6G architecture as well as a functional view is described. Thereafter, a description on how the enablers fit into the system view is given, which is also an overview of the content in this book. The chapter dives into the security and privacy area in a bit more detail and gives an overview of the 6G security and privacy architectural components. Finally, the main capabilities needed for a future 6G M&O systems are discussed.

References

[1] United Nations (UN), "Transforming our world: the 2030 Agenda for Sustainable Development," 21 October 2015. Accessed: April 6, 2023. [Online]. Available: https://sdgs.un.org/2030agenda.

[2] Ericsson, "6G – Connecting a cyber-physical world," February 2022. Accessed: April 6, 2023. [Online]. Available: https://www.ericsson.com/4927de/assets/local/reports-papers/white-papers/6g--connecting-a-cyber-physical-world.pdf.

[3] Nokia, "Communications in the 6G era," 9 September 2020. Accessed: April 6, 2023 [Online]. Available: https://onestore.nokia.com/asset/207766.

[4] GSM Association, "2022 Mobile Industry Impact Report: Sustainable Development Goals," September 2022. Accessed: April 6, 2023. [Online]. Available: https://www.gsma.com/betterfuture/wp-content/uploads/2022/11/2022-SDG-Impact-Report.pdf.

[5] M. Matinmikko-Blue, S. Aalto, M. I. Asghar, H. Berndt, Y. Chen, S. Dixit, R. Jurva, P. Karppinen, M. Kekkonen, M. Kinnula, and P. Kostakos, *White paper on 6G drivers and the UN SDGs*, University of Oulu, June 2020. Accessed: April 6, 2023. [Online]. Available: http://urn.fi/urn:isbn:9789526226699.

[6] G. Fettweis, "6G – Just a better 5G?: 2020 5G World Forum keynote series," 18 November 2020. Accessed: April 6, 2023. [Online]. Available: https://ieeetv.ieee.org/2020-5g-world-forum-keynote-gerhard-fettweis.

[7] N. Demassieux, "6G. Why?," In *EuCNC 2020*, Dubrovnik, Croatia, 2020. Accessed: April 6, 2023. [Online]. Available: https://www.eucnc.eu/wp-content/uploads/2020/07/2020-05-29-6G.-Why-Nicolas-Demassieux-EUCNC-VDEF.pdf.

[8] Samsung, "The Next Hyber-Connected Experience for All," 14 July 2020. Accessed: April 6, 2023. [Online]. Available: https://cdn.codeground.org/nsr/downloads/researchareas/20201201_6G_Vision_web.pdf.

[9] NTT DOCOMO, "White paper 5G Evolution and 6G, version 5.0," January 2023. Accessed: April 6, 2023. [Online]. Available: https://www.docomo.ne.jp/english/binary/pdf/corporate/technology/whitepaper_6g/DOCOMO_6G_White_PaperEN_v5.0.pdf.

[10] Raconteur, "The economic impact of 5G," 2020. Accessed: April 6, 2023. [Online]. Available: https://www.raconteur.net/infographics/the-economic-impact-of-5g/.

[11] European Commission, "A New Industrial Strategy for Europe," 10 March 2020. Accessed: April 6, 2023. [Online]. Available: https://eur-lex.europa.eu/legal-content/EN/TXT/?uri=CELEX:52020DC0102.

[12] Arcep, "Networks and the Environment," 8 September 2020. Accessed: April 6, 2023. [Online]. Available: https://en.arcep.fr/fileadmin/cru-1677573101/user_upload/37-20-english-version.pdf.

[13] The Connexion, "Lille issues 'moratorium' on 5G technology," 11 October 2020. Accessed: April 6, 2023. [Online]. Available: https://www.connexionfrance.com/article/French-news/Lille-issues-moratorium-on-controversial-5G-technology-pending-2021-Anses-report.

[14] International Commission on Non-Ionizing Radiation Protection, 2023. Accessed: April 6, 2023. [Online]. Available: https://www.icnirp.org/.

[15] V. Ziegler, H. Viswanathan, H. Flinck, M. Hoffmann, V. Räisänen, and K. Hätönen, "6G architecture to connect the worlds," In *IEEE Access*, vol. 8, pp. 173508–173520. Sep. 2020.

[16] Orange, "Orange inaugura tienda en el metaverso," 13 September 2022. Accessed: April 6, 2023. [Online]. Available: https://blog.orange.es/innovacion/orange-inaugura-tienda-en-el-metaverso/.

[17] R. Minerva, G. M. Lee, and N. Crespi, "Digital twin in the IoT context: A survey on technical features, scenarios, and architectural models," In *Proceedings of the IEEE*, vol. 108, no. 10, pp. 1785–1824, 2020.

[18] European Telecommunications Standards Institute (ETSI), "About ETSI," 2023. Accessed: April 6, 2023. [Online]. Available: https://www.etsi.org/about.

[19] M. Rasti, S.K. Taskou, H. Tabassum, and E. Hossain, "Evolution toward 6g multi-band wireless networks: A resource management perspective," In *IEEE Wireless Communications*, vol. 29, no. 4, pp. 118–125.

[20] Z. Zhang, Y. Xiao, Z. Ma, M. Xiao, Z. Ding, X. Lei, G. K. Karagiannidis, and P. Fan P, "6G wireless networks: Vision, requirements, architecture, and key technologies," In *IEEE Vehicular Technology Magazine*, vol. 14, no. 3, pp. 28–41, Jul 2019.

[21] Hexa-X, "D1.1 – 6G Vision, use cases and key societal values," 28 February 2021. Accessed: April 6, 2023. [Online]. Available: https://ec.europa.eu/research/participants/documents/downloadPublic?documentIds=080166e5d9f611f1&appId=PPGMS.

[22] Hexa-X, "D1.2 – Expanded 6G vision, use cases and societal values – including aspects of sustainability, security and spectrum," 30 April 2020. Accessed: April 6, 2023. [Online]. Available: https://ec.europa.eu/research/participants/documents/downloadPublic?documentIds=080166e5dc8b611b&appId=PPGMS.

[23] Hexa-X, "D1.3 – Targets and requirements for 6G – initial E2E architecture," 28 February 2022. Accessed: April 6, 2023. [Online]. Available: https://ec.europa.eu/research/participants/documents/downloadPublic?documentIds=080166e5e90431aa&appId=PPGMS.

[24] 6G Infrastructure Association, Vision and Societal Challenges Working Group, Societal Needs and Value Creation Sub-Group, "What societal values will 6G address?," May 2022. Accessed: April 6, 2023. [Online]. Available: https://5g-ppp.eu/wp-content/uploads/2022/05/What-societal-values-will-6G-address-White-Paper-v1.0-final.pdf.

[25] Hexa-X, "D5.1 – Initial 6G architectural components and enablers," December 2021, Accessed: April 6, 2023. [Online]. Available: https://ec.europa.eu/research/participants/documents/downloadPublic?documentIds=080166e5e652c1f2&appId=PPGMS.

[26] M. Ericson, S. Wänstedt, M. Saimler, H. Flinck, G. Kunzmann, P. Vlacheas, P. Demestichas, D. Rapone, A. De La Oliva, C. J. Bernardos, and R. Bassoli, "Setting 6g architecture in motion–the hexa-x approach," In *2022 Joint European Conference on Networks and Communications & 6G Summit (EuCNC/6G Summit)*, June 2022.

[27] Hexa-X, "D3.2 – Localisation and sensing use cases and gap analysis," September 2022. Accessed: April 6, 2023. [Online]. Available: https://ec.europa.eu/research/participants/documents/downloadPublic?documentIds=080166e5e648d02a&appId=PPGMS.

[28] Hexa-X Deliverable "D5.2 – Analysis of 6G architectural enablers' applicability and initial technological solutions," October 2022, Accessed: April 6, 2023. [Online]. Available: to appear in https://cordis.europa.eu/project/id/101015956/results.

[29] Hexa-X Deliverable "D2.1 – Towards Tbps Communications in 6G: Use Cases and Gap Analysis" June 2021. Accessed: April 6, 2023. [Online]. Available: https://ec.europa.eu/research/participants/documents/downloadPublic?documentIds=080166e5df2529fd&appId=PPGMS.

[30] E. U. Soykan, L. Karaçay, F. Karakoç, and E. Tomur, "A Survey and Guideline on Privacy Enhancing Technologies for Collaborative Machine Learning," In *IEEE Access*, vol. 10, pp. 97495–97519, Sep 2022.

[31] N. Rajatheva, I. Atzeni, E. Björnson, A. Bourdoux, S. Buzzi, J. B. Dore, S. Erkucuk, M. Fuentes, K. Guan, Y. Hu, and X. Huang, J. Hulkkonen, J. M. Jornet, M. Katz, R. Nilsson, E. Panayirci, K. Rabie, N. Rajapaksha, M. J. Salehi, H. Sarieddeen, T. Svensson, O. Tervo, A. Tolli, Q. Wu, and W. Xu, "White paper on broadband connectivity in 6G," In *arXiv preprint* arXiv:2004.14247. April 2020.

[32] H. Q. Ngo, A. Ashikhmin, H. Yang, E. G. Larsson, and T. L. Marzetta, "Cell-free massive MIMO versus small cells," In *IEEE Transactions on Wireless Communications*, vol. 16, no. 3, pp. 1834–1850, Jan 2017.

[33] M. B. Khorsandi, R. Bassoli, G. Bernini, M. Ericson, H. F. Fitzek, A. Gati, H. Harkous, H. Hoffmann, I. L. Pavon, G. Landi, and D. Lopez, "6G E2E Architecture Framework with Sustainability and Security Considerations," In *2022 IEEE Globecom Workshops (GC Wkshps)* Dec 4, 2022.

[34] Hexa-X Deliverable "D6.2 – Design of service management and orchestration functionalities", Apr. 2022. Accessed: April 6, 2023. [Online]. Available: to appear in https://cordis.europa.eu/project/id/101015956/results.

[35] 5GPPP, "Architecture Working Group – View on 5G Architecture", Version 4.0, August 2021. Accessed: April 6, 2023. [Online]. Available: https://5g-ppp.eu/wp-content/uploads/2021/11/Architecture-WP-V4.0-final.pdf.

DOI: 10.1561/9781638282396.ch3

Chapter 3

Towards Versatile Access Networks

By Mir Ghoraishi, et al.[1]

3.1 Introduction

Compared to its previous generations, the 5th generation (5G) cellular network features an additional type of densification, i.e., a large number of active antennas per access point (AP) can be deployed. This technique is known as massive multiple-input multiple-output (mMIMO) [1]. Meanwhile, multiple-input multiple-output (MIMO) evolution, e.g., in channel state information (CSI) enhancement, and also on the study of a larger number of orthogonal demodulation reference signal (DMRS) ports for MU-MIMO, was one of the Release 18 of 3rd generation partnership project (3GPP Rel-18) work item [2]. This release (3GPP Rel-18) package approval, in the fourth quarter of 2021, marked the start of the 5G Advanced evolution in 3GPP [3]. The other items in 3GPP Rel-18 are to study and add functionality in the areas of network energy savings, coverage, mobility support, multicast broadcast services, and positioning [2].

The 6th generation (6G) cellular network will continue to modernize the existing radio access networks (RANs) towards satisfying 6G network's key performance

1. The full list of chapter authors is provided in the Contributing Authors section of the book.

indicators (KPIs) [4]. Future networks will likely combine a range of RAN technologies from macro cells to small cells with very high-capacity, short-range links. This calls for dense small cell deployments, especially for throughput-demanding use cases required simultaneously by a high proportion of people in populated areas such as dense cities [5]. On the other hand, the technology's evolution towards cost reduction and improved efficiency requires all future deployment scenarios to rely on a superior transport network and network fabric that is flexible, scalable, and reliable to support demanding use cases and novel deployment options, such as a mixture of distributed RAN and centralized/cloud RAN enabled by artificial intelligence (AI)-powered programmability. The benefit of cell densification is to achieve a certain area capacity using less complicated hardware at the expense of using more APs (adding to the infrastructure requirement), which then means more interference. Distributed MIMO (D-MIMO) is a technology that combines the best aspects of ultra-dense cellular networks with MIMO technology to enjoy the strengths of both technologies [6].

From a technical perspective, an increased number of cooperating APs will pose new challenges for providing front/back-haul access to all nodes. On the other hand, at mmWave and sub-THz frequencies, a lot of bandwidth will be available (a total of 17.25 GHz of spectrum has been identified between 24 and 86 GHz [6, 8]), and links will be supported by very directive antennas limiting the amount of interference. This opens the door for integrated access and backhaul (IAB) solutions, sharing the same resources, but requiring potentially new beamforming and scheduling concepts [9]. It is foreseen that the wireless backhaul capacity will be enhanced in 6G by orders of magnitude, enabling the cost-efficient provision of 6G services using IAB even to remote areas that are currently cost prohibitive [5]. One cost-effective and backward-compatible method in densifying the cellular network is by combining different radio access technologies via multi-connectivity technology [10]. This technology was introduced originally as dual-connectivity in 3GPP Rel-11; nevertheless, it is still viable beyond the 5G era [11]. Another technology considered for 6G networks is reconfigurable intelligent surface (RIS), comprising an array of reflecting elements for reconfiguring the incident signals. Owing to their capability of proactively modifying the wireless communication environment, RISs have become a focal point of research in wireless communications to mitigate a wide range of challenges encountered in diverse wireless networks [12]. RIS technology is attractive due to its several advantages, namely, ease of deployment, spectral efficiency enhancement, and environmental affability [13].

There are yet several envisioned 6G use cases that require an extreme data rate in the order of 100 Gb/s even up to 1 Tb/s in specific scenarios. This cannot be fulfilled with current communication standards or realized by a simple evolution of available wireless technologies, rather it requires new technologies beyond the capabilities of existing systems [14]. One approach for achieving such an extreme data rate relies on exploiting ultra-wide bandwidth of multiple GHz, which is available in the frequency range above 100 GHz, especially in the range 100–300 GHz, which is denoted as sub-THz [15].

The current chapter, "Towards Versatile Access Networks," provides an overview of the technologies introduced above, namely, D-MIMO, IAB, RIS, multi-connectivity, and sub-THz, in some more detail. The reader will notice that the chapter provides a non-exhaustive list of candidate technologies for the 6G RAN.

3.2 Distributed MIMO

3.2.1 What is D-MIMO for 6G?

Multi-antenna system technologies have evolved during every generation of wireless networks. After single-input single-output (SISO) systems, point-to-point MIMO has been introduced in the 3rd generation mobile communications (3G), and subsequently, the 4th generation (4G, or *long-term-evolution advanced*, LTE-Advanced) system is based on multi-user MIMO. Densification, i.e., increasing the number of base stations (BSs) per unit of space, is one of the main techniques that resulted in improving spectral efficiency in 4G and 5G cellular networks [1]. The drawback of this solution is high interference that negatively affects the performance of cell-edge users [15]. Networks of next generation will have to deal with even higher density of infrastructure to provide the expected performance [14]. This requires a rethinking of the underlying architecture to eliminate the cell boundaries [15, 17].

Heterogeneous networks (HetNets) have been utilized as an alternative to the homogenous cellular network architecture containing BS and possessing similar properties, where various transmission nodes (e.g., pico-cell, femto-cell, and remote-radio-head (RRH)) are installed within the same macro-cell area to improve the quality of service of cell-edge users as well as the system-wide service quality. These transmission nodes are network components that work in coordination with the central BS, but also differ from the BS in various aspects, such as transmit power levels and hardware. For example, RRHs are composed of antennas and RF amplifiers, and are connected to the BS via fibre cables or radio links using baseband signals, which transform the macro-cell deployment into a distributed antenna system concept.

The 5G NR standard introduces mMIMO, where each BS is equipped with many antenna elements, enabling it to serve numerous user equipment (UE) simultaneously by means of highly directional beamforming techniques. This approach is benefiting from channel hardening and favourable propagation utilizing the deterministic channel and, hence, potentially eliminates the need for combating small-scale fading [18–21].

Joint Transmission Coordinated Multi-Point (JT-CoMP), which enables coherent transmission from clusters of BSs to overcome the inter-cell interference within each cluster [20, 22], has also attracted the attention during the past 10 years; however, it did not become part of the 5G NR standard, as LTE standardization [23] did not deliver significant gains in practical deployments. This can be mainly attributed to the considerable amount of backhaul signalling for CSI and data sharing resulting from a network-centric approach to coherent transmission [24], whereby the BSs in a cluster cooperate to serve the UE in their joint coverage region. The practical implementation of JT-CoMP was also hindered by other attributes of LTE, such as frequency division duplex (FDD) operations and a rigid frame/slot structure, which did not allow for effective channel estimation.

One question for the 6G is what the next evolution of MIMO is going to be, and what the demands and needs of 6G networks from multi-antenna systems will be. 6G should provide limitless connectivity with both functional and deployment values [25]. Among other licenced bands, e.g., mid-bands, it will likely utilize mmWave frequencies and is expected to provide high spectral efficiency and predictable quality of service (QoS) to the UE. To ensure more consistent quality and non-intrusive, flexible, and robust networks, multi-point transmission is expected to become common [26]. It is envisioned that joint transmission and reception via spatially separated transceivers are going to be vital in upcoming systems [27].

Cell-free (CF) mMIMO [24, 28] combines the elements of small cells [20], mMIMO [21], and UE-centric JT-CoMP [20, 29], as illustrated in Figure 3.1. In a CF context, the mMIMO regime is achieved by spreading many antenna elements across the network (even in the form of single-antenna BSs [17, 30]), which provide enhanced coverage and reduced path loss. Moreover, a UE-centric coherent transmission extended to the whole network, where each UE is served jointly by several BSs, allows to practically eliminate the interference, as shown in [31].

Such a large-scale D-MIMO system, which can be thought of as the ultimate embodiment of concepts, such as network MIMO [31], multi-cell MIMO cooperative networks [22], virtual MIMO [32], ultra-dense networks, and JT-CoMP, is now regarded as a potential physical-layer paradigm shift for 6G networks [33].

A set of transmission nodes (APs) is assumed to be connected to a central processing unit (CPU) via fronthaul links, e.g., high-capacity fibre-optic cables, which convey both the UE-specific data and processing weights that enable network-wide processing for the computation of the AP-specific precoding strategies [24], see Figure 3.1. Since APs can perform channel estimation and distributed precoding locally, D-MIMO constitutes a scalable way to implement the network MIMO concept. Moreover, precoding can be designed by using channel estimates acquired

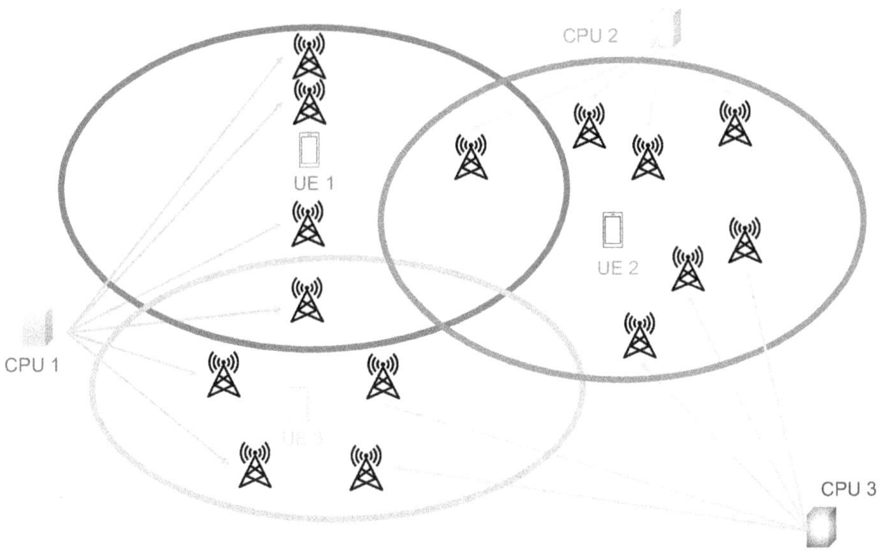

Figure 3.1. Illustration of distributed MIMO.

via uplink pilots by leveraging the channel reciprocity of time division duplex (TDD) operation; therefore, the overhead due to the channel estimation is independent of the number of APs and scales with the number of transmission layers. APs pertaining to different CPUs may serve a given UE.

As depicted in Figure 3.1, APs pertaining to different CPUs may serve a given UE, resulting in a distributed communication computation platform. Such a platform offers several features that are very well suited to support the diverse novel interactive use cases envisioned for 6G [34]:

1. Local connectivity-computational resources: many "mixed reality" applications rely to a significant extent on local content, and distributed processing can increase efficiency and reduce bottlenecks both regarding bandwidth and energy.
2. Proximity: excellent connectivity can be achieved with drastically lower transmit power. Applications can benefit from the distributed, local computational capabilities in the access infrastructure. The expenditure of overall network bandwidth can be drastically reduced as the dependence on medium- and long-distance connections will be minimized, resulting in very efficient usage of network bandwidth and energy. Interaction with energy-neutral devices essentially and realistically needs charging features to be close to these devices.
3. Redundancy: retransmissions can be avoided to achieve unperceivable latency and ultra-reliable connections.

4. Diversity: they can offer consistently excellent service levels that are essential for robust, innovative interactive applications, requiring imperceptible latency and zero outages. Precise and accurate indoor positioning can benefit from hyper-diversity, and favourable propagation conditions can be created to achieve consistent good communication quality and extensive spatial multiplexing.

The terminology for the transmission nodes in D-MIMO needs to be clarified, where both BS and AP are used in the literature; however, without expressing the functions placed in the units, any term like BS or AP does not provide sufficient information. Since BS can be interpreted with different functionalities, throughout this context, it is preferred to use the term AP as a distributed antenna or transmission node. A gNodeB (gNB) is introduced with functional splits in 5G era, e.g., radio unit (RU), distributed unit (DU), and centralized unit (CU), and the splits in D-MIMO have not been defined yet. One interpretation is that the APs may have little processing capability, and CPU may be the DU as it involves low-level functions. Accordingly, in such an interpretation, multiple CPUs may connect to CU. Herein, DU may be further split by centralizing the medium access control (MAC) and radio link control (RLC) layers with reference to the physical layer (PHY). In investigations, where the functional split is not fundamental, BS/AP and AP can be used.

3.2.2 D-MIMO Potential

At lower carrier frequencies (sub-6 GHz), where coherent transmission is possible, D-MIMO can be used to increase the spectral efficiency of the system and, in principle, to avoid inter-cell interference. Moving up in frequency, available bandwidth becomes larger, and spectral efficiency is not necessarily the main concern anymore. Instead, the reliability of the communication links becomes a primary concern. Reliability is impacted by the higher path loss, lower available output power of semiconductors, narrower antenna beams, and most importantly, a higher level of signal blockage. On the other hand, the feasibility of practical implementation highly depends on the RF hardware capabilities and other constraints, such as size, power source, and mobility. Moreover, the responses of different hardware components are influenced by the centre frequency, bandwidth, and waveform. Furthermore, the beamforming architecture and the possibility of exploiting spatial multiplexing depend on the radio channel characteristics, which need extensive measurement and modelling.

As, with D-MIMO, a link between the network and UE is provided by multiple APs, the likelihood that a link or combination of links with minimal blockage is available increases with *macro-diversity*. Hence, D-MIMO offers great potential to exploit the spatial multiplexing gain of the channel and, at the same time, mitigate both unreliable links due to blockage and increased path loss. D-MIMO also allows for the densification of APs with, in the ideal case, no increase in interference. An increased density of operating APs will further reduce the likelihood of blockage and will also be necessary to have sufficient link margin due to output power limitations and increased path loss at mmWave and sub-THz frequencies.

3.2.3 D-MIMO: Roll-out Considerations

The appeal of D-MIMO as a 6G solution is its high degree of macro-diversity that results in a predictable service quality over the entire service area. The line-of-sight (LoS) probability for D-MIMO is very high, which makes it suitable for deployment in very high-frequency bands where radio propagation makes it challenging to provide a robust access link for mobile users. In theory, under the assumption of full CSI, there is no upper limit on the capacity of a D-MIMO system while densifying the distribution of nodes [31]. However, the capacity will be limited by practical constraints such as cost, power, and hardware impairments, as discussed in Section 3.2.4.

The main challenge for large-scale D-MIMO roll-outs is arguably the cost of installing many nodes in different places, each requiring fast and high-speed fronthaul connections. D-MIMO installations need to be easy to deploy, have a small and non-intrusive visual footprint, and be flexible to scale and extend. Clearly, there are seemingly conflicting requirements that D-MIMO needs to fulfil, as shown in Figure 3.2.

If every D-MIMO AP requires a dedicated fronthaul cable, then there is no way to make a large installation economically feasible. That is, in case every AP requires separate cables to each of its neighbouring APs, the installation would become complex, labour-intensive, and costly. Therefore, it must be possible to cascade multiple APs on the same fronthaul connection. Moreover, cascading or serializing the fronthaul may not be sufficient.

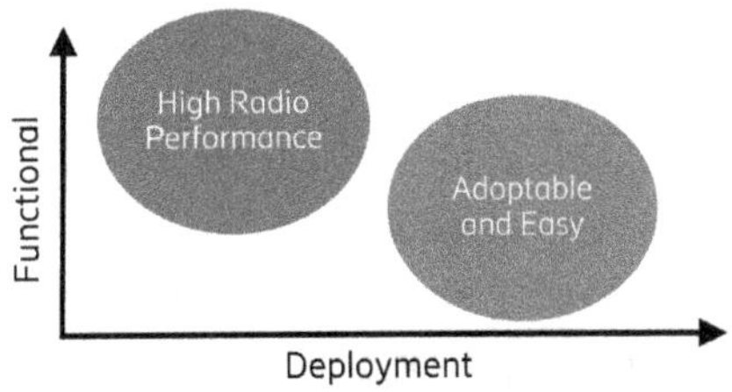

Figure 3.2. D-MIMO needs to fulfil seemingly conflicting requirements on functional complexity and deployment costs.

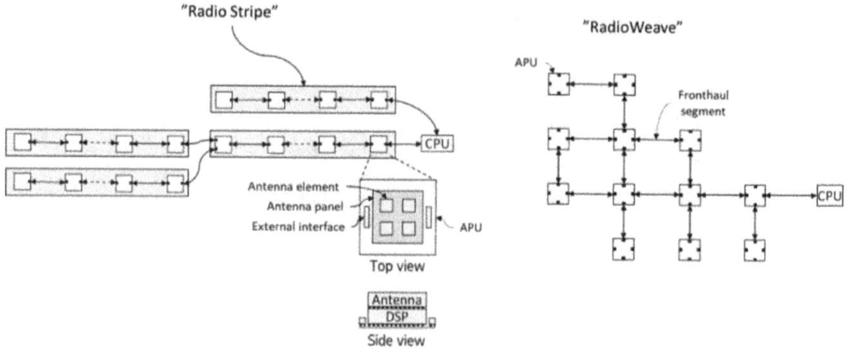

Figure 3.3. D-MIMO installations: (left) with radio stripe; (right) using both serialized (or cascaded) fronthaul connections plus parallel (or meshed) fronthaul connections. The fronthaul segments may be wired or wireless.

There are two different ways to provide fronthaul/backhaul to different APs, i.e., through fibre or wirelessly. For D-MIMO deployments, both ways need to be highly efficient. An optimized wired fronthaul can be realized through the so-called "radio stripes" [35], while wireless fronthaul/backhaul can be realized efficiently by integration with the access network. In this scenario, APs are integrated into a *radio stripe* as depicted in the left part of Figure 3.3. The radio stripe provides the fronthaul data and power supply for each AP as well as physical protection and encapsulation. With similar low-power techniques that are also used in mobile phones, the entire AP can be implemented in a system-on-chip (SoC) package that is denoted as antenna processing unit (APU) in Figure 3.3. An APU may contain some digital processing function (e.g., a digital signal processing (DSP) unit), an antenna panel consisting of one or more antenna elements, and one or more external interfaces that can connect to other APUs or to CPU. D-MIMO may also make use of a meshed fronthaul, sometimes denoted a RadioWeaves [36], as depicted in the right part of Figure 3.3. In a meshed fronthaul, the actual fronthaul segments may be implemented using wires, but the fronthaul segments may also be wireless. Further analysis of RadioWeaves is presented in Section 3.2.5.1. These architectures open possibilities to realize ultra-reliable links and interact with low-power devices. Moreover, capacity can be scaled up quite smoothly by adding more distributed components in the network.

Outdoor D-MIMO installations need to have a low visual footprint. With radio stripes, it is possible to hide the installation in existing construction elements in the environment, see Figure 3.4. To enable a small form factor of the APs or the APUs integrated inside a radio stripe, the power consumption needs to be very low. Fronthaul connections between APs or APUs can sometimes, but not always, be realized very easily with a cable. It is, however, not realistic to assume that every

Figure 3.4. Outdoor D-MIMO is well suited for dense urban areas, where it can provide invisible and high-capacity deployments.

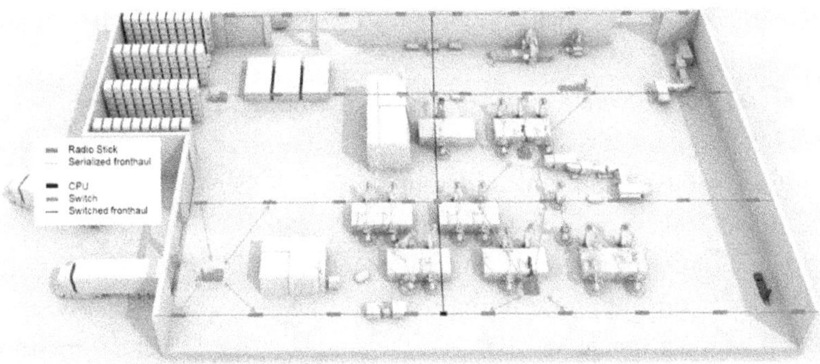

Figure 3.5. In indoor deployments, such as in a factory, D-MIMO promises to provide extremely reliable services with high capacity.

CPU can have a fibre backhaul connection. In dense urban areas, the backhaul connection can then be provided wirelessly by an existing macro BS.

Cascading or serializing the fronthaul connections is important for indoor installations as well, such as in a factory environment, as depicted in Figure 3.5. The concept can be compared to the electrical lights that are also installed in large volume inside a factory, where it is noted that this would not be possible if each electrical light required a separate cable installation. Just as one power cable can power up several lights, one fronthaul cable can connect several APs or APUs in a D-MIMO installation.

3.2.4 D-MIMO Deployment Considerations

3.2.4.1 Hardware constraints

As it will be outlined in Section 3.6, signals transmitted at mmWave and sub-THz frequencies are subject to distortion by the transceiver hardware. This is even a more important issue in D-MIMO scenarios, as the received (or transmitted) signal is collected (produced) from several hardware in separate locations experiencing different environmental conditions that may result in different effects on part of the received signal. In this section, the main hardware effects that are particularly relevant for D-MIMO systems are discussed, that is:

- Phase and frequency coherency
- Output power
- Reciprocity

To utilize distributed antennas to their maximum extent, coherent transmission and reception would be desirable. This requires all antennas participating in coherent transmission to be tightly synchronized. Going to sub-THz frequencies, coherent schemes will be very challenging to realize due to phase noise and frequency errors. Robust techniques against coherency errors must be developed, e.g., using antenna selection.

Another key hardware effect of current semiconductor technology at mmWave and sub-THz frequencies is that it is challenging to produce sufficient output power to maintain a robust communication link over a larger distance. As a result of increased power, more non-linear distortion is created. D-MIMO is a possibility to address this limited output power challenge: APs will have challenges to reach end-points due to output power limitations, but the inherent macro-diversity in D-MIMO and support for dense deployments enable more APs close to end-points. The effect of lower output power will result in transceiver designs with more parallel power amplifiers (PAs) combined with a higher number of antenna elements, yielding higher equivalent isotropic radiation power (EIRP). At lower mmWave frequencies where the EIRP, rather than AP output power, might be the limiting factor at certain deployment scenarios, D-MIMO has the capability to allow for lowering the EIRP by dense AP deployment and still to provide sufficient coverage.

To perform beamforming in a D-MIMO system, the channel information has to be received and known. One way to do this is to assume channel reciprocity, which means uplink and downlink channel are identical (in TDD mode). The channels estimated by a device contain the air-channel and also the analogue transceiver chains. Reduced coherence time (of channel and hardware) and components that are driven more non-linearly will make it more challenging to build D-MIMO systems based on channel reciprocity.

3.2.4.2 Architectural considerations

Among the desirable features, techniques for enabling scalable D-MIMO systems with converged access-backhaul-and-fronthaul, delivering extreme spectral efficiency, reliable access and robust mobility in scenarios ranging from low-bands to sub-THz bands, taking hardware impairments and deployment options into account can be mentioned.

As a first step, it would be necessary to understand how much distribution is required and where is the sweet spot in terms of complexity versus robustness and performance considering the trade-off between distributed and centralized processing. Then, one needs to deal with the practical approaches to non-coherent operation in higher bands and transport solutions satisfying the requirements. Optimum solution would be phase-coherent transmission and one centralized processor, but it will be difficult to build and meet the feasibility requirements. The other extreme would be phase non-coherent transmission with duplicating every data in each AP and relying on single-frequency network, but it will be inefficient. Further research on finding the balance, in terms of complexity versus robustness and performance, is necessary. In addition, problems such as beam management aspects, practical approaches to non-coherent operation in higher bands, and transport solutions, e.g., wired/wireless, optical/electrical, and analogue/digital, satisfying the requirements need to be addressed.

At low-frequency bands (e.g., sub-6 GHz), the spectrum is scarce, thus there is a need to grow the spectral efficiency to increase capacity for the available limited bandwidth (e.g., 10–100 MHz). One way is to have a distributed coherent antenna system. In theory, D-MIMO and coherent multipoint transmission can enable higher capacity everywhere by adding additional coordinated APs. For these bands, the digital processing is feasible in the APs and can be carried out locally. It is only sort of impairments and practical limitations that will eventually limit the performance. Furthermore, distributed processing means that there will be less information, e.g., channel estimation and precoding, exchange between the APs and CPU. Since baseband data will be transferred, there will be less load (compared to RF modulated data) on the fronthaul that can be handled without high-capacity fronthaul links, e.g., a 10 Gb/s Ethernet digital fronthaul may be enough.

At higher frequencies, i.e., mmWave and sub-THz, relatively large system bandwidth, e.g., several GHz, are available, and there is not much traffic in early phases of the utilization of these bands, even the UE is limited in how much bandwidth it can process. That means there will be available spectrum that can be utilized for the fronthauling, and non-coherent transmission can be enough for early phases. Since APs are generally not nomadic, robust fronthaul links will be easy to maintain. Moreover, as frequency increases, antenna elements shrink, the antenna array will be relatively small, and beams become narrower. The main challenge in these bands is to realize and maintain a robust access link that supports mobility where the propagation environment is more challenging. Macro-diversity brings an advantage to achieve the high reliability, especially in NLoS, and a D-MIMO can benefit from a much higher degree of macro-diversity to overcome radio blocking in case of narrow beams and weak signal penetration. Random blockage due to moving objects can be handled by redundant links. Nevertheless, digital processing, analogue-to-digital converters (ADCs) and digital-to-analogue converters (DACs) in these bands are power consuming and increase the chip cost and size of the processing unit. Small APs are essential in terms of deployment values, e.g., invisible, and aesthetic impacts, which is why it is required to move the digital processing from APs to the CPU, which may, as an option, mean that fronthaul will become analogue and needs to have a capacity on the level of fibre connection.

3.2.4.3 D-MIMO support in the ORAN

The D-MIMO network architecture intends to disaggregate the traditional CPU in multiple DUs, in line with 3GPP's 5G architecture [37] and to propose novel solutions based on fully distributed (aligned with ORAN Alliance specifications [38]), data-driven processing and local coordination. The disaggregation is vital for creating scalable versions of D-MIMO architectures [39], which will unlock the potential of deploying D-MIMO networking in future 6G networks with massive RU deployments.

In addition, in case of user-centric approaches, e.g., CF architecture, an important issue is cluster formation (selection of serving RUs). In contrast with the available simplistic distance-based solutions [29], it is important to dynamically allocate a sub-set (or cluster) of RUs to each UE based on (i) the radio propagation environment; (ii) quality of CSI estimates; (iii) constraints introduced by computation requirements; (iv) fronthaul links capacity; and (v) user mobility. Going a step further, innovative machine learning (ML) algorithms could be adopted for optimal cluster formation focusing on real-time operation by using historic data coming from the network, as well as for advanced modulation schemes and/or channel estimation/equalization. Finally, clustering RUs served by multiple DUs (in contrast to disjoint clusters in [39]) could provide an inter-DU coordination algorithm for decoding the actual signal. Moreover, exploring inter-DU coordination requirements and their effect in the spectral efficiency performance as well as dynamic adaptability of the coordination levels jointly addressing RU-DU and DU-DU coordination are some of the main research trends in the D-MIMO domain.

The D-MIMO architectures require fine granularity of the processing options (processing at AP versus CPU, inter-CPU coordination), which can be offered by hardware and software supporting the ORAN disaggregation concept [40]. In the following, the D-MIMO terminology is mapped onto the ORAN architecture, and a possible deployment is presented.

In ORAN architecture, the next-generation radio access network (NG-RAN) is disaggregated into ORAN CU (O-CU), ORAN DU (O-DU), and ORAN RU (O-RU). The O-CU is further split into O-CU control plane (O-CU-CP) and

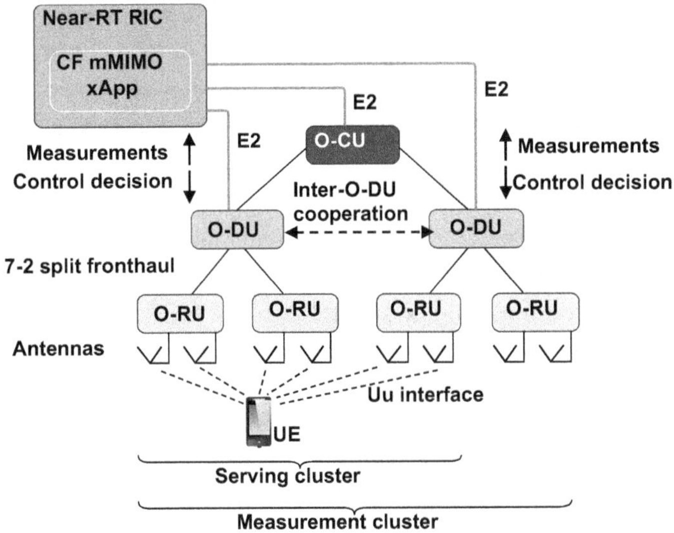

Figure 3.6. D-MIMO based on ORAN architecture, when a UE is served by O-RUs connected to several O-DUs, inter-O-DU cooperation may be enabled to boost the performance.

O-CU user plane (O-CU-UP). Multiple O-CUs and O-DUs are connected to the near-real-time RAN intelligent controller (near-RT RIC) for centralized NG-RAN performance control. Consequently, D-MIMO network components, particularly the CPU, can be associated with the ORAN nodes. The CPU is represented by the O-DU, while the AP is represented by O-RU.

The O-DU hosts the scheduler, which allocates radio resources for UE. Upon scheduler's decision, the user data are sent to the O-RU for transmission, or data are received at the O-RU. In the CF mMIMO, transmission/reception takes place simultaneously over multiple O-RUs, as shown in Figure 3.6. Thus, from the O-DU's point of view, UE can be divided into two categories as follows:

- **Local users**, who can be satisfyingly served by O-RUs connected to one O-DU.
- **Edge users**, who should be served collaboratively by O-RUs connected to two (or more) O-DUs. Serving of the edge users requires inter-DU cooperation (not yet in spec, but a studied interface [40]). It is also shown that serving edge users with O-RUs of multiple O-DUs improves the user spectral efficiency.

The main purpose of the inter-DU interface is to serve the user by having O-RUs connected to more than one O-DU, which is especially advantageous for edge users. One of the cooperating O-DUs should be considered a serving O-DU. The inter-O-DU cooperation interface can be used to both exchange user data between O-DUs as well as send necessary signalling required by the multi-antenna processing (MAP). To avoid any confusion, the MAP terminology is used for practical deployment to describe the application of precoding and combining to the signals sent in downlink and received in uplink, respectively. Based on the recent studies, five D-MIMO deployment options with the support from ORAN architecture are identified [40]. The deployment options of the proposed approaches are discussed very briefly in Table 3.1.

Another challenge is how to apply the current state-of-the-art in user clustering for D-MIMO systems (e.g., in CF mMIMO literature) in an ORAN-oriented framework, which is done using Near-RT RIC. The ORAN specification defines the Near-RT RIC, which is responsible for optimizing the RAN performance. The RAN should work in the absence of Near-RT RIC. However, with the Near-RT RIC, the system performance improves, as it can be connected to multiple RAN nodes, where based on telemetry data from multiple E2 nodes, Near-RT RIC can optimize the RAN configuration to decrease interference and increase spectral efficiency.

The Near-RT RIC plays an important role in moving RAN away from cell-based design towards a D-MIMO architecture, in particular in a CF mMIMO scenario. In an mMIMO architecture, the number of antennas in an O-RU is high, but the number of O-RUs connected to O-DU is limited. In the CF approach, the number of antennas in an O-RU should be limited, but the number of RUs connected to an O-DU should be large. Each UE will be served by more than one O-RU, whereas O-RUs serving the UE can be connected to different O-DUs based on the location of the UE.

The Near-RT RIC based on E2 link capacity is connected to multiple E2 nodes, as shown in Figure 3.6. In the CF case, multiple O-DUs are connected to a Near-RT RIC and export the telemetry data of UE measured from multiple O-RUs to Near-RT RIC via E2 connection. As Near-RT RIC receives data from multiple O-RUs that can be connected to different O-DUs, it can cluster the O-RUs which should transmit to a particular UE. Two clustering approaches are considered for the CF approach:

- **Network-centric clustering (NCC)**, which consists in deploying fixed disjoint clusters of APs where the APs in a cluster serve only the UE residing in their joint coverage area.
- **User-centric clustering (UCC)**, which consists in deploying dynamic (possibly partially overlapped) clusters of APs based on the requirements of each connected UE in the system.

By analysing the performances, it is understood that UCC is superior compared to the NCC. In UCC, O-RUs serving the UE are grouped together based on signal strength of pilots received by the RU. Each O-RU measures the signal strength

Table 3.1. Deployment options.

Deployment Option	Description of the Option	Inter-DU Coordination	ORAN Support
Option 1	The MAP vector is computed separately for each O-RU neglecting the channel between the user and antennas of other O-RUs	Absent (all O-RUs serve only the users located within the same O-DU)	Yes (Option 1 deployment is supported by current ORAN architecture)
Option 2	Same as Option 1, only difference here is that the edge users are served through the inter-DU coordination process	Present (all O-RUs of the O-DU serve local users. Additionally, the O-RUs serve edge users of neighbouring O-DUs)	No (Option 2 deployment is not supported by ORAN architecture)
Option 3	The MAP vector used by each O-RU for a particular user is computed jointly for all O-RUs of one particular O-DU (O-DU level information)	Absent (all O-RUs serve only the users located within the same O-DU)	Yes (Option 3 deployment is supported by current ORAN architecture)
Option 4	Same as Option 3, but only difference here is with inter-DU coordination interface present in the architecture: (i) all O-RUs of the O-DU serve local users; additionally (ii) the O-RUs serve edge users of the neighbouring O-DUs	Present (all O-RUs of the O-DU serve local users. Additionally, the O-RUs serve edge users of neighbouring O-DUs)	No (Option 2 deployment is not supported by current ORAN architecture)
Option 5	Option 5 may be implemented by connecting all O-RUs to one global O-DU where the global processing is performed	No (inter-DU coordination is not required where all the UE serve by a single DU)	Yes (ORAN can support this, but it is impractical due to high requirements on the O-DU hardware resources (processing, storage, and network)

of pilot for a particular UE and sends it to O-DU, which in turn forwards it to the Near-RT RIC over E2 interface. The Near-RT RIC collects all the telemetry data from multiple O-RUs that can be from more than one DU. If inter-DU communication is supported, RU clusters can be formed independently of O-RU to O-DU connection for a particular UE. The cluster information then can be transmitted to the primary O-DU of the UE. This primary O-DU is then responsible for transmitting and receiving UE data from the clustered O-RUs.

The Near-RT RIC uses signal-to-interference-plus-noise ratio (SINR) of the pilot from a particular O-RU, O-RU load, and O-DU load to cluster O-RUs for a particular UE. Location service can also be used to determine the location of the UE in the network and be considered as a factor for clustering of O-RUs. As pilot measurement is periodic, the clustering is also done in a periodic manner. ML can be used to predict the UE movement based on its current SINR and previous SINR values and, in general, change in the SINR over time. This can help in predicting the UE movement towards a particular direction which, in turn, can help to cluster O-RUs serving the UE.

3.2.4.4 Analogue centralized beamforming

The type of D-MIMO system architecture and functional split that can be used depends very much on the overall system bandwidth. Therefore, on lower bands, where the bandwidth is limited, it can be feasible to use an electrical-based fronthaul technology such as Ethernet. Ethernet provides up to 10 Gb/s rate and can also provide up to 100 W or electrical power using PoE 802.3 bt type 4 [41]. On lower frequency bands, the digital processing requirements of each APU are also reasonably limited.

However, on sub-THz frequency bands, the very wide system bandwidth prohibits the use of electrical cables for the fronthaul, and we must instead use optical fronthaul connections. It is also not feasible, at least within the early 6G timeframe, to assume that the APUs will be capable of performing any meaningful digital processing on such high bandwidth signals. The APUs need to use low power to remain small enough for large volume installation, and this is not compatible with extreme requirements on digital processing.

Therefore, one promising architecture for high-frequency and high-bandwidth D-MIMO systems is to utilize analogue-radio-over-fibre and is depicted in Figure 3.7. To keep the APUs operating with low power usage (to reduce the heat dissipation and the resulting size and weight), the APU consists of an optical-add-and-drop-multiplexer, a photo-detector, a directly-modulated-laser, and some analogue components related to transmitter and receiver. Using wavelength-division multiplexing (WDM), it is still possible to serialize the fronthaul and connect many APUs to the same physical fibre cable. All digital processing (e.g., precoding and channel

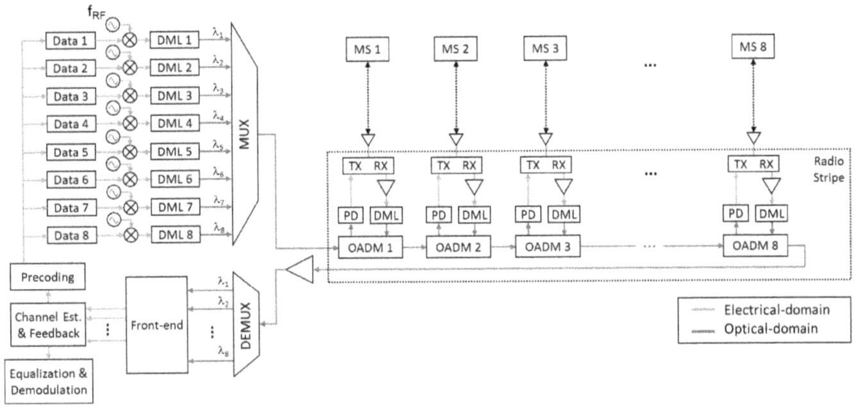

Figure 3.7. A wide bandwidth D-MIMO system can be implemented using analogue-radio-over-fibre and WDM.

estimation) is performed in the CPU where the constraints on size and weight are more relaxed, and a somewhat higher power consumption can be acceptable.

3.2.5 Some Recent Analysis of D-MIMO Scenarios

This section presents some recent results as a first step towards understanding how to develop a common scalable and frequency-agile architecture for practical D-MIMO deployments. In the following subsections, the theoretical analysis of a RadioWeaves deployment followed by the performance comparison of a distributed versus centralized zero-forcing as well as an analysis of a hybrid precoding is discussed.

3.2.5.1 RadioWeaves deployment analysis

A preliminary study compared the connectivity in a RadioWeaves, as illustrated in Figure 3.3, infrastructure with a deployment based on one central "candelabrum" array with an equal number of antennas [42, 43]. The two deployment scenarios are shown in Figure 3.8. This analysis has considered 200 simultaneous, randomly located, users, and different topologies for the RadioWeaves infrastructure. It has assessed the transmit power requirement needed to guarantee that all users simultaneously receive 4 Mbit/s (200 × 4 b/s/Hz spectral efficiency).

The results are reported in [42]. The randomness in the distribution stems entirely from the randomness of the user locations. A significant gain can be achieved, that is, with a total transmit power of −20 dBm, the distributed RadioWeaves deployment on four walls will decrease the probability of not achieving the targeted throughput on a random location with a factor of 1000 compared to the candelabrum-based infrastructure operating with 0 dBm total output power.

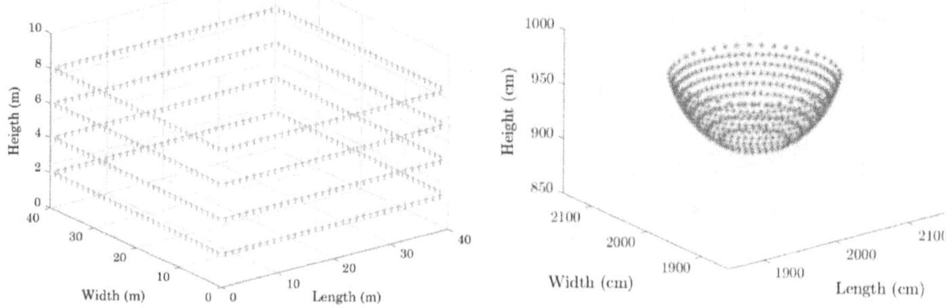

Figure 3.8. Distributed RadioWeaves infrastructure and central "candelabrum" deployment scenarios [42].

The RadioWeaves deployment achieves this 1000 times increase in QoS with 100 times less transmit power. This is shown in particular with respect to a central array in a candelabrum topology, which in itself is a quite good case and does create favourable conditions in a deployment with a central array.

The aim of RadioWeaves technology is to realize its great potential both in terms of performance metrics and energy efficiency [43]. Algorithm-architecture co-design is pursued, with specific attention for the bottlenecks in interconnects and mixed fronthaul-backhaul requirements [34].

3.2.5.2 Beamforming computations and weight distribution

When enough bandwidth is available at the fronthaul, a centralized digital beamforming can be applied. The fully digital beamforming is advantageous (compared to the analogue beamforming) since all the signal processing is done digitally. This can facilitate a more flexible design by using more degrees of freedom compared to the analogue beamforming introduced in Section 3.2.4 [44]. This section presents reciprocity-based (UE-centric) interference-aware distributed zero-forcing (IADZF) method and compares with centralized zero-forcing (CZF). Each distributed precoding done in clusters/subsets is interference-aware, i.e., it considers minimizing the power interfering to other clusters.

Let N APs coherently serve K randomly distributed UE in a UE-centric way. For UE-centric AP clustering ("subset"), the method in [41] has been set by grouping the APs that provide sufficiently high signal-to-noise ratio (SNR) values with the best channel quality that contribute at least $\alpha\%$, e.g., 95%, of large-scale fading coefficients towards the k-th UE.

For the CZF, all APs form a single cluster. This means APs can transmit signals to any UE, and ZF is done in the central entity. In the distributed zero-forcing (DZF) case, multiple clusters are formed in a UE-centric way where APs in each cluster

Table 3.2. Simulation parameters.

Parameter	Model Specification	
Frequency range (GHz)	28	100
Bandwidth (MHz)	200	5500
Maximum AP transmit power (dBm)	13	
Number of subarrays	1	
Vertical/horizontal antenna elements	4/8	8/16
Number of UE	{20, 40}	
Propagation Model	3GPP InH [45]	
Area size	100 m × 100 m	
Number of blockers	1000	
Blocker size	Max: 2 m × 3 m, Min: 0.5 m × 1 m	
Duplexing	TDD with 50% downlink	
AP noise figure (dB)	7	10
Overhead ratio	1:3	

only serve the UE inside the given cluster. The precoding coefficients are calculated for each cluster separately. Performance has been evaluated in an indoor scenario with randomly distributed blockers, for different number of regularly deployed APs each with M antenna elements and K uniformly distributed single-antenna UE. The configuration parameters used in the simulations are given in Table 3.2. Perfect channel estimation and lossless, high capacity fronthaul is assumed. RF imperfections, hardware impairments, phase noise, and PA non-linearities are omitted from the scope of this work.

Figure 3.9 demonstrates that IADZF method applied over UE-centric clusters with $\alpha = 95\%$ clustering method can achieve the performance of CZF method, at which all APs serve each UE for different deployment sizes on sub-THz band. Two methods converge in relatively dense deployments. Due to the higher path-loss and weaker signal penetration, SE decreases as frequency increases.

IADZF has been compared with interference unaware TDZF in Figure 3.10 for different cluster sizes, $N_k = 1$, $N_k = 4$, $\alpha = 95\%$ clustering approach, and $N_k = N$, i.e., all AP simultaneously transmit to all UE, operating at 28 GHz. It has been shown that larger subsets using IADZF method bring more precoding gain and increase the spectral efficiency (SE). However, if distributed precoders are interference unaware, increased subset size degrades the performance.

Cumulative distribution functions (CDFs) of SINR values of 20 UE served by 49 APs for different serving cluster sizes, e.g., $N_k = 1$, $N_k = 4$, $\alpha = 95\%$ clustering approach, and $N_k = 49$ are shown in Figure 3.9. It has been shown that more

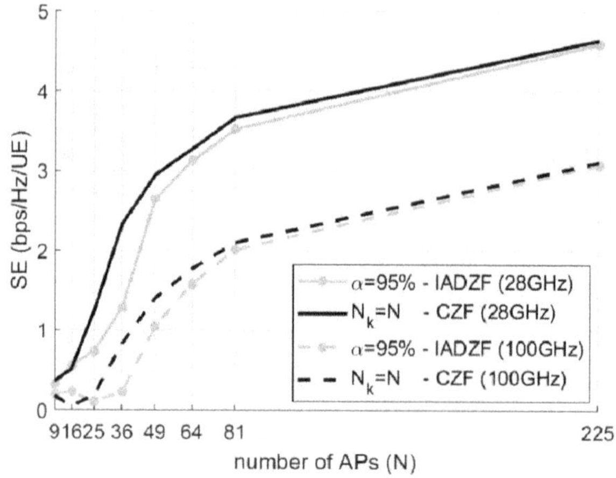

Figure 3.9. SE performance comparison of IADZF and CZF for different cluster sizes for 20 UE operating at 28 and 100 GHz.

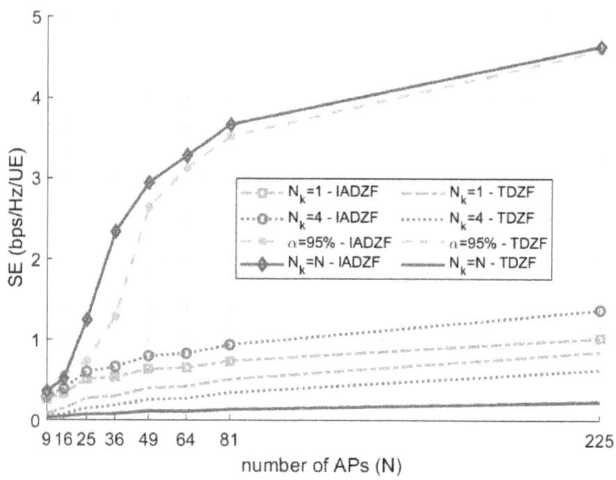

Figure 3.10. SE Performance comparison of IADZF and interference unaware TDZF for different cluster sizes for 20 UE operating at 28 GHz.

APs involved in coherent transmission bring higher degree of diversity, hence better interference cancellation and performance. Effect of the operating frequency is negligible for small size of serving AP subsets. Nonetheless, as frequency increases, APs close to the UE become dominant while distant APs do not contribute much, which decreases the received signal strength, eventually the SINR. Small clusters can still provide good enough SE performance on the average, and particularly can utilize the high bandwidth at high-frequency bands. Another highlight is that more APs can enable the cell-edge UE, e.g., 5% of the UE, having positive SINR.

Figure 3.12 demonstrates the average SE performance over 40 UE with reference to different antenna distributions, e.g., $N = 64 \times M$ antennas are distributed over

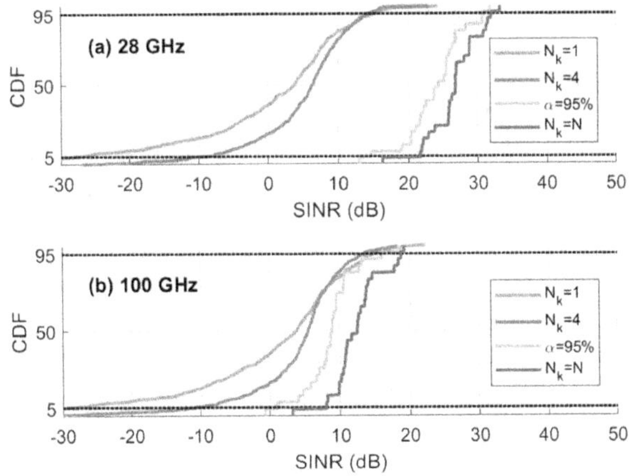

Figure 3.11. SINR performance for IADZF for $N = 49$ APs and $K = 20$ UE evaluated at (a) 28 GHz and (b) 100 GHz.

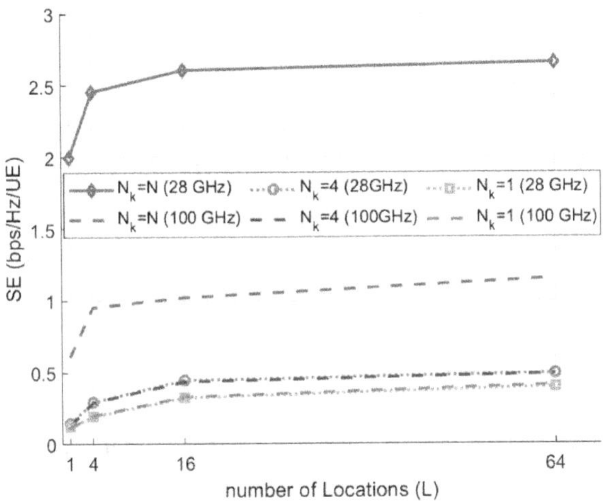

Figure 3.12. Performance comparison for collocated ($L = 1$), partially ($L = \{4, 16\}$), and fully distributed ($L = 64$) deployments for different operating frequencies for $N_k = \{1, 4, 64\}$.

$L = \{1, 4, 16, 64\}$ locations. It has been shown that semi-distributed ($L = 4$) or fully distributed ($L = 64$) deployments perform similar for single cluster case. Herein, there is a trade-off between implementation and deployment complexities. One cluster including all APs in the network can still achieve high SE values by collocating the antennas ($L = 1$), but it has higher implementation complexity. Many but smaller subsets ($N_k = \{1, 4\}$) are easier to implement joint processing; however, more scattered deployment is necessary for good enough performance.

3.2.5.3 Hybrid precoding in cooperative mmWave scenarios

The BS coordination schemes in mmWave networks that allow multiple streams to be transmitted jointly from multiple BSs are investigated by taking hardware and channel characteristics into account. Users are being served using spatial multi-flow and implements successive interference cancellation to decode the data streams sequentially.

In such a scenario, a downlink multi-cell and multi-user mmWave network with M multi-antenna BSs and K single antenna users are considered. It is assumed that hybrid precoding architecture is used at the BSs, to achieve high spectral efficiency with reduced hardware power consumption compared to that of the fully digital precoding, due to the reduced number of RF chains [46]. The number of antennas at BS is denoted by N_m and the number of RF chains L_m. Depending on the number of phase shifters, the fully connected hybrid precoding architecture requires each RF chain connected to all antennas and the partially connected hybrid precoding architecture connects each RF chain to a subset of the antennas [47]. In addition, it is assumed that the sum power consumption of all BSs is divided into the hardware power consumption, including the phase shifters and the DACs, and the RF transmit power.

The objective is to minimize the sum power consumption such that the per-user minimum spectral efficiency, the per-BS maximum power constraint, and the hybrid precoding constraint are guaranteed. A sub-optimal algorithm by decoupling the optimization problem into an analogue precoding problem is proposed that only depends on the channel information, and a digital precoding problem that minimizes the sum power consumption by solving a semi-definite program. The proposed hybrid precoding algorithm jointly associates users to the BSs, finds the optimal BS silence strategy with minimum power, and enables us to jointly serve a user by multiple BSs. For the detailed system model and algorithm, cf. [48].

In Figure 3.13, the sum RF transmit power and the sum power consumption against the number of BSs are shown. In the scenario of this simulation, for each architecture, results based on all BSs being active are compared to the optimal case when the silent mode is enabled. Network parameters are as follows:

Description	**Parameter**	**Value**
Number of antennas	N_m	64
Number of users	K	4
Number of RF chains per BS	L_m	4
PS power consumption	P_{PS}	40 mW
DAC power consumption	P_{DAC}	200 mW
RF chain power consumption	P_{RF}	40 mW
Target spectral efficiency per user	τ_k	4 bit/s/Hz

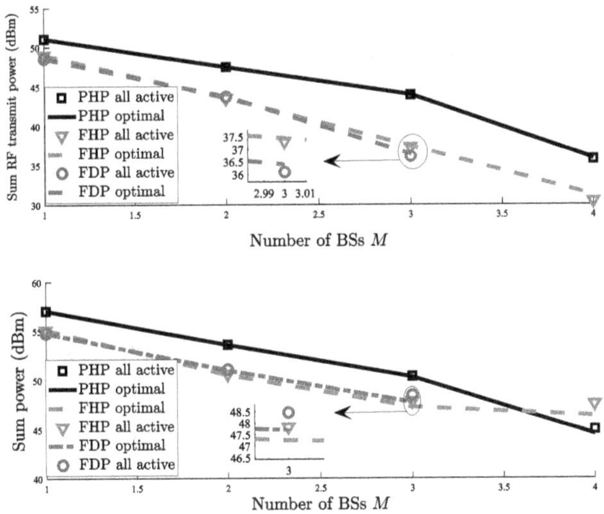

Figure 3.13. (a) Sum RF transmit power and (b) sum power consumption versus the number of BSs.

In Figure 3.13(a), the result shows that network densification and cooperative transmissions lead to a better chance for a user to be served by BSs with good channel conditions, thus, requiring less RF transmit power to achieve a target spectral efficiency. In Figure 3.13(b), for $M > 4$, the partially hybrid precoding starts to consume less sum power than the fully hybrid precoding, since the hardware power increase has less effect on the sum power consumption than that of the fully hybrid precoding. Also, Figure 3.13 shows that the power consumption difference between the optimal case and the sub-optimal case is small. It is worth noting that the total power consumption depends on the power consumption of the phase shifters, RF chains and the DACs, and the power scaling in the silence mode. Setting a low value for the hardware power consumption will give advantages to the all-active case and, thus, reducing the difference between the case of all-active and the case of the optimal silence mode. This parameters setting is based on the carrier frequency at 28 GHz. When higher frequencies are considered, the available transmit power will be further limited. Thus, the proportion of the hardware power will most likely increase, and the advantage of silence mode may be more prominent.

3.3 Integrated Access and Backhauling

The main aim in the IAB network architecture is to facilitate the dense deployment of the modern network architectures without the need for a fibre connection to each BS, by using the same spectral resources and infrastructures to serve both UE in access as well as the BSs in backhaul [50]. IAB networks reduce deployment costs by replacing expensive wired backhaul to each cellular BS and do not cause massive degradation compared to all-wire network in realistic scenarios, as validated in [49]. Thus, it is expected that IAB architecture will be part of any advanced cellular RAN infrastructure lacking fibre connectivity to the RUs, but in particular for a D-MIMO architecture, IAB could be the efficient solution to the backhaul/fronthaul challenge. The D-MIMO infrastructure is expected to have many links connecting the CU and DU to multiple RUs covering the designated area. For such a case, IAB can provide a cost-effective connectivity.

Specifically, if bandwidth is not the primary constraint, as for example in early deployment and at sub-THz frequencies, IAB can be an efficient approach to maintain flexibility and efficient deployment. Moreover, IAB offers an advanced and flexible solution with multi-hop communications, dynamic resource multiplexing, and a plug-and-play design for low-complexity deployments. Additionally, the higher spatial reuse due to directivity in mmWave band and multi-beam systems and MIMO reduces cross-link interference between backhaul and access links allowing higher densification [49, 51].

Figure 3.14 shows a general IAB scenario, where the access and the backhaul of each BS need to share the same resources. BS1, master BS (MBS), is directly linked to the core network, CN, (e.g., by fibre backhaul), while BS2 and BS3 (secondary BS, SBS) use in-band backhauling [50]. In the network, there are constraints applied to all BSs and UE, notably, UE can see at least one BS. Besides, BS2 and BS3 can communicate with each other, while all BSs work in the TDD manner, i.e., either in transmit or receive mode in each time slot. In any case, one UE is served by only one BS in receive mode in each time slot, i.e., antenna/AP selection is used as the most primitive form of D-MIMO. One BS allows multiple access links in one time slot using different frequency resources. Note that this will increase the cost of link when scheduled and routed.

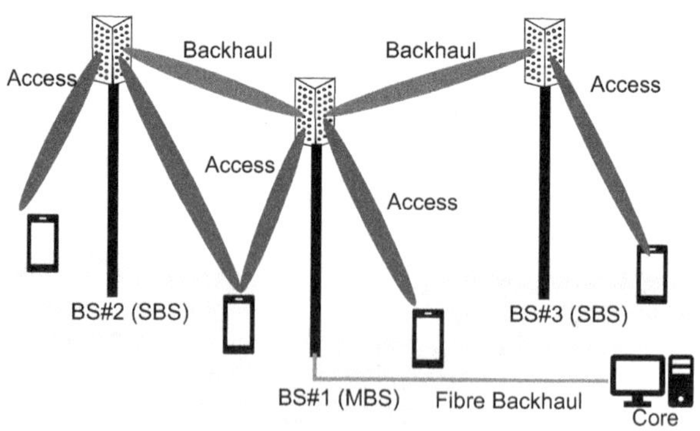

Figure 3.14. 5G Self backhaul-integrated access and the backhaul concept.

In this section, the efficient AP selection combined with access and backhaul scheduling in a TDD system where backhaul/fronthaul share resources is investigated.

3.3.1 IAB in 3GPP

To provide high data-rate requirements for backhauling a 6G BS, mmWave and sub-THz bands are expected to be used. In this case, backhaul links could be vulnerable to blockage, e.g., due to moving objects, seasonal changes (foliage), or infrastructure changes (new buildings). Thus, from a resilience perspective, it is important to ensure that an IAB node can continue to operate (e.g., provide coverage and end-user service continuity) even if an active backhaul path is degraded, lost, or even overloaded and failed. For this purpose, 3GPP has agreed on dynamic topology for IAB networks to autonomously reconfigure the backhaul network to achieve optimal backhaul performance under the above-mentioned circumstances [52], by supporting the autoconnection of an IAB node, network topology adaptation, and redundant connectivity. Moreover, the architecture is scalable, so that the number of backhaul hops is only limited by network performance (single and multi-hop backhauling are supported) [53].

In 5G NR IAB architecture, the network nodes are split into two types, namely: IAB-nodes and IAB-donors. Multiple IAB-nodes use wireless backhaul, and IAB-donors have fibre connectivity towards the CN. IAB-nodes and IAB-donors can serve UE and other IAB-nodes. Each IAB-node hosts two NR functions [49]:

1. Mobile termination (MT): for the wireless backhaul connection towards an upstream IAB-node or IAB-donor.
2. DU: for the access connection to the UE or the downstream MTs of other IAB-nodes.

Depending on using an AP or a BS (according to the definition provided in the previous section), a functional split of the radio protocol stack could be implemented with the control and upper layers in the IAB-donor CU, and the lower layers in the DUs of the IAB-nodes.

There is ongoing research regarding path selection techniques for 5G NR IAB networks. Different path selection techniques using a distributed approach are presented in [54], and the investigation of the performance in terms of hop count and the bottleneck SNR is analysed in [54, 55].

IAB adds new challenges in the task of scheduling to allocate resource between the access and the backhaul in IAB-node. Moreover, in a dense city where uplink/downlink traffics are constantly changing, scheduler with a fixed resource allocation could prevent the 5G network to realize its full potential.

Features of the scheduler task in the 5G IAB network is reviewed in [55], where the scheduling problem is split into two sub-problems and the ML task is defined for each sub-problem, namely, for the network access and the network backhaul. Finally, a smart scheduling solution based on deep learning is presented to address the problems above.

3.3.2 IAB Versus Fibre

In this section, the IAB network performance is discussed, and the evaluation of its performance in comparison with those achieved by hybrid IAB/fibre-connected networks using a finite homogeneous Poisson point process (FHPPP)-based stochastic geometry model, i.e., a Poisson point process (PPP) with a constant density, with random distributions of the IAB nodes as well as the UE inside a finite region is presented [56–58].

Figure 3.15 shows the service coverage probability of the IAB networks with those obtained by the scenarios having a fraction of fibre-connected SBSs, as

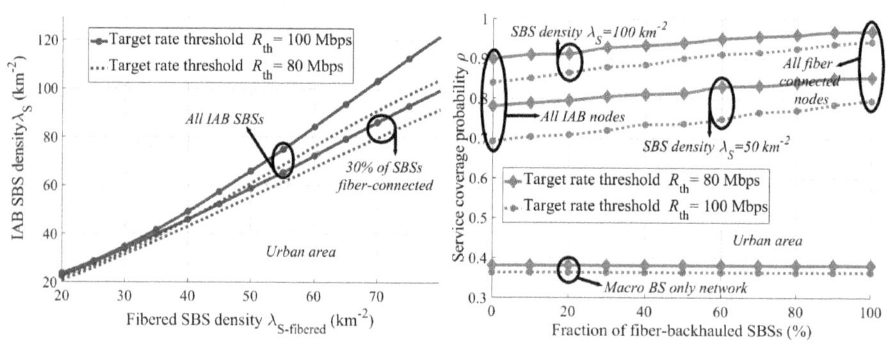

Figure 3.15. (left) Fibre-backhauled networks. (Right) Service coverage probability as a function of percentage of fibre-backhauled SBSs.

Table 3.3. Simulation parameters.

Parameter	Value
Carrier frequency	28 GHz
Bandwidth	1 GHz
Path loss exponents	[LoS, NLoS] = (2, 3)
Main lobe antenna gains	[MBS, SBS, UE] = (24, 24, 0) dBi
Side lobe antenna gains	[MBS, SBS, UE] = (−2, −2, 0) dBi
Antenna powers	[MBS, SBS, UE] = (40, 24, 0) dBm

well as the cases without SBSs (refer to Table 3.3 for main simulation parameters). Figure 3.15-left compares the performance of IAB and fibre-connected networks, in terms of service coverage probability, i.e., the probability of the event that the minimum target rate requirements of the UE are satisfied. Moreover, Figure 3.15-right shows the network service coverage rate as a function of the fraction of fibre-connected SBSs and characterizes the system performance with the cases without SBSs.

This is motivated by the fact that some of the SBSs may have easy access to fibre. Here, it is assumed fibre-connected SBSs to be randomly distributed in the considered network area. It is observed that for a wide range of parameter settings, the IAB network can effectively provide the same levels of network service coverage probability as that of fibre-backhauled network with relatively small increase in the number of deployed IAB nodes.

Such a small increment in the number of IAB nodes leads to several advantages:

- **Increased network flexibility:** In contrast to fibre-backhauled networks, where the APs can be installed only in the places with fibre connection, the IAB nodes can be installed in different places if they have an acceptable connection with their parent nodes. This increases the network flexibility and the possibility for topology optimization remarkably.
- **Reduction in network cost:** An SBS is much lower in cost than laying fibre. Also, different evaluations reveal that, for dense urban/suburban areas, even in the presence of dark fibre, the IAB network deployment reduces the total cost of ownership.
- **Reduction of time-to-market:** Due to the required regulatory permissions, digging, and construction time, laying fibre typically takes a long time. In such cases, IAB can help to install new BSs/radio sites quickly.

3.3.3 Coordinated Mesh-based IAB

In a more general form, all BSs together form a backhaul/fronthaul mesh. Note that at this point, no assumption is made on the traffic to be routed to the core, i.e., the functional split between BS and core. Up to this point, this is abstracted in a cost-metric, which can be changed depending on beamforming capability, channel model, and deployment.

The overall problems to be solved are:

- Which AP should a UE connect to?
- In which time-slot should it be scheduled?
- Which route should the traffic be routed through fronthaul/backhaul mesh?

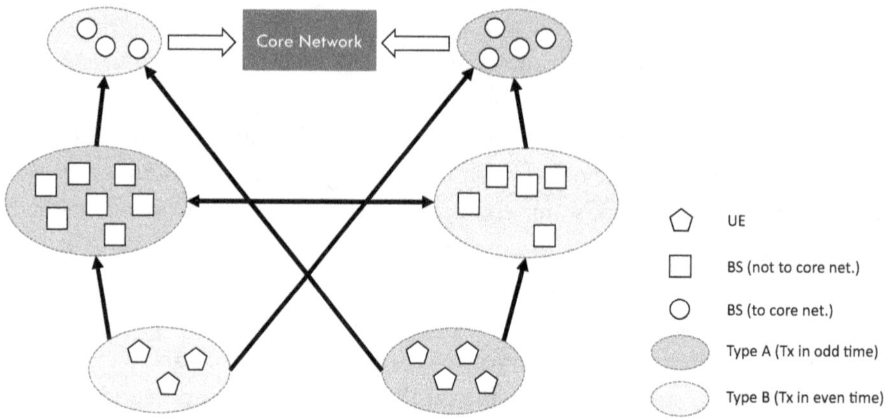

Figure 3.16. UE and BS/AP grouping in odd and even time slots, prior to routing optimization.

Given a cost metric (e.g., number of hops to core and throughput), the problem can be divided into time slot allocation and shortest-path optimization. Time slot allocation is illustrated in Figure 3.16, which shows how BSs and UE are grouped as part of optimization process.

For a given grouping of nodes (odd and even time slots), the second problem to be solved is between which nodes traffic should be sent. From which UE to which AP, and if an AP needs to transport traffic to the core, which backhaul/fronthaul links should be used?

Both problems need to be solved jointly, and exhaustive search becomes already prohibitively complex for small networks. Thus, a greedy algorithm was developed, which has a complexity that is linear in the number of UE.

3.3.3.1 IAB coverage analysis

As discussed in Section 3.2, the traditional cellular network will probably evolve to a kind of D-MIMO scenario, where it is likely to use APs with large distributed antenna arrays. The potential offered by IAB techniques can be very helpful in this regard by alleviating the fronthaul problem and by reducing the deployment cost in massive BS densification scenarios compared to the fibre deployment, which requires a noteworthy initial investment for installation.

Motivated by the presence of very wide bandwidths at mmWave carrier frequencies and above, IAB networks allow the operator to use part of the spectrum resources for wireless backhauling [53]. In 3GPP NR, IAB network configurations allow to provide flexible low-cost wireless backhaul using 3GPP NR technology in international mobile telecommunication (IMT) bands and provide not only backhaul, but also the cellular services in the same node. This will be a complement to existing microwave point-to-point backhauling in suburban and urban areas.

3.3.3.2 Genetic algorithm-based topology optimization for IAB

Due to the increase of network size in dense areas, which is the main point of interest in IAB networks, finding solutions for optimal network topology/routing is important. Since such optimization problem is very complex, an exhaustive search over all possible deployment options quickly becomes infeasible. This motivates a potentially suboptimal ML approach, which gives effective (sub)optimal solutions with reasonably satisfactory implementation complexity. Particularly, a genetic algorithm (GA)-based scheme to optimize the BS locations and non-IAB backhaul link placement is proposed. The details of the proposed scheme can be found in [59].

Unlike the non-IAB backhaul-connected networks, IAB networks may be prone to environmental effects, especially due to the blockage in dense urban environments and the tree foliage in sub-urban environments. It should also be noted that although IABs' main point of interest is dense urban areas, it has the capability to be deployed in suburban areas as well.

The evaluation of the blockage effect in urban areas and tree foliage on the network performance of an IAB network in suburban areas with random deployment and GA-optimized non-IAB backhaul link distribution is presented in Figure 3.17. Here, the results are presented for different rate requirements of the UE for the access links with path loss exponents 3 and 4 for LoS and NLoS propagations, and main lobe antenna gains of 18, 18, and 0 dBm for MBS, SBS, and UE, respectively. Particularly, Figure 3.17 (left) shows the service coverage probability considering the PPP-based germ-grain blocking model, while in Figure 3.17 (right) presents the results for the average hop distance of 450 m which corresponds to SBSs density of 8 km^{-2} in a suburban area. In Figure 3.17 (left), the MBS, SBS, and UE transmit powers are as Pm, Ps, Pu = (40, 24, 0) dBm, and in Figure 3.17 (right) the tree density are $\lambda_M, \lambda_S, \lambda_U = (2, 50, 500)$ km^{-2}.

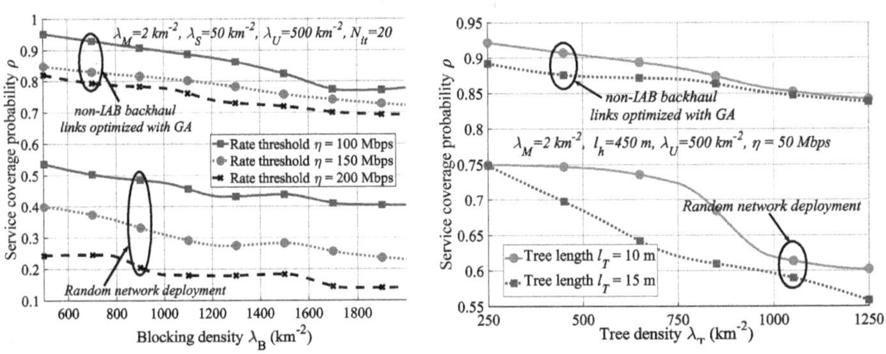

Figure 3.17. Service coverage probability of the IAB network; (left): as a function of the blocking density; (right) as a function of tree density.

Based on Figure 3.17, the following conclusions can be derived:

- The GA-based optimized network deployment shows considerable resilience to blockage and tree foliage, compared to the unoptimized random network.
- The service coverage probability in suburban area with a random IAB network deployment is considerably affected by the tree foliage loss. This is especially witnessed when the tree density is increased. However, with the introduction of GA-based optimization on selecting the appropriate non-IAB backhaul link distribution, a resilience to the tree foliage is observed.

In general, IAB robustness is hard to predict in the presence of tree foliage due to the influence of characteristics of the trees/vegetation on the link quality. Particularly, the link quality may vary due to the number of wet trees, snow on the trees, leaf percentage in different seasonal changes, and wind. However, it can be concluded that proper network planning can reduce the adverse effects on IAB, even though it is prone to medium/highly densified tree foliage in suburban areas. Moreover, mmWave IAB in areas with low/moderate amounts of tree foliage is expected to perform well.

3.4 Reconfigurable Intelligent Surfaces

Research in MIMO and D-MIMO context can be linked with another technology, i.e., RIS adoption. RIS adoption in telecommunications is a novel technology foreseen as one of the key enablers of the 6G mobile networks.

RIS is a two-dimensional surface of engineered material whose properties are reconfigurable rather than static, and with RIS, it is possible to shape how the surface interacts with wireless signals, enabling the wireless propagation environment to be fine-tuned [60]. The surface can include mainly passive elements without doing digital processing or any signal amplification. Nonetheless, some RIS surfaces can have relaying capability providing signal amplification. Moreover, RIS can have active elements that can enable digital signal generation, i.e., it can be interpreted as a sort of extra-large mMIMO.

RISs are low-cost and limited-power devices that can redirect, in a programmable fashion, the impinging waves towards the desired directions. Hence, providing control on the propagation environment in turn becomes an optimization variable to enhance communication link performance [13, 61]. RIS is suitable for a number of use cases, such as performance boosting, electromagnetic field exposure minimization, localization, and sensing [62]. Moreover, proposed RIS applications in the recent literature cover several application scenarios under different assumptions. These scenarios can be categorized as [13]:

- RIS-enhanced cellular network, where RIS is used to establish the link between the users located in a blind spot and the BS. Thus, the QoS in HetNets and the latency performance in mobile edge computing (MEC) networks are improved [63] or can act as a signal reflection to support massive connectivity via interference mitigation in device-to-device (D2D) communication networks, or to strengthen the received signal power of cell-edge users and mitigating the interference from neighbour cells [64].
- RIS-enhanced unmanned systems, where RIS can be leveraged for enhancing the performance of unmanned aerial vehicle (UAV)-enabled wireless networks [65], cellular-connected UAV networks [66], autonomous vehicular networks, autonomous underwater vehicle (AUV) networks, and intelligent robotic networks by fully reaping the aforementioned RIS benefits.
- RIS-enhanced Internet of Things (IoT), where RIS is exploited for assisting intelligent wireless sensor networks, e.g., in intelligent agriculture and intelligent factory scenarios [67, 68].

In this section, first a proposed architecture based on introducing logical components for the control and orchestration of the RIS operation is discussed. Then, an analysis on the influence of the RIS position and orientation on its overall performance is presented, followed by discussions on two specific RIS scenarios, i.e., cascaded multi-RIS scenario and RIS-assisted UAV system.

3.4.1 Proposed Architecture for Efficient RIS Deployment

RIS can be organized for adaptive network configuration and orchestration depending on scenarios and application needs [13, 61]. The integration of RIS in the RAN foresees additional logical components, interacting with each other and with the legacy components, to enable integration and orchestration of such devices. A proposed architecture and corresponding components for the RIS-aided control in the RAN are depicted in Figure 3.18, where the following components are represented [68]:

- **RIS:** the RIS itself is the reflecting intelligent surface based on technologies such as reflect-array or meta-material. RIS devices are typically constituted by a grid of discrete unit cells spaced at sub-wavelength distance. The electromagnetic (EM) response of each unit cell can be controlled in a programmable manner by altering several EM response parameters, such as phase, amplitude, polarization, and frequency. Moreover, recently proposed RIS hardware designs are providing the surface with channel sensing capabilities [69, 70]. Each RIS deployed in the network is associated with a RIS actuator that controls the RIS surface within a time granularity in the order of 10–100 ms.

- **RISA** (RIS actuator): the logical entity in charge of actuating the commands received from the RISC (RIS controller, refer to the definition below) and translate them in the configurations of the RIS, i.e., reflection properties of the surface. RISA can provide feedback to the RISC, e.g., in case of sensing capabilities at the RIS to provide the network with additional context information. The envisioned time granularity of the RISA is in the order of 1–20 ms.
- **RISC** (RIS controller): the controller associated to a RIS actuator that generates the logical commands associated with the switching between different states/configurations of the RIS elements (e.g., predefined or custom phase shifts configuration at the RIS). The RISC can embed third-party smart algorithms to derive the desired RIS configuration to be applied. Alternatively, the configuration can be setup based on received commands from other elements of the network. In the former case, the RISC takes care of the optimization of the RIS surface autonomously to form other network optimization operations, whereas in the latter case, the RISC behaves as an interface that controls the RIS based on external instructions, e.g., if the RIS configuration is optimized jointly with other network optimization operation. A RISC time granularity in the order of 20–100 ms is envisioned.
- **RISO** (RIS orchestrator): a logical component placed at a higher hierarchical level, to serve the role of orchestrating multiple RISCs in the network. Depending on the application, the time granularity of the RISO is expected to be in the order of 100–1000 ms.

As it is presented in Figure 3.18, there is a one-to-one correspondence between RIS and RISA. The RISA, RISC, and RISO can be virtualized, abstracted, and deployed into edge or central clouds, while physical devices (RF) shall be placed on site. The RISA operates directly on the RIS devices through the general interface called Open Environment that would accommodate heterogeneous RIS technology. Still depending on the RIS technology, different functions can be placed in the RISA to accommodate different RIS technologies, e.g., CSI feedback in the case of sensing capabilities at the RIS, etc. Therefore, the RISA provides an open interface to integrate heterogeneous RIS devices capabilities.

3.4.2 RIS Position and Orientation Influence on the Performance

As a challenge in the RIS-assisted scenarios, the quality of the communication is expected to depend strongly on the RIS placement in a RIS-aided link. This is

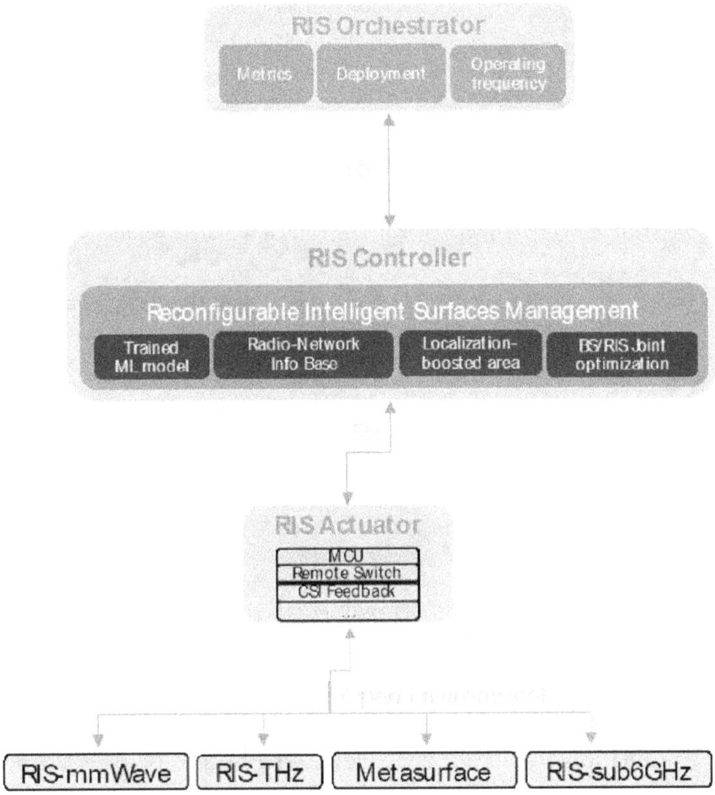

Figure 3.18. Proposed components and architecture for RIS control and orchestration.

because the path-loss changes with the relative distance of the RIS from the BS, the user, and their relative orientations. Therefore, the position and orientation of the RIS, with respect to the location of the BS and the user, need to be taken into account [71, 72].

Figure 3.19 shows an indoor scenario where the direct LoS link between the BS and the user is interrupted. The angular relation between the orientation of the RIS, the transmitted signal to the RIS, and the reflected signal is also clarified. The optimal RIS placement refers to the topology that guarantees a minimum threshold for the received power, for every possible location of a mobile user within the area of interest. To understand the impact of the RIS position and orientation on the received power, it would be instructive to investigate how the RIS captures the incident power and how it redistributes it in space. With respect to the incident beam, the power captured by the RIS depends on the illumination conditions, i.e., whether it is fully illuminated (therefore capturing only a part of the incident beam) or partially illuminated (capturing the entire incident beam). With respect to the reflected beam, assuming that the RIS is lossless, the total power of the beam

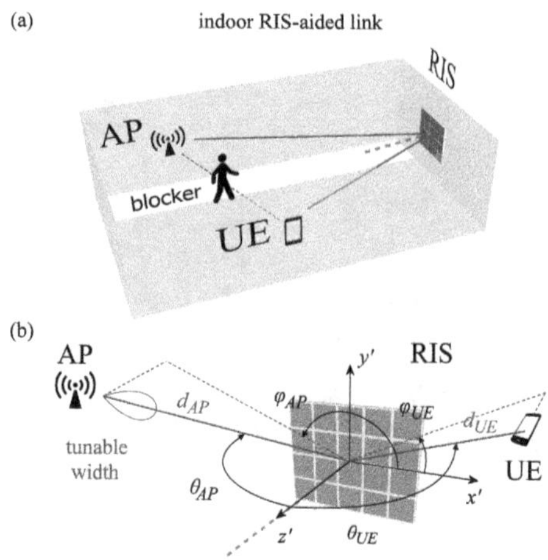

Figure 3.19. Indoor RIS-aided link system model: (a) LoS communication link interruption; (b) relative angles between the BS, RIS, and the user equipment.

reflected by the RIS is the same for all possible UE positions. However, the received power is determined by the local power density of the reflected beam (not the total beam power) and, therefore, beam spreading that causes the beam peak power to drop will have a direct impact on the received power.

A set of simulation results for the spatial distribution of received power as a function of user equipment position in ideal alignment is presented in Figure 3.20. Here, Figure 3.20(a) and (c) shows the received power as a function of the UE position in the absence of misalignment, while Figure 3.20(b) and (d) shows cross section at $z = 3$ m for the cases marked at Figure 3.20(a) and (c) with steering angles at $0°$, $20°$, and $40°$. Furthermore, the cases in Figure 3.20(a) and (b) represent the fully illuminated case, whereas Figure 3.20(c) and (d) shows the partially illuminated case. The contour lines represent the locus of user equipment positions, where the same received power can be achieved for each individual case. This is shown in Figure 3.20(a) and (c) for all possible user equipment positions within the illustrated area. The solid lines represent three chosen steering angles, namely steering angles at $0°$ (blue), $20°$ (orange), and $40°$ (yellow), and the dashed line marks their values at distance $z = 3$ m from the RIS (denoted with the short black line in the top). The contour lines represent the locus of user equipment positions at which the same received power can be achieved for each individual case. The spatial distribution of the received power for the chosen angles along the dash-line is shown in detail in Figure 3.20(b),(d). It is observed that, with increasing RIS to UE elevation angle, the beam spreading becomes more severe, resulting in the reduction of the beam's peak power and, consequently, of the received power (in case of ideal alignment the user equipment is located at the peak of each distribution).

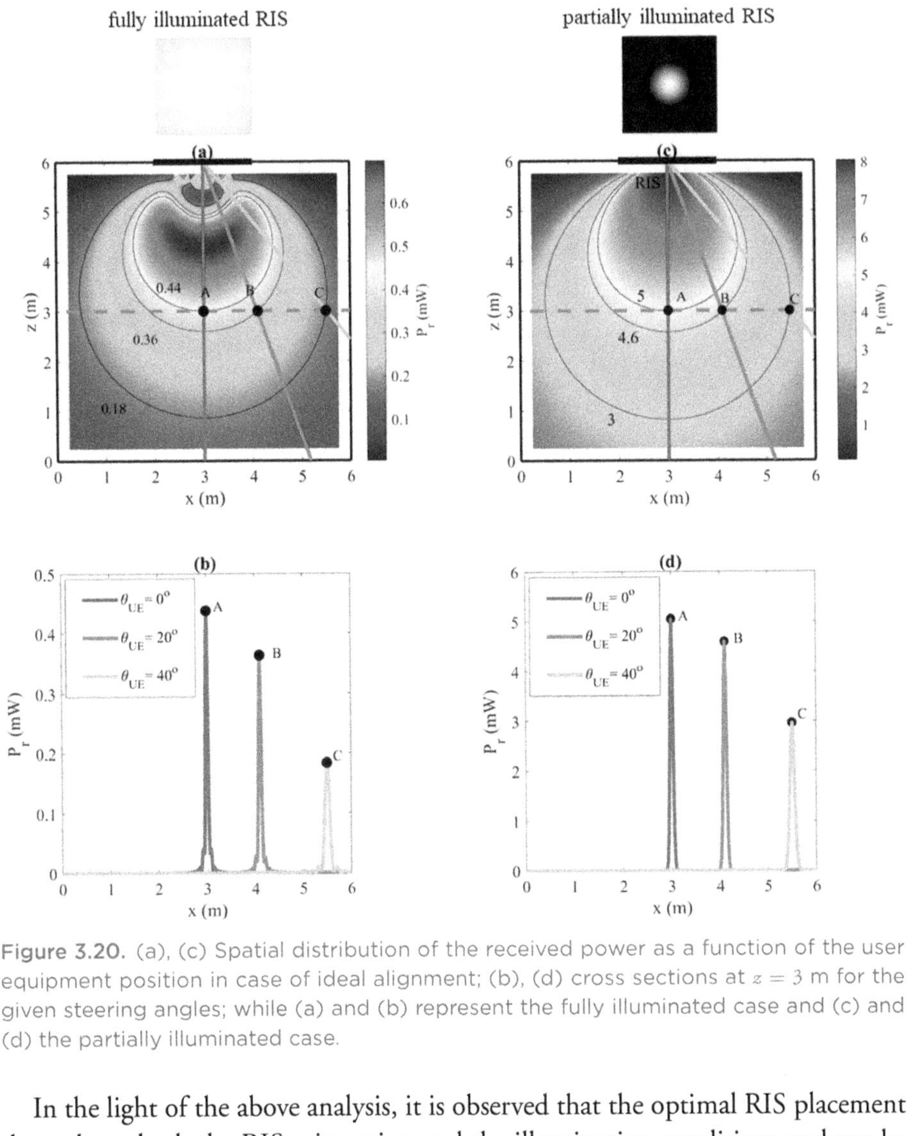

Figure 3.20. (a), (c) Spatial distribution of the received power as a function of the user equipment position in case of ideal alignment; (b), (d) cross sections at $z = 3$ m for the given steering angles; while (a) and (b) represent the fully illuminated case and (c) and (d) the partially illuminated case.

In the light of the above analysis, it is observed that the optimal RIS placement depends on both the RIS orientation and the illumination conditions and can be formulated in terms of the coverage provided by the RIS (or RIS efficiency).

As an example, the RIS orientation efficiency is shown in Figure 3.21 in terms of coverage and as a function of the orientation angle, i.e., the angle between the top wall and the RIS normal, as shown in the inset. While both BSs are equidistant from the RIS, the RIS orientation angle θ_o modifies the angle θ_{AP}, with direct consequences on the room coverage. Figure 3.21(a) represents the fully illuminated case with BS antenna gain equal to 35 dB, whereas Figure 3.21(b) shows the partially illuminated case with AP antenna gain equal to 55 dB. For the fully illuminated case shown in Figure 3.21(a), the received power is proportional to $\cos(\theta_{AP})$. As a

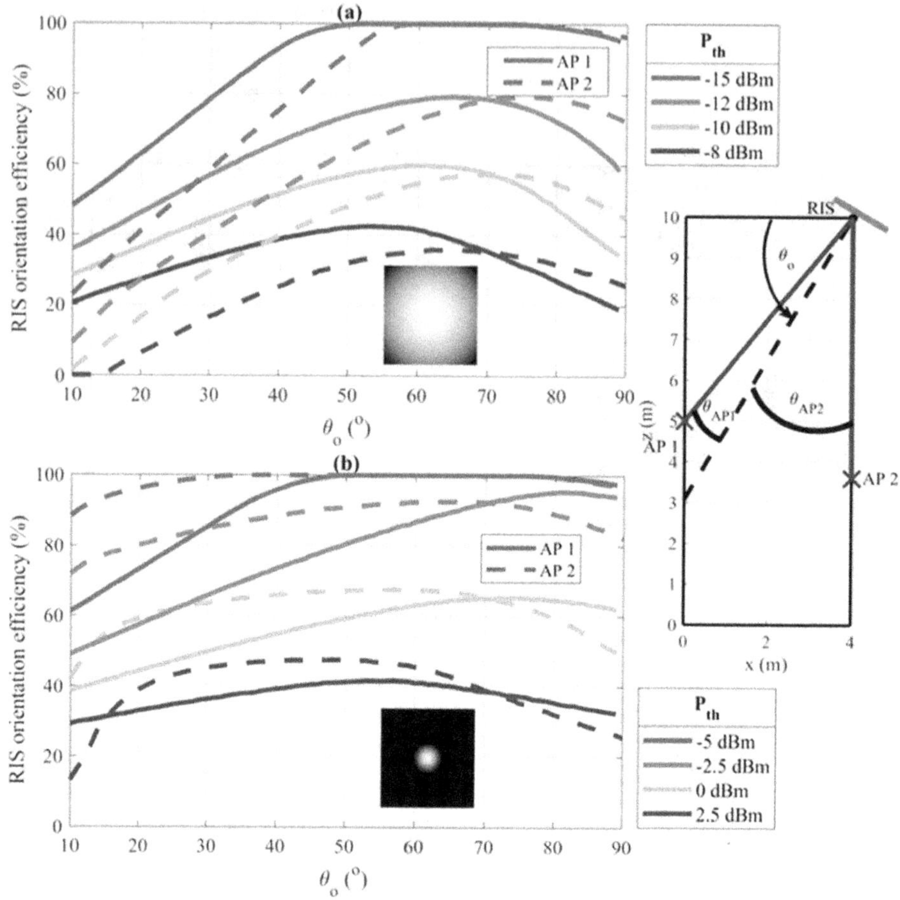

Figure 3.21. RIS orientation efficiency versus its orientation angle θ_o. (a) Fully illuminated RIS (35 dB BS antenna gain) and (b) partially illuminated RIS (55 dB BS antenna gain), for the indoor scenario shown in (c).

result, the efficiency of the first BS is higher than the efficiency of the second BS when $\theta_{AP1} < \theta_{AP2}$ (i.e., for $\theta_o \rightarrow 10°$), and vice versa. The two efficiencies become equal when $\theta_{AP1} = \theta_{AP2}$, which occurs when the RIS normal points towards the opposite corner of the room (bottom left corner). On the other hand, for the partial illumination case shown in Figure 3.21(b), the efficiency increases with the increase of θ_{AP}, because the beam footprint on the RIS becomes larger (acquires elliptical shape), in turn inducing weaker beam spreading and stronger peak power. In this case, the efficiency of AP1 is higher than the efficiency of AP2 when $\theta_{AP1} > \theta_{AP2}$ (i.e., for $\theta_o \rightarrow 90°$), and vice versa. Again, when $\theta_{AP1} = \theta_{AP2}$, the efficiency for both AP positions is equal. The solid and dashed lines illustrate the room coverage for the first and second BSs, respectively ($d_{AP} = 6.4$ m for both cases). The power thresholds examined in each case are depicted with the color-coded lines.

In conclusion, the RIS efficiency depends on its (a) position, (b) orientation, and (c) size relative to the size of the incident beam's footprint. In case of a fully illuminated RIS, the received power is proportional to cosine of the incident signal, which translates in decreased efficiency when the incident angle increases, and vice versa. The opposite occurs in the partially illuminated case, where increasing the incident angle leads to wider incident beam-footprint, in turn to weaker beam spreading, higher peak power and, consequently, to higher received power at the user equipment and therefore higher efficiency.

3.4.3 Cascaded Multi-RIS Scenarios

In case, the blind spot problem cannot be resolved using a single RIS (as, e.g., the LoS from the RIS to any of BS or the user is missing), then multiple-RISs can be used to provide uninterrupted connectivity between the BS and the user [73–75].

Multi-RIS can enable several novel functionalities, including, but not limited to, blockage avoidance, routing, coverage expansion, and beam splitting. An example scenario is a multi-RIS-empowered outdoor THz wireless systems discussed in [74] (Figure 3.22).

The result of a statistical characterization of the multi-RIS outdoor THz links and the associated performance gains, e.g., outage probability of the link as a function of RISs sizes and distances, are discussed in [73, 74]. These references analyse the impact of turbulence on the outage performance of multi-RIS-empowered THz

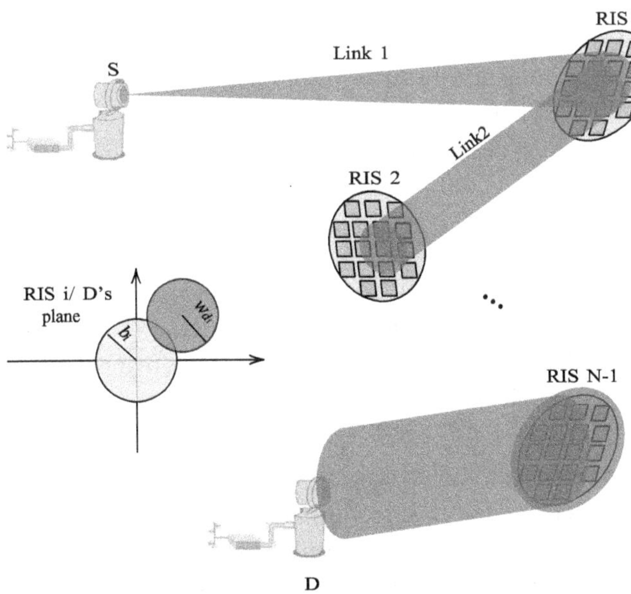

Figure 3.22. Cascaded multi-RIS-empowered THz wireless model.

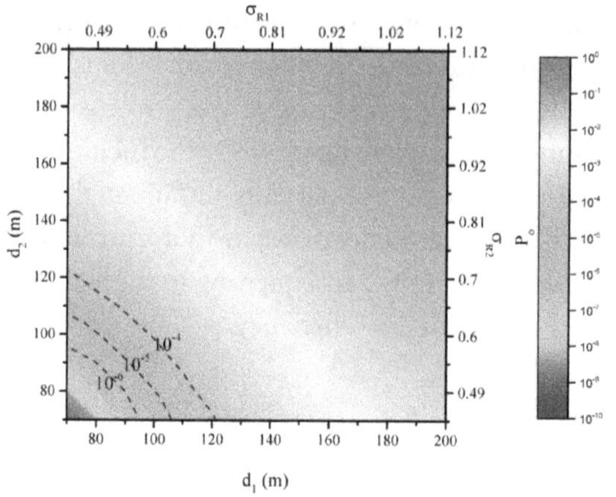

Figure 3.23. Outage probability for a link consisted of two RIS at distances d_1 and d_2.

wireless systems in the absence of misalignment and hardware imperfections for a double RIS link setup. Figure 3.23 shows the outage probability of the link as a function of d_1 and d_2, which are the distances to the first and from the first to the second RIS. Moreover, σ_{R1} and σ_{R2} are defined as

$$\sigma_{R_i}^2 = 1.23 C_n^2 \left(\frac{2\pi}{\lambda}\right)^{7/6} d_i^{11/6}$$

with λ is the wavelength and C_n^2 is the reflection index structure parameter [73, 74].

As expected, for a fixed d_1, as d_2 increases, σ_{R2} increases; thus, an outage performance degradation is observed. Similarly, for a given d_2, as d_1 increases, σ_{R1} also increases, i.e., turbulence intensity increases, in turn, the outage probability increases. Finally, from Figure 3.23, we observe that for a given transmission distance, $d_1 + d_2$, the worst outage performance is observed for $d_1 = d_2$.

3.4.4 RIS-assisted UAV Systems and Performance Analysis

As it was mentioned in the introduction of Section 3.4, one potential application of RIS is in enhancing the performance of UAV networks. Most existing analysis of the RIS-assisted UAV wireless systems, however, assumes that the RIS-UAV link is not directional. This results in neglecting the adverse effects of UAV disorientation and/or misalignment of the RIS-UAV beam. However, as the operating frequency and thus the directionality of links increase, even small disorientations and/or misalignments may adversely affect the performance of the RIS-assisted UAV wireless system. Aside from the joint effects of disorientation and misalignment, another important performance-limiting factor in high-frequency communications is the effect of transceiver hardware imperfections.

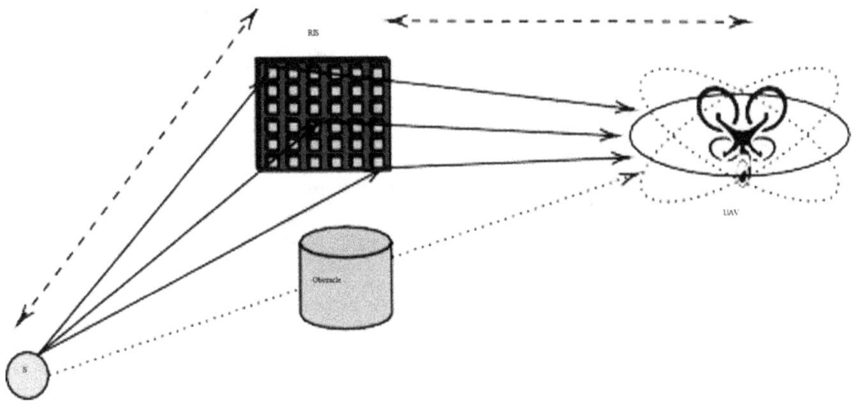

Figure 3.24. RIS assisted UAV scenario.

The performance analysis of RIS-assisted UAV wireless systems that accounts for the impact of different propagation environments, UAV disorientation and/or misalignment of the RIS-UAV beam, and the transceivers RF front-end imperfections is presented in [73]. Motivated by this, in this section, a contribution in this area is presented [76].

Figure 3.24 shows a RIS-assisted UAV scenario in which a BS (or a UE) communicates with a UAV via a RIS. For the sake of analysis, it is assumed that no direct link can be established between the BS and the UAV, due to blockage. It is assumed that the UAV is in a hover state, where both the position and orientation of the UAV are not completely fixed. In this case, both the BS and the UAV are equipped with single antennas, while the RIS is equipped with multiple antennas.

Figure 3.25 demonstrates the joint impact of hardware imperfections and fading on the outage performance of RIS-assisted UAV wireless systems, where the outage is plotted as a function of the transmitter and UAV receiver error vector magnitudes, κ_s and κ_d, for different values of threshold [76]. In this case, $\kappa_s = 0$ indicates that the BS is equipped with ideal RF front-end. Similarly, $\kappa_d = 0$ means that the UAV is equipped with ideal RF front-end. Hence, the $(\kappa_s, \kappa_d) = (0, 0)$ point represents the case in which both the S transmitter and UAV receiver are ideal. As expected, for given threshold and κ_s, as κ_d increases, the outage also increases. Similarly, for fixed threshold and κ_d, as κ_s increases the outage performance degrades.

3.5 Multi-Access Connectivity

The 5G CN supports integration of non-3GPP access networks via interworking functions that provide a secure connection for the UE accessing the 5G CN over non-3GPP access networks. Such integration enables a UE to establish multiple

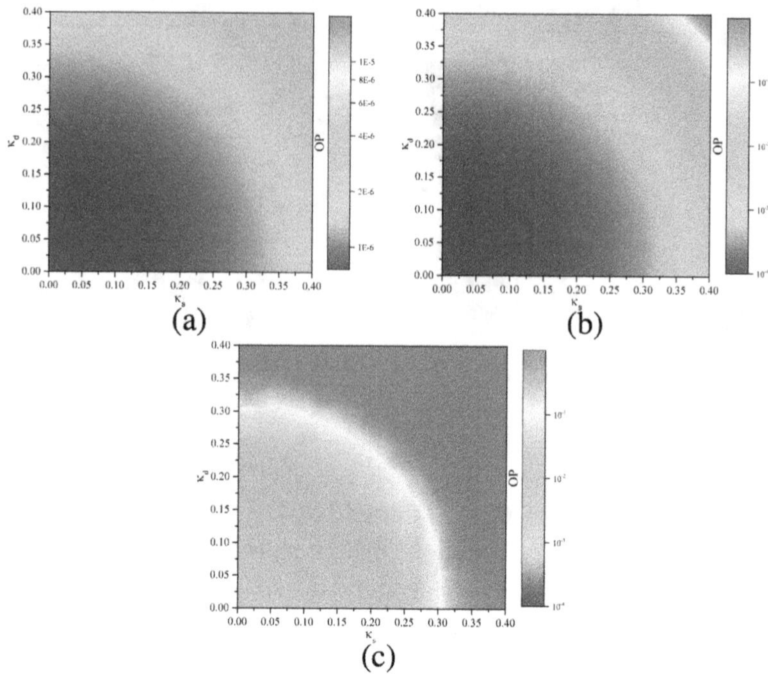

Figure 3.25. RIS-assisted UAV link outage probability against imperfect link parameters.

sessions to receive and send data traffic over 3GPP access and/or non-3GPP access. In other words, the UE can have simultaneous connectivity over 3GPP and non-3GPP access networks.

The procedure that steers a UE traffic onto available access networks and enables multi-access connectivity for a single UE is termed as Access Traffic Steering, Switching and Splitting (ATSSS) aka AT3S, by 3GPP [77]. AT3S framework promises to have a great importance for beyond 5G and 6G networks to enable ubiquitous connectivity and service continuity aspects by utilizing all available networks including 3GPP and non-3GPP access networks.

Current framework defined for AT3S by 3GPP has several steering modes, such as active-standby, smallest delay, throughput aggregation, load balancing, redundant, and priority based, and can use multi-path transmission control protocol (MPTCP) to send data through different available network interfaces. The widespread availability of HetNets and the enablement of multi-access connectivity from a single UE to the available access networks increase the complexity level of access traffic routing and resource management for beyond 5G and 6G networks.

An improved version of the 3GPP AT3S framework, called enhanced AT3S (eAT3S), is proposed that uses multi-wireless access technology (multi-WAT) telemetry in an xApp within the ORAN RIC framework to add near-RT RAN control of the AT3S policies behind the user plane function (UPF) [78].

Accordingly, the telemetry and performance measurement values may be used to configure how the traffic splitting shall be performed, e.g., to estimate the proportion of traffic that shall be sent over the different access traffic flows to achieve a particular objective. The implementations presented in this section are done on a 5G Wi-Fi 6 constellation; however, the learnings are relevant for beyond 5G evolution towards 6G.

The 3GPP system architecture specifies that the AT3S can support two steering functions: a high-layer steering function, based on the MPTCP protocol, and a low-layer (LL) steering function based on the AT3S function (AT3S-LL). Each steering functionality in the UE enables traffic steering, switching, and splitting across 3GPP access and non-3GPP access, in accordance with the AT3S rules provided by the network. Regarding MPTCP, the UPF may support MPTCP Proxy functionality on the network side. This capability interacts with the MPTCP functionality in the UE by using the MPTCP protocol.

Standard TCP connections are identified by a four tuple, which includes the source and destination IP addresses and source and destination ports, whose packets are sent through a single link. In the case of MPTCP, defined in RFC 6824 [79], several paths (called sub-flows) are aggregated to create one connection. The protocol includes operations to handle when and how to add or remove paths, to be compatible with legacy TCP hardware (e.g., firewalls may reject TCP connections, if the sequence numbers are not successive) and to define a fair congestion control strategy between the different links and the different hosts.

MPTCP allows the use of different packet scheduling schemes, which select the sub-flow that will forward the next packet. These schedulers may have been designed for different purposes, such as throughput aggregation, reliability improvement, and latency reduction. In particular, the MPTCP implementation available at [80] includes three main packet schedulers:

- A default MPTCP scheduler, which selects the sub-flow with the shortest smooth round-trip time (SRTT) estimated delay, which may be useful for time critical services.
- A redundant scheduler, which forwards each packet through all the sub-flows to achieve redundancy and, thus, increase reliability. Since the packet received first is processed, it also achieves low latency.
- A round-robin scheduler, which transmits packets over the different sub-flows according to a fixed cyclic schedule. Since the different sub-flows forward different data, it achieves throughput aggregation.

These concepts are implemented and can be tested in a virtualized testbed [81, 82]. A framework built on virtual machines (VMs) that incorporate numerous network interfaces as though they were the various radio access technologies has

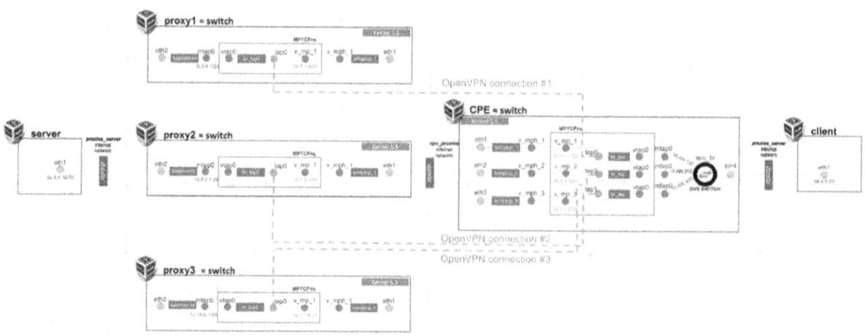

Figure 3.26. Virtualized multi-connectivity testbed [82].

been developed to evaluate these goals. This virtual testbed is publicly available at [82] and is depicted in Figure 3.26. In this framework, one of the VMs acts as customer premises equipment (CPE) that connects to another VM that acts as a proxy, which is located behind the 5G CN. Both the CPE and the proxy use an MPTCP-enabled kernel [80]. One (or more) client VMs attached to the CPE connect to a server behind the proxy, and a tunnel is used between the CPE and the proxy so that all traffic is transmitted using MPTCP. The main results of the evaluation of the three different schedulers are included in [78, 81]. The repository also includes a version of the testbed which connects non-3GPP networks (such as Wi-Fi 6 and LiFi) through the non-3GPP interworking function (N3IWF) using the *free5gc* platform.

Between the CPE and the MPTCP proxy, an L2 tunnel is created that enables transparent L2 traffic transit over various WATs. Since the traffic from this L2 tunnel is sent via MPTCP, all traffic passing through the CPE (such as that sent by clients connected to the CPE) is also done and is therefore scheduled using the MPTCP scheduler that has been chosen. Additionally, by establishing many L2 tunnels to various proxies, the CPE can change the packet scheduler in real time without disrupting the TCP (MPTCP) connection.

Based on this testbed, several demonstrators have been implemented [81]. Figure 3.27 shows one of these multi-connectivity testbeds that include 5G NR and Wi-Fi 6 as wireless access technologies. The adoption of MPTCP for the multi-access framework allows the performance of seamless vertical handover, since the schedulers will consider only the links that are available. The tests have shown that the data transportation also is resumed when an interface is recovered. Using these demonstrators, three aggregation modes have been tested, i.e., capacity aggregation mode, delay sensitive aggregation mode, and reliability aggregation mode [81]. These modes are described in the following subsections.

Capacity aggregation mode: For use cases requiring the highest user data rate, like enhanced mobile broadband (eMBB), the capacity aggregation mode might be

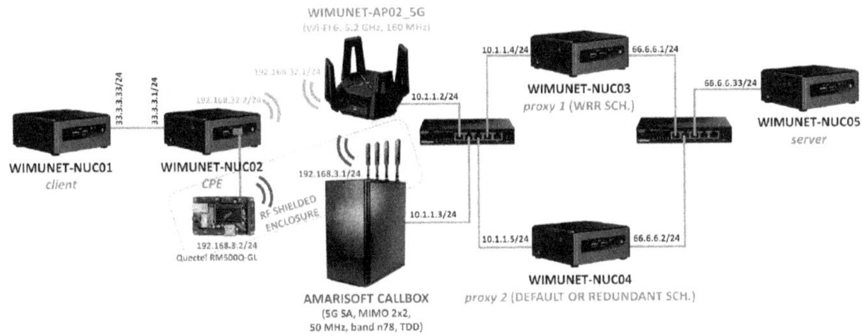

Figure 3.27. 5G NR, Wi-Fi multi-connectivity testbed.

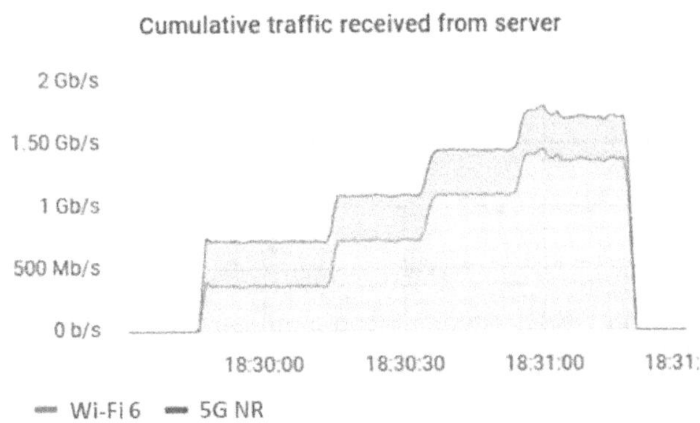

Figure 3.28. Sample results from the scheduler implemented in [81] using 5G NR and Wi-Fi 6.

beneficial. To achieve the maximum aggregate throughput, a variant of the weighted round-robin (WRR) scheduler (available at [79]) is implemented [81]. This WRR implementation allows to dynamically assign the number of turns (or weights) for each available network interface within a round.

The benefit of using this approach is twofold. On the one hand, the operator is able to adjust what proportion of traffic will be served by each radio technology. The proper assignment of the weights can lead to achieve the maximum aggregate throughput. Figure 3.28 shows an example in which the testbed reached 350 Mb/s with 5G and 1.4 Gb/s with Wi-Fi 6. As it can be observed, by using the proposed scheduler, it is possible to adjust the percentage of traffic served by each technology (varying the weights between 1–1, 1–2, 1–3, and 1–4 in this example). Moreover, with this approach, the implementation reaches the maximum aggregate throughput (around 1.75 Gb/s in this case).

Delay sensitive aggregation mode: For ultra-reliable and low-latency communication (URLLC) service cases, the delay sensitive aggregation mode can be beneficial.

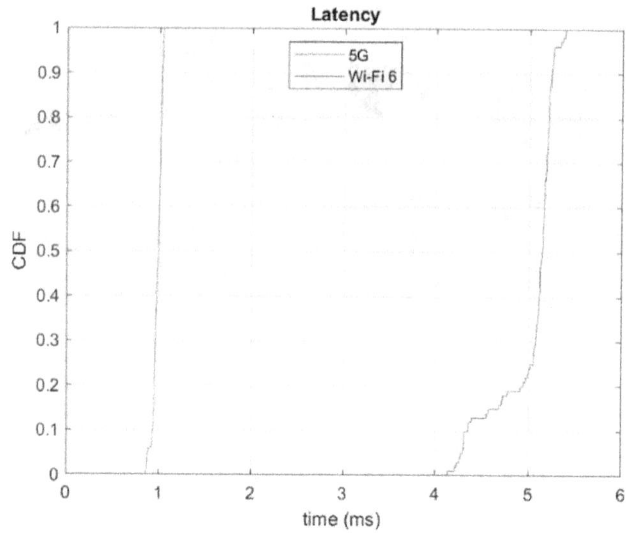

Figure 3.29. Latency of 5G and Wi-Fi 6 in the proposed testbed [81].

To achieve the minimum latency, the terminal shall employ an MPTCP scheduler such as the redundant (which sends the same information through all the different wireless technologies) or the default (which selects the network interface with lowest round trip time (RTT)). As shown in Figure 3.29, Wi-Fi 6 achieves a much lower latency (around 1 ms) compared to 5G (around 5 ms), mainly due to the complexity of the 5G core and the gNB (BS in 3GPP terminology), which increases the processing time.

Figure 3.30 shows the latency achieved by using multi-connectivity with both 5G and Wi-Fi 6. As expected, the latency obtained with the redundant and the default schedulers is the one achieved by the Wi-Fi 6 interface (approx. 1 ms). Even for the round-robin scheduler, the latency is reduced to approximately the average between the latency of 5G and Wi-Fi 6 (slightly higher than 2 ms). Thus, the redundant and the default MPTCP schedulers are suitable for time critical services.

These measurements have been performed using the *netperf* tool from NMAP in order to measure the latency in TCP (i.e., using MPTCP). Latency is estimated as the one-way delay, i.e., half of the RTT.

Reliability aggregation mode: Regarding the air interface reliability requirement, it is achieved by design because MPTCP ensures a reliable packet delivery. The trade-off introduced by MPTCP is that packets could experience too much latency, e.g., due to TCP retransmissions, thus not being useful for the application. We can, therefore, reinterpret the reliability requirement in terms of the latency requirement, which was already commented in this section.

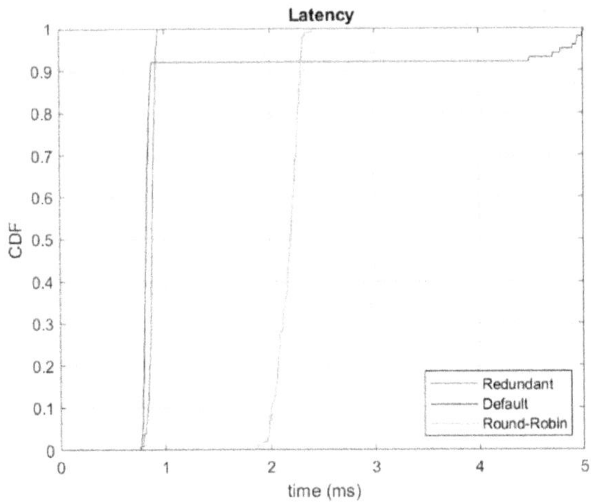

Figure 3.30. Latency of MPTCP using different schedulers [81].

3.5.1 Vertical Handover

AT3S enables simultaneous access to 3GPP and non-3GPP networks. The handover can be considered as horizontal handover or vertical handover. In horizontal handover, the UE moves within the same access network. In vertical handover, the UE moves between the 3GPP and non-3GPP access networks. When UE moves within the same access network, AT3S can be used to enable service continuity by steering the access traffic onto the other access network. Therefore, AT3S can be utilized in network performance optimizations, such as load balancing, service continuity, or seamless handover. The noted three optimization problems can also be considered as part of one single problem that focuses on the overall mobility optimization for the UE.

Within the AT3S framework, there is technically no vertical handover between access networks. In other words, there is no break-before-make approach in multi-access connectivity scenarios. Instead, when an access link is enabled, MPTCP detects the link as an available network interface and performs data traffic routing/scheduling accordingly. Therefore, the vertical handover time between available access networks can be considered as an elapsed time for an interface transition from disabled to enabled where MPTCP can start sending data through the recently enabled interface. This timer can be termed as "transmission resume delay" to mitigate any confusion on the conventionally used handover timers.

In order to measure transmission resume delay for multi-access connectivity scenarios, a controlled experiment environment is created. Accordingly, two machines are virtualized with VirtualBox version 6.1 and connected to two virtual networks via two network interfaces. Each machine runs an Ubuntu 20.4 image with the modified kernel, which includes a specific MPTCP module designed in [81]. A test traffic is generated using the *iperf* tool. Traffic is sent from one VM (VM1) to the other VM (VM2). VM1 runs a script which detects when the network interfaces become enabled or disabled. To this end, this script uses *NETLink* sockets. This way, as soon as an interface changes, the event is captured and notified. VM1 also runs the *tcpdump* tool to capture traffic from the interface that will be enabled and disabled. Since *detect-interfaces* and *tcpdump* run in the same machine, they both use the same clock. This way, the timestamp of the event of a new available interface generated by *detect-interfaces* and the one from the *tcpdump* network capture can be used to determine the transmission resume delay. The delay is estimated by an AWK script which processes the output of *detect-interfaces* and *tcpdump*.

To simulate the outage of a link, one of the network interfaces of VM1 is disabled and enabled from the VirtualBox user interface several times during the test. A total of 100 temporal outages were simulated. As it can be seen from Figure 3.31, more than the 60% of the outages resume the transmission after 0.065 s since the interface is re-enabled. Two allocation and retention priority (ARP) resolution requests are made before sending a TCP segment, one for each destination IP address. Nevertheless, this ARP resolution lasts 0.6 ms in the worst case. The results of these experiments demonstrate that the MPTCP approach for the multi-access connectivity allows to provide continuous transmission even when a link fails. This "always-on" type of connection replaces the need for a vertical handover across access networks. Therefore, the time with no connectivity due to a handover becomes zero, and service continuity can be provided during the handover process.

Figure 3.31 shows that the measured time to resume of the transmission through a new link is nearly 60 ms. To identify the source of this delay, several experiments are conducted. To start with, Ethernet frames are forged with the source and destination MAC addresses of some of the MPTCP packets captured during the previous experiments. The payload of these frames is randomly created. A Python script that sends these frames continuously is also created. The time to resume is measured as in previous experiments. The resulting measurements show that the average delay is near to 5 ms.

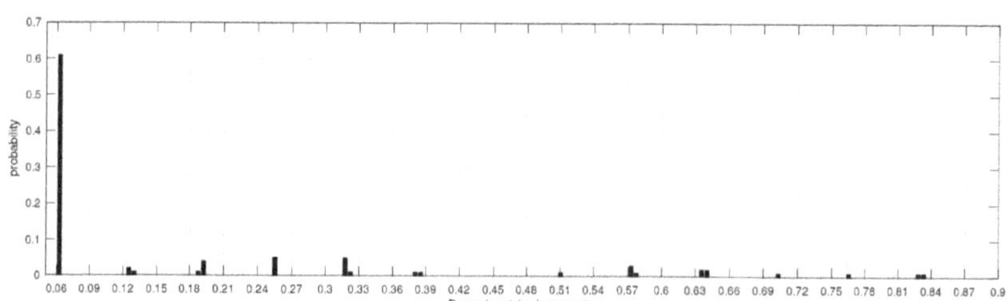

Figure 3.31. Time to recover from link outage [81].

3.6 Sub-THz for Ultra-High Data Rate

As discussed in Section 3.2.4.2, the availability of several GHz bandwidths at mmWave and Sub-THz bands makes them an attractive technology for future throughput demanding 6G use cases. Besides its potential for communication systems, the wide sub-THz spectrum is beneficial for radio sensing to increase the time, angular, and frequency resolution, in addition to providing the ability to explore the physical properties with spectroscopy [83]. However, these advantages come with challenging link budget, because of the propagation loss caused by blockage and atmospheric absorption, the decreased antenna aperture at high carrier frequencies, the increased noise associated with bandwidth, in addition to the RF hardware non-idealities and limitations. Therefore, multi-antenna implementation with high-gain beamforming is necessary to increase the link range. As a result, it is not feasible to develop one solution that fits all scenarios, but the radio design needs to be analysed for specific link requirements, considering the characteristics of the radio channel and the physical properties of the hardware components. Starting from the link requirements, in terms of data rate, range, and mobility, which are determined from the relevant use cases, the technical requirements and parameters for radio design can be analysed [84].

This section provides an overview of the use case families subjected to the radio access in the sub-THz range (100–300 GHz), and discusses technical aspects related to link modelling, RF impairment, hardware modelling, radio, and beamforming architecture, in addition to the impact of radio channel and waveform design.

3.6.1 Use Cases and Technical Requirements

The relevant communication use cases require ultra-high data rate, and they can be mapped to two scenarios based on the range, which highly influence the underlying wireless technologies [5, 85]:

Short-range wireless connectivity: corresponds to small-cell scenario with typical cell size blow 100 m and typical peak data rate of 100 Gb/s. This scenario enables wireless access for device to infrastructure communication at a short range, for general purpose access through a hotspot deployed in indoor environment, such as home for augmented reality/virtual reality (AR/VR) applications, industrial campus for digital twin and industrial control applications, and outdoor for the applications of smart cities. This scenario can also be considered for D2D communication in the applications of digital immersion and telepresence, such as for wirelessly connecting displays and docking stations. Due to the short range of this scenario, the target mobility is low (<10 km/h).

Long-range wireless connectivity: can be exploited to improve coverage by providing long-range fixed wireless access to sparsely populated or hardly accessible areas, which cannot be covered by short-range infrastructure. Additionally, they are deemed a cost-effective alternative of optical fibres for the interconnection between small cells in network densification, and for backhaul. The communication range is from 200 m up to 2 km in predominantly fixed outdoor LoS environment, but it is also possible for interconnecting indoor RUs, and for providing backhaul to flying BS in non-terrestrial networks (NTN). The data rate depends on the specific scenario and can reach 1 Tb/s for infrastructure backhaul.

The basic technical requirements to achieve the peak data rate include the bandwidth, number of RF chains, modulation and coding scheme (MCS), in addition to the corresponding minimum SNR. The required SNR is derived from the link budget, which imposes transceiver design requirements including transmit power, antenna array gain, while considering the hardware impairment models, channel characteristics and mobility requirements. Moreover, the link budget is directly related to the link range [84].

3.6.1.1 Bandwidth and number of RF chains

The RF bandwidth B (i.e., the passband bandwidth) required for supporting a peak data rate R_p depends on the number of orthogonal spatial and frequency channels M_{ch}, the modulation order $Q_{c,m}$, and code rate $r_{c,m}$ in each chain. MCS corresponds to a number bit/symbol $L_{c,m} = r_{c,m}\log_2 Q_{c,m}$. Without considering the impact of filtering and the required guard band, and assuming the symbol rate is $1/B_m$, the relation between these parameters can be expressed as $R_p = \sum_{m=1}^{M_{ch}} L_{c,m}B_m$. Using multiple channels reduces the bandwidth requirements for the RF chain, and it is a solution when a contiguous wideband is not available. In addition, the availability of spatial channels depends on the properties of the wireless channel and the number of serving radio nodes in the case of distributed MIMO deployments. When M_{ch} corresponds to the number of spatial streams, a smaller bandwidth $B = B_m$ can be used leading to an increase in the spectral efficiency, $\mathrm{SE} = R_p/B = \sum_{m=1}^{M_{ch}} L_{c,m}$. Whereas in the case of multiple frequency channels, the aggregated bandwidth is $B_{\mathrm{agg}} = \sum_{m=1}^{M_{ch}} B_m$, and the required spectral efficiency per channel is $\mathrm{SE}_m = L_{c,m}$, with average $\mathrm{SE} = \frac{1}{M_{ch}} \sum_{m=1}^{M_{ch}} \mathrm{SE}_m$. Note that the frequency channelization does not increase the spectral efficiency in comparison to spatial multiplexing.

LoS scenario: assuming $B_m = B$, $r_{c,m} = r_c$, and $Q_{c,m} = Q_c$, then, $B_{\mathrm{agg}} = M_{ch}B$, such that $\mathrm{SE} = L_c = r_c\log_2(Q_c)$, and therefore,

$$R_p = M_{ch}B\, r_C\log_2 Q_c \Rightarrow B = \frac{R_p}{M_{ch}r_C\log_2 Q_c}. \tag{3.1}$$

As can be seen from (3.1), there are different possibilities to choose the bandwidth and MCS parameters. For instance, when employing 16-QAM, code rate 5/6, and a single channel, the required RF bandwidth for achieving 100 Gb/s is 30 GHz, and the baseband bandwidth (i.e., the cut-off frequency of low-pass filter) is 15 GHz. This is less or equal to 10% of the carrier frequencies in the range 150–300 GHz, which is seen as the target for RF circuity design. Each MCS requires sufficient SNR at the receiver to fulfil predefined performance measure, such as bit error rate (BER) and packet error rate (PER) [LJT+19] or based on information theoretical analysis with additional margin for the implementation algorithms, as discussed in the next subsection.

3.6.1.2 SNR requirements

Using the additive noise model after equalization, $y = d + v$, the SNR of the symbol is defined by $\mathrm{SNR} = \frac{E[|d|^2]}{E[|v|^2]}$. Shannon formula assumes additive white Gaussian noise (AWGN) channel and Gaussian symbols, and thus, the minimum SNR can be derived from $\log_2(1 + \mathrm{SNR}) \geq L_c$. However, considering discrete constellation $\mathcal{M}_c = d_0, \ldots, d_{Q_c-1}$ and uniform distribution of the symbols, the mutual information for a noise variance $\sigma^2 = E[|v|^2]$ can be computed from the formula

$$I(\mathcal{M}_c, \sigma^2) = \log_2 Q_c - \frac{1}{Q_c \pi \sigma^2} \int \sum_{0}^{Q_c-1} e^{-\frac{|y-d_m|^2}{\sigma^2}} \log\left(\sum_{0}^{Q_c-1} e^{-\frac{|y-d_m|^2}{\sigma^2}}\right) dy. \quad (3.2)$$

Figure 3.32 shows the theoretical achievable mutual information for different QAM order at different SNRs. As can be seen from the figure, for uncorded modulation, the minimum SNR for uncorded 16-QAM is about 17.5 dB so that $L_c \approx 4$ bits/symbol. By giving sufficient margin for processing, a reliable transmission can be achieved. Moreover, by employing channel coding with code rate $r_c = 1/2$, $L_c = 2$ bits/symbol. In theory, the minimum SNR for the latter case is 5 dB. However, by allowing a margin for decoding algorithms, the required SNR is higher. Practically, the SNR requirements in the coded case compared to the uncoded one can be reduced by $10 \log_{10}(r_c)$ dB. Thus, in the case of 16-QAM and $r_c = 1/2$, a minimum SNR of 14.5 is sufficient.

The SNR is calculated from the link model, and it is proportional to the ratio between the transmitted power and bandwidth, $\frac{P_{TX}}{B}$, per chain. Based on that, the required SNR in case of employing multiple frequency channel for the same MCS is identical. Since the bandwidth $B = \frac{B_{\text{agg}}}{M_{ch}}$, and the total power $P_{TX,\text{tot}} = M_{ch} P_{TX}$, the required SNR does not change with frequency channelization. Nevertheless,

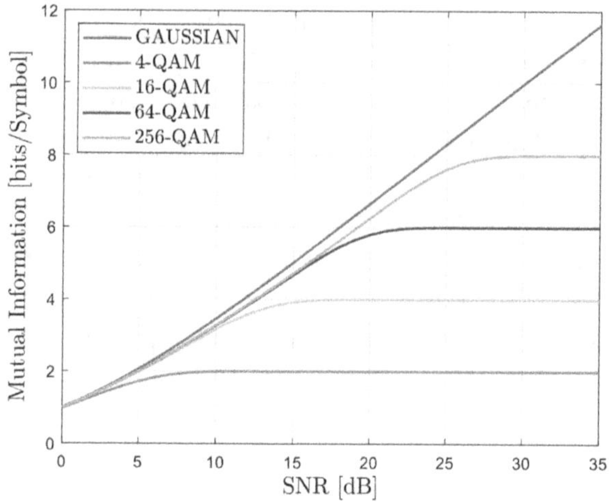

Figure 3.32. Mutual information for QAM.

the realization of the hardware architecture to achieve such SNR can be relaxed for smaller bandwidth.

3.6.1.3 Link model

The abstracted link model of the digital baseband signal relates the transmitted signal $x[n]$ to the received signal $y[n]$. In LoS, the link model can be expressed in the form

$$y[n] = hx[n] + z[n] + v[n], \tag{3.3}$$

where h is the channel gain, $v[n]$ the additive noise that is independent of the signal, and the additional term $z[n]$ denotes the self-interference term because of the hardware non-idealities. The channel gain is related to the link budget, the noise power is determined from the thermal noise, the noise figure, and the bandwidth, and the interference power is computed from the hardware models. In the low SNR region, the receiver performance is inevitably noise limited, while in the higher SNRs the performance is limited by the non-linearity. Note that $x[n]$ can be the modulated signal, and thus, after demodulation, additional processing gain can be achieved. In fact, (3.3) considers the linearization of the hardware response, where the RF non-idealities are treaded as additive noise. In general, a non-linear model can be used such that

$$y[n] = f(h, x[n]) + v[n], \tag{3.4}$$

where $f(h, x[n])$ is a non-linear function that depends on the hardware models.

As illustrated in Figure 3.33, which focuses on one signal chain, the link budget, is computed from the relation between the received power P_{RX} and the transmitted

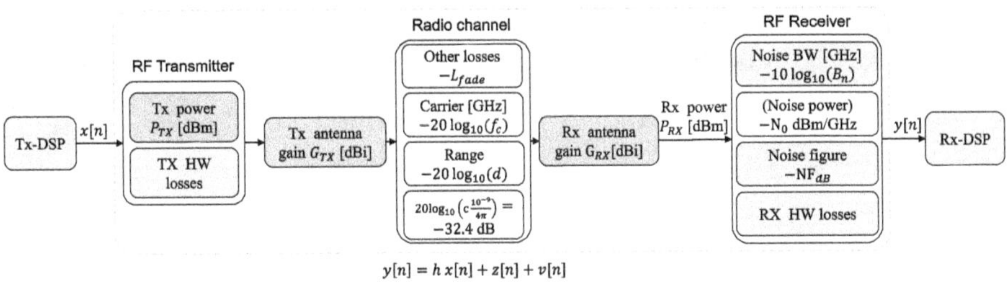

Figure 3.33. Link budget model.

power P_{TX}, given the total antenna gains $G_{a,TX}$ and $G_{a,RX}$ at the transmitter and receiver, respectively, in addition to the propagation losses, such that [85, 86]:

$$\begin{aligned} P_{RX}[dBm] = {} & P_{TX}[dBm] + G_{a,TX}[dBi] + G_{a,RX}[dBi] \\ & - 20\log_{10}(f_c[GHz]) - 32.4 - 20\log_{10}(d) - L_{fade}. \end{aligned} \tag{3.5}$$

Here, f_c is the carrier frequency, d is the range, and L_{fade} denotes other path losses, such absorption or fading in the case of non-LoS. The link budget decreases by 6 dB by doubling the range or doubling the carrier frequency. This can be compensated, for instance, by doubling the antenna gains at the transmitter and receiver for the same transmit power.

Considering the hardware imperfections, the SNR can be expressed in the form

$$\begin{aligned} \text{SNR[dB]} = {} & P_{RX}[\text{dBm}] - \text{N}_0[\text{dBm/Hz}] \\ & - 10\log_{10}(B_n[Hz]) - \text{NF[dB]} - L_{\text{HW}}. \end{aligned} \tag{3.6}$$

The term $\text{N}_0 = 10\log_{10}(1000\ \text{kT})$ is the noise spectral density, where k is Boltzmann's constant, T temperature in Kelvin (for typical temperature $T = 300$ K, $\text{N}_0 = -174$ dBm/Hz). Moreover, B_n is the noise bandwidth, which can be higher than the signal bandwidth B. Furthermore, the term L_{HW} represents the overall losses in the hardware. The SNR can be increased by increasing the link budget, which depends on the RF transceiver architecture. In particular, the antenna gains represent the overall antenna element gains and array gain, and P_{TX} corresponds to the total power generated by the PAs. Thus, the EIRP ($P_{TX}[dBm] + G_{TX}[dBi]$) can result from the power radiated by different number of antenna elements and PAs, and the behaviour of highly directive antennas such as horn, lenses, and reflectors. The sensitivity of the receiver P_{sens} is computed according to the minimum SNR_{min} requirements for the targeted MCS,

$$\begin{aligned} P_{\text{sens}}[\text{dBm}] = {} & \text{SNR}_{\text{min}}[\text{dB}] + \text{N}_0[\text{dBm/Hz}] \\ & + 10\log_{10}(B_n[Hz]) + \text{NF[dB]} + L_{\text{HW}}. \end{aligned} \tag{3.7}$$

From system level perspective, the link budget parameters highly depend on the technology, bandwidth, carrier frequency, and waveforms. For instance, the transmitted power results form

$$P_{TX} = P_{\text{sat}} - P_{\text{BO}} - L_{\text{ant-PA}}, \tag{3.8}$$

where P_{sat} is the saturated (maximum) output power, P_{BO} is the back off needed, and $L_{\text{ant-PA}}$ is the loss between PA and antenna. Moreover, the hardware imperfections, which contribute to L_{HW}, depend on parameters related to RF, and analogue and digital constraints that tend to worsen by increasing the frequency. These parameters include

- Transition frequency of the transistor f_T of low-noise amplifier (LNA), which impacts the receiver noise.
- Frequency of unity unilateral gain f_{max} of PA, which impacts the transmitted power.
- Parameters impact the detection including the phase noise N_{ph}, digital gate delay (t_{Gate}), clock frequency (f_{clk}), clock jitter Δt_{jitter}, quantization noise N_q, and maximum signal level V_{max}.
- Other signal-dependent non-idealities or distortion components D_{other}, and cross-coupling between channels I_{mutual}.

3.6.2 Radio Design Consideration

A functional block diagram of potential single phased array transmitter and receiver models is shown in Figure 3.34, whereas the overall system may consist of several chains. At the transmitter, the complex-valued discrete complex signal $x[n]$ is converted to baseband signals, $x_I(t)$ and $x_Q(t)$, using DAC and LP. The baseband is upconverted to RF signal $s(t)$ using mixers assuming direct conversion. The signal is spitted to parallel signals at the beamforming module that pass through the gain and phase control blocks, then to the PA, and finally to the antenna elements. The overall transmitter signal model considers the impacts of DAC quantization, I/Q imbalance, phase noise, analogue beamforming, PA non-linearity, and the antennas. The received signal at each antenna is amplified by an LNA connected to gain and phase control at the receiver beamforming block, then the output signals are compinged in the received signal $r(t)$. The baseband signals $y_I(t)$ and $y_Q(t)$ are generated by the IQ mixer and LP filter, which are converted by the ADCs to obtain the discrete complex signal $y[n]$. Similarly, the transmitter signal model contains the impacts of ADC quantization, I/Q imbalance, phase noise, and analogue beamforming. The LNA can be non-linear as well, which impacts the analogue gain control (AGC) functionality.

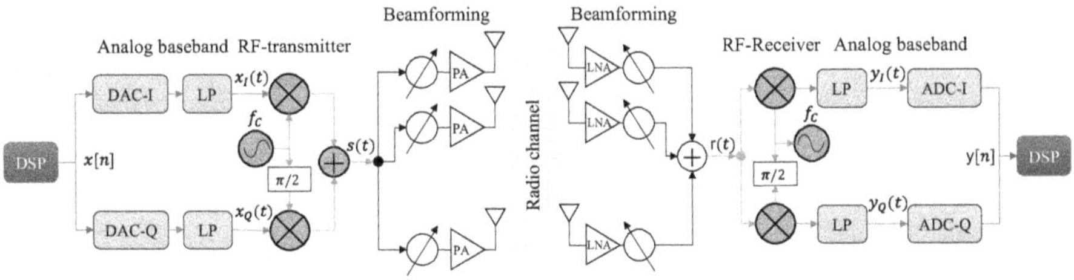

Figure 3.34. Functional RF blocks of a single phased-array transmitter and receiver.

Although the transceiver architecture looks similar to the one used in 5G mmWave, the key difference is in the non-ideal behaviour of the RF components at the frequency range above 100 GHz and ultra-wideband, as well as the increased technical challenges in the hardware implementation. Moreover, to achieve data rate larger than 100 Gb/s at a specified range with affordable complexity, form factor, and energy consumption, implementation with multiple RF chains is required either in the form of multiple aggregated frequency channels, in addition to the consideration of hybrid beamforming architecture depending on the availability of spatial beams. Therefore, several crucial aspects need to be considered in the radio design:

Duplex mode: TDD operation is envisioned because of the highly integrated modules of the antennas and the related LNAs and PAs as well as a large number of antennas. TDD allows also exploiting radio channel reciprocity as the same band is used for transmission and reception. However, to exploit this feature, the transmitter and receiver need to be calibrated to account for different frontend hardware responses. In addition, the antenna can be used for transition and reception, which requires a T/R switch at the cost of additional losses between PA and the antennas [87]. To avoid such losses, another option is to use separate antenna arrays for the transmitter and receiver. In this case, the channel reciprocity is not fulfilled, but still the TDD operation is favourable to avoid costly diplexers [88].

Transmit power: Generating sufficient power is one of the most important and well-known challenges in wireless communications, which become more challenging with the increase of frequency. In RF performance, the power is limited by the capabilities of transistors implemented using different semiconductor technologies, and as a general trend the output saturated power decades severely as a function of frequency, especially, above 100 GHz. As this power corresponds to non-linear operation, sufficient back-off needs to be considered, which depends on the waveform and increases consumption [89].

Receiver noise: Like any other electronic devices, both active and passive receiver components contribute to the receiver noise, which consequently degrades the signal quality in terms of, e.g., SNR. These include insertion losses (ILs) in connected ports of different components, losses caused by interconnection and routing, in addition to the noise factor of LNA and other blocks. ILs increase at higher frequencies, which require high level of integration to mitigate. Also, the noise figure of LNA, which is the dominant active component, generally increases with the frequencies above 100 GHz, but can be less severe in some technologies [90]. In addition, as the noise is proportional to the bandwidth, doubling the bandwidth results in 3 dB losses of SNR.

Phase noise: It causes significant degradation in the performance of high data rate by causing random rotation of the received signal constellation and causes interference to adjacent channel in case of frequency division systems. This effect highly depends on the waveform, bandwidth, and oscillator source selection. The larger is the bandwidth, the higher is the amount of accumulated phase error, and such effect cannot simply be compensated by increasing SNR, as the noise floor becomes dominant. The phase noise depends on the technology of local oscillators (LOs). For instance, one common technique generates high frequency by multiplication of low-frequency oscillator. Each doubling of the frequency results in 6 dB higher phase noise level. Thus, it is critical to afford LO with low phase noise in order to allow operation at ultra-wideband [91].

DAC/ADC: The speed of converters is proportional to the bandwidth assuming Nyquist sampling. The essential challenge for high-speed converters is the increase of energy consumption with the sampling frequency and resolution. Nevertheless, the implementation of high speed DAC is relatively easier in comparison to ADC, and its power consumption at the transmitter is not significant when compared to the PA. In relation to ADC, the sampling frequency f_s needs to be at least twice the baseband bandwidth to avoid aliasing, and up-sampling might be required for digital synchronization. The quantization noise should not have a big impact on the SNR if the used amount of bits is sufficient. In particular, after quantization $\mathrm{SNR_d} = \frac{P_S}{\sigma^2+\sigma_q^2} = \frac{P_S}{\sigma^2}\frac{1}{1+\sigma_q^2/\sigma^2}$, where σ_q^2 is the quantization noise power, if $\sigma_q^2/\sigma^2 < 0.1$, the reduction of the SNR $\frac{P_S}{\sigma^2}$ is less than 0.4 dB. Therefor, the ADC dynamic range $\mathrm{SNR_{ADC}}$ should be considered according to the minimum SNR, such that, $\mathrm{SNR_{ADC}[dB]} > \mathrm{SNR_{min}[dB]} + 10$. In uniform quantization, $\mathrm{SNR_{ADC}}[dB] = 6.02Q_d$, where Q_d is the resolution in number of bits. Moreover, to avoid clipping, compensate for AGC, and tackle the near-far problem at the BS, additional margins are required. All that increases the required resolution, which increases the ADC power consumption P_{ADC} according to Walden's figure-of-merit

(FoM), which is used to evaluate ADC performances and given by [92].

$$\mathrm{FOM_W} = \frac{P_{\mathrm{ADC}}}{2^{Q_d} \cdot f_s}. \tag{3.9}$$

This parameter depends on the technology, and it is shown to be constant up to 100 MHz, where doubling the sampling rate or increasing the resolution by 1 bit double the power consumption. In contrast, for sampling rate above 1 GHz, $\mathrm{FOM_W}$ significantly increases as a function of the sampling rate, and therefore the power consumption grows more than double by doubling the sampling rate.

DSP: The implementation of real-time signal processing algorithms to achieve reliable throughput larger than 100 Gb/s within low-latency constrained is challenging and requires processing papalism, which leads to high energy consumption as well. Therefore, the DSP architecture needs to be optimized, such as considering the design of low-complexity algorithms without significant performance degradation and implementation with low-resolution operations. This might require adding an additional SNR margin, which impacts the RF design, and thus, a joint optimization of RF hardware and DSP is important [93].

Antenna: As the wavelength decreases, the antenna sizes also decrease, which create challenges for the transceiver radio-frequency integrated circuit (RFIC) design to match the form factor of the electronics and the antenna matrix [94]. For instance, to increase the transmit power, the size of PA consumes larger area than the antenna, which limits the practical size of the antenna array. Nevertheless, an array of large number of elements is required to increase the aperture and compensate for the reduced physical size of a single antenna. The array gain, assuming $\lambda/2$ element spacing, is proportional to the number of elements N, $G_{a,x} = NG_x$, where G_x is the antenna element gain. Following the link budget model (3.5), doubling the frequency or doubling the range requires doubling the number of elements at the transmitter and receiver sides, or to increase by factor 4 at one end of the link. Thus, employing antenna array at frequencies above 100 GHz, for sufficient range, is challenging. Besides the design RFIC design and packaging limitations, antenna arrays are impacted by non-idealities such as coupling between antenna elements. An alternative option of large antenna array is using high directivity antenna such as horn and lenses, but this is not steering-friendly choice [95].

Beamforming: The high-gain antenna arrays required to improve the link budget produce narrow beams, which requires strict alignment between the transmitter and receiver. Moreover, upon mobility, switching the beam needs to be fast enough when analogue beamforming is employed. However, because of the limited angular resolution because of the limitation of the gain and phase control, a successful beam alignment might not be guaranteed in all positions. Although digital beamforming is fully flexible, it is not practically feasible, and thus, hybrid beamforming arises as a reasonable approach. In hybrid beamforming, the antenna array is partitioned into subarrays driven by multiple RF chains. This allows also serving multiple users or exploiting MIMO spatial multiplexing. The size and number of subarrays are design parameters to be considered in the radio design based on the scenario and channel characteristics. Other challenges for the beamforming design are inherited from the non-idealities of other RF components, the constrained on the transmitted power limits the exploitation of gain control, the violation of the narrow band approximation, violation of the far-field assumption with large size of the array relative to the wavelength, and the antenna coupling.

3.6.3 RF Hardware Modelling

The modelling of individual block is essential for the computation of the link budget and the design of the waveform to mitigate the non-linearity in the system, in addition to be able to estimate the performance of realistic system. Three modelling approaches are required based on the studied problem, namely

- **Second order statistics:** this model considers the second-order statistics only, to compute the SNR ($P_{RX} = P_{TX} + P_N$) for link-level range and coverage analysis. This modelling can be carried out to determine the contribution of each block and identify the dominant source of non-ideality.
- **Additive non-linearity model:** in this approach, the RF impairment is considered as an additive term after linearization according the signal model $y[n] = hx[n] + z[n] + v[n]$, and h is the linearization gain, and $z[n]$ is signal-dependent term corresponds to the overall all RF block non-idealities such as quantization noise, phase noise, non-linear distortion of PA, IQ imbalance, etc. Thus, $z[n]$ can be approximately modelled by simplified Gaussian noise considering the sum of different random non-idealities. This model is useful to perform simulation in affordable time to evaluate the link performance in terms of BER, FER, etc., but it is not necessarily realistic due to the non-Gaussian nature of the signal-dependent RF non-idealities.
- **Response function:** where the goal is to obtain a function representing the relation between the input and output of blocks such as PA and LO to study compensation techniques. These models are mostly memory-dependent, such that the current response depends on the previous state. The simulation using this model is time consuming, as it emulates the actual hardware. The overall response of the system can be expressed as $y[n] = \mathrm{f}(x[n], \mathrm{x[n-1)}, \ldots) + v[n]$.

The following subsections provide an overview of modelling examples for different components.

3.6.3.1 ADC/DAC models

DAC and ADC add impairments to the signal by non-linear clipping and quantization noise, in addition to the effect of anti-aliasing and reconstruction filtering, as illustrated in Figure 3.35.

The filtering effects can be neglected by designing the signal to occupy a spectrum in the flat region of these filters [96]. Thus, it is required to model the impact of the quantization, which involves proper scaling of the signal at the ADC input, and proper mapping of floating point or fixed-point representation at the DAC input. The scaling of the input signal is achieved by AGC at the receiver, whereas the scaling of discrete signal is achieved digitally. Two models are shown in Figure 3.36, the first corresponds to the response function, where the signal is scaled and quantized, whereas the other model performs scaling and adds quantization noise, which can be modelled as uniformly or Gaussian-distributed random signal.

3.6.3.2 Power amplifier

In addition to the maximum saturated power of PA, modelling the non-linearities and memory effects are important. Measured data are used to model the PA response in the discrete-time domain. Following memory polynomial model, which

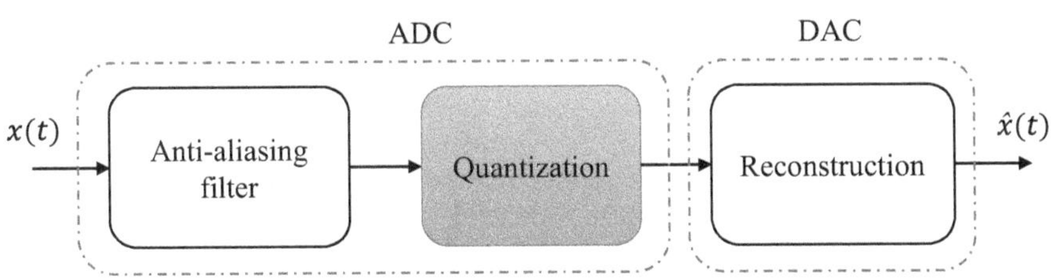

Figure 3.35. ADC/DAC functional model.

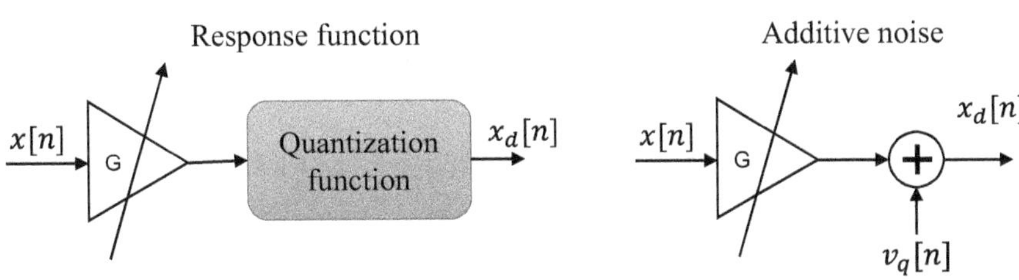

Figure 3.36. ADC/DAC signal models.

can be expressed as [97]:

$$Vout(s) = \sum_{q=1}^{Q} \int_{k=1}^{K} \tilde{a}_{kq} Vin(s-q)|Vin(s-q)|^{2(k-1)}, \tag{3.10}$$

where s is the number of samples, K polynomial order, Q memory length, and $\tilde{a}_{kq}$ are the coefficient to be estimated by fitting. The signals V_{in} (s) and V_{out} (s) are the measured discrete input and output complex envelope signals of the s-th sample, respectively. Note that the selection of the proper PA model depends highly on the scenario that is investigated. For example, for linearization studies, very accurate models are required, while for generic link-budget investigations, rather simplified behavioural approaches may be sufficient to achieve the correct order of magnitude for the modelling.

3.6.3.3 LNA noise figure

The losses and noise contribution from components after the LNA are mitigated by the power gain of the LNA, and the major noise impact is introduced by LNA. The noise-figure of LNA depends on the frequency of operation, which can be estimated based on measured data of certain semiconductor technology or based on circuit simulation. Based on empirical results and curve fitting in the mmWave range and sub-THz range (30–300 GHz), a common noticeable trend among the investigated technologies is the exponential increase of the noise figure with frequency [98], which is defined by exponential function parametrized by two parameters α and β, which depends on the technology, such that the minimum noise figure, NF, is given by

$$NF_{\text{min}} = \alpha exp\left(\frac{f_c}{\beta}\right) \tag{3.11}$$

where exemplary values are provided in Table 3.4.

Table 3.4. LNA noise minimum noise figure parameters of (3.11).

	CMOS	SiGe	GaAS	InP
α	1.50	1.75	0.70	1.50
β	112.4	130.7	129.9	188.7

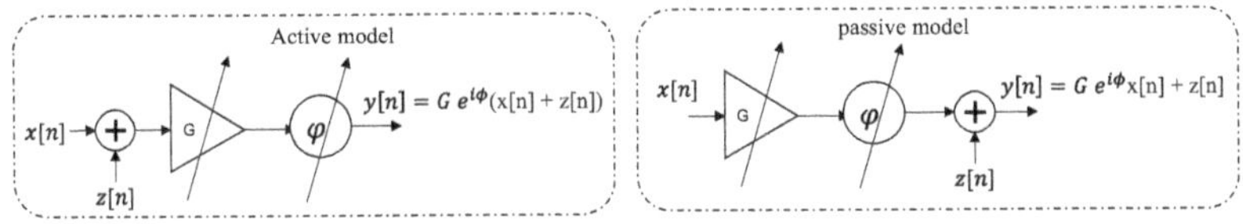

Figure 3.37. Analogue phase shifter models.

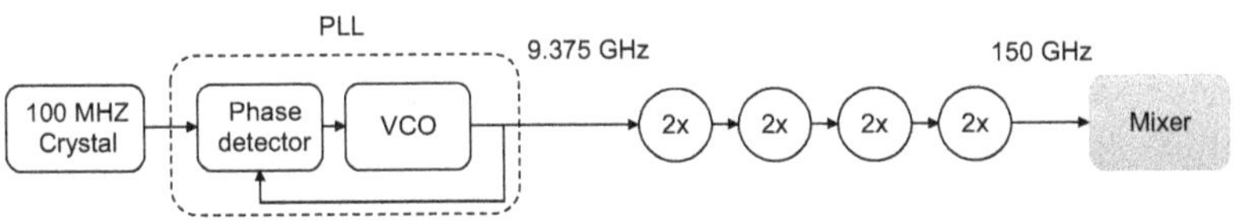

Figure 3.38. Example of LO generation architecture to support 150 GHz frequency.

3.6.3.4 Analogue beamforming non-linearities

The gain and phase control block, which can be implemented by different passive and active technologies, is inserted before the PA at the transmitter and after the LNA at the receiver. The non-idealities of this RF block result from the quantized beamforming weights and potential noise that has impact, especially if the amplitude is controlled. In most cases, this impact is neglected at the receiver compared to the impact of LNA noise, as sufficient gain is assumed from the amplifier. However, when the amplitude is controlled for beamforming purpose, modelling the noise of the VGA/phase shifter should be considered at least to analyse a realistic amplitude control dynamic range. A simplified modelling considers additive noise, as shown in Figure 3.37, which can be added at the input in the case of active component, or after in the case of passive components. At the transmitter side, the noise can be neglected in the overall link model. The transfer function model can be derived from the employed components for the evaluation of the beamforming.

3.6.3.5 Phase noise

The total phase noise highly depends on the frequency synthesis approach of the transceiver. By utilizing multiple multiplications by two or three of the voltage-controlled oscillator (VCO) signal in series, as shown in Figure 3.38, where the operation frequency of the VCO can be lowered for practical implementation [99]. However, doubling the LO frequency effectively raises the respective phase noise by 6 dB since the same signal is in intermediate frequency (IF) and LO port of the mixer and correlated noises are multiplied into the RF port of the mixer.

The single-side phase noise power spectrum can be defined by [100], Section 4.2.3.1.

$$S(f_o) = \mathrm{PSD0}\frac{\prod_{n=1}^{N}\left(1+\left(\frac{f_o}{f_{z,n}}\right)^{\alpha_{z,n}}\right)}{\prod_{m=1}^{M}\left(1+\left(\frac{f_o}{f_{p,m}}\right)^{\alpha_{p,m}}\right)}, \tag{3.12}$$

where f_o is the offset frequency, $\{f_{z,n}\}$ zeroes of orders $\{\alpha_{z,n}\}$, $\{f_{p,m}\}$ poles of orders $\{\alpha_{p,n}\}$. The average root mean square phase jitter, which can be further converted to an equivalent SNR or error vector magnitude (EVM), can be used in the simulation, and it can be defined as

$$\sigma_{\text{rms}} = \sqrt{2\int_{f_1}^{f_2} S(f)df}. \tag{3.13}$$

The integration goes from minimum to maximum frequencies that matters, such as the frequency range of the signal over which the phase compensation is performed. Accordingly, a signal model using random phase $\varphi_{pn}(n) \sim N(0,\ \sigma_{\text{rms}}^2)$ can be used, such that

$$y(n) = x(n)\exp(j\varphi_{pn}(n)). \tag{3.14}$$

This is suitable for short symbols, but to elaborate the impact of longer signal, a better model can be obtained by filtering Gaussian noise whose frequency response is $\varphi_w(f)$ using the filter $H(f) = \frac{1}{\sqrt{2}}\sqrt{S(f)}$, such that the generated phase noise is given by

$$\varphi_{LO}(t) = \int_{f_1}^{f_2} H(f)\varphi_w(f)\exp(2\pi ft)df. \tag{3.15}$$

3.6.4 Radio Architecture

A generic block diagram of the radio architecture that employs hybrid beamforming is illustrated in Figure 3.39. The transmitter is equipped with M RF chains and a phased antenna array with $P > M$ antenna elements, which are partitioned into M subarrays. A total number of antenna elements is $Q > N$ split into N subarrays connected to N receive RF chains at the receiver. A single RF chain follows the diagram presented in Figure 3.34. This generic architecture shows various degrees of

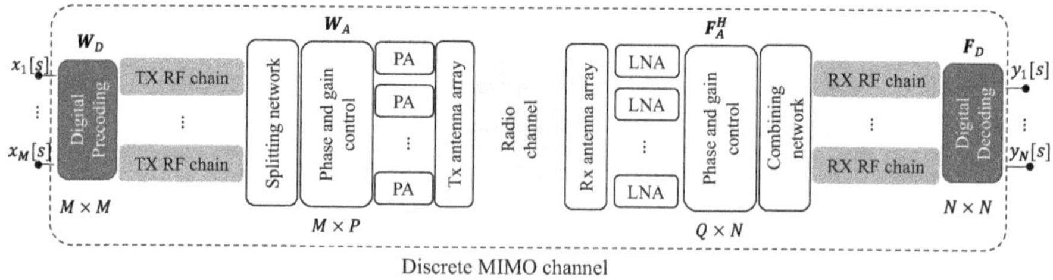

Figure 3.39. Generic hybrid beamforming radio architecture.

freedom that need to be determined at the design stage, and also at the operation, depending on the scenario and requirements. The defined configuration should provide the suitable beamforming architecture, based on the radio channel characteristics, in addition to the waveform properties before and after digital precoding, if applicable, to cope with the RF hardware limitations. The following subsections introduce an overview of the relevant radio modules.

3.6.4.1 Beamforming

The beamforming is achieved by means of digital precoding and analogue beamforming. The beamforming at the transmitter is achieved by a digital precoding matrix $\boldsymbol{W}_D$ of size $M \times M$ and analogue beamforming matrix $\boldsymbol{W}_A$ of size $M \times P$. Similarly, the receiver employs analogue beamforming matrix $\boldsymbol{F}_A$ of size $N \times Q$ and digital decoding matrix $\boldsymbol{F}_D$ of size $N \times N$.

The effective MIMO signal model, assuming ideal hardware is given in the frequency domain by

$$\tilde{\boldsymbol{y}}(f) = \boldsymbol{F}^H \tilde{\boldsymbol{H}}(f) \boldsymbol{W} \tilde{\boldsymbol{x}}(f) + \boldsymbol{F}^T \tilde{\boldsymbol{v}}(f), \tag{3.16}$$

where $\tilde{\boldsymbol{H}}(f)$ is the $M \times M$ MIMO channel in the frequency domain, and $\boldsymbol{W} = \boldsymbol{W}_A \boldsymbol{W}_D$ and $\boldsymbol{F} = \boldsymbol{F}_A \boldsymbol{F}_D$. This model can be used for initial assessment of the theoretical limits w.r.t. the channel characteristics. For instance, to select the number of RF chains based on the rank of the channel, evaluate the achievable rate and compare the performance of initial beam access techniques. Moreover, from implementation perspective, the structures of $\boldsymbol{W}_A$ and $\boldsymbol{F}_A$ are constrained by the implementation of the beamforming. For instance, as in practice, each antenna element is connected to phase and gain control, $\boldsymbol{W}_A$ is sparse, where non-zero values appear only once in each column and row. Moreover, the non-zero entries are limited by the resolution of the beamforming control.

3.6.4.2 Waveform and precoding

The design of the waveform and precoding needs to consider mitigating the hardware limitations at frequency above 100 GHz. This includes robustness to the phase noise as a major impairment that cannot be solved without compensation even at high SNR. To reduce the back-off and, thus, allowing higher transmit power, the waveform needs to be chosen to have smaller peak-to-average power ratio (PAPR). Moreover, to reduce the required ADC power, it is important to consider waveforms that work with low-resolution quantization. Furthermore, the processing complexity, such as equalization and detection, needs to be minimized to reduce the power consumption by DSP, especially when targeting data rate above 100 Gb/s. In addition, considering the channel characteristics, and the impact of beamforming architecture, it is foreseen that 6G waveform might need to go beyond orthogonal frequency-division multiplexing (OFDM). For example:

Zero-Crossing Modulation (ZXM): which is based on temporal oversampling and 1-bit quantization. The information bits are encoded in the zero crossing of the signal. Although it offers low complexity and power efficiency in AWGN, its performance under phase noise and in selective channel needs further investigation [101].

Analogue Multicarrier: is based on dividing the wideband to multiple narrower bands, where each band is modulated independently or considering overlap with other bands, in similar fashion to analogue OFDM. This allows using conventional wideband waveforms per sub-band at lower frequencies, then aggregating the bands at the high frequency. This approach allows using high-resolution ADC without linear scale of the ADC power consumption, but the impact of the high-frequency phase noise needs to be evaluated [102].

DFTS-OFDM and SC-FDE: are preceded version of OFDM using discrete Fourier transform (DFT) matrix, which is smaller than the DFT size in the first, and the same as the DFT in the later, with the goal of achieving a trade-off between reducing the PAPR and complexity. The impact of phase noise and low-resolution ADC should be considered.

Other waveforms include constant phase modulation (CPM) and its constrained envelope with a main focus on reducing the PAPR by employing precoding on OFDM [103].

3.6.5 Radio Channel

The radio channel characteristic affects several KPIs, and the path loss impacts the SNR and thus the data rate. Rich multi-path environment and polarization allow exploiting spatial multiplexing and thus increasing the spectral efficiency. The spatial correlation affects the area traffic capacity and connection density. The channel dynamics such as fading in addition to the path loss has a significant impact on the energy efficiency, in terms of transmitted power and signal processing for channel estimation and equalization. This dynamicity has also an impact on the latency and reliability as going in deep fade results in lost packets and a need for retransmissions. The impact of the channel needs to be studied in terms of wave-material interaction, atmospheric losses, and multi-path characteristics. In addition, to the propagation models, which are independent of the hardware, the channel model should consider the number of antennas and the RF design parameters, in order to provide a discrete channel model suitable for designing the signal processing algorithms. This methodology is conventional in all wireless communications systems. However, it becomes challenging for frequencies above 100 GHz, because of the complexity of designing a channel sounder and obtaining measurements in different environments.

3.6.5.1 Path loss, angular, and delay dispersions

The path loss, delay, and angular spread models of a wireless channel are useful information to predict the characteristics of the channel. These parameters are estimated for the LoS and NLoS links in the entrance hall and three outdoor scenarios, namely suburban, residential, and city centre, described in [86].

Two path loss models, such as the close-in (CI) reference free space reference and the alpha-beta-gamma (ABG) models, are considered here. The path loss of a signal with frequency f_c at distance d based on the CI model is expressed as

$$PL^{\mathrm{CI}} = \mathrm{FSPL}(f_c, 1\ \mathrm{m}) + 10n\log_{10}\left(\frac{d}{d_0}\right) + \chi_{\sigma}^{\mathrm{CI}} \tag{3.17}$$

where $\mathrm{FSPL}(f_c, 1\ \mathrm{m})$ is the free space path loss of the signal with frequency f_c at 1 m distance, n is the path loss exponent, and $\chi_{\sigma}^{\mathrm{CI}}$ is the CI shadow fading. Meanwhile, the ABG model is given by

$$PL^{\mathrm{ABG}} = 10\alpha\log_{10}(d) + \beta + 10\gamma\log_{10}\left(\frac{f}{1\ \mathrm{GHz}}\right) + \chi_{\sigma}^{\mathrm{ABG}} \tag{3.18}$$

where α is the distance-dependent loss coefficient, β is an offset coefficient, γ is the frequency-dependent loss coefficient, and $\chi_{\sigma}^{\mathrm{ABG}}$ is the ABG shadow fading. The omnidirectional path losses for the indoor and outdoor scenarios, together with the fitted path loss models, are plotted in Figures 3.40 and 3.41.

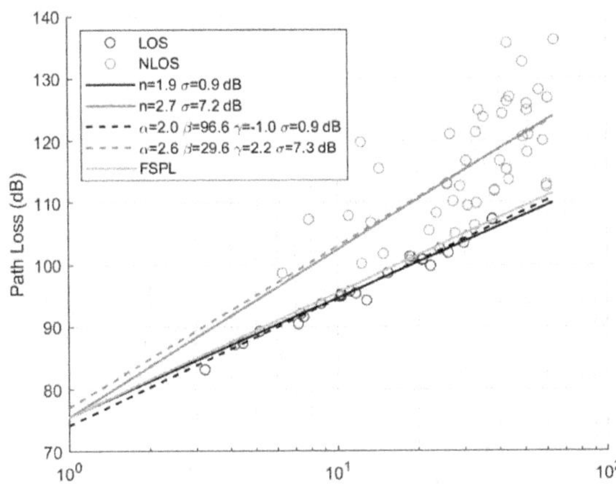

Figure 3.40. Path loss model for indoor scenario.

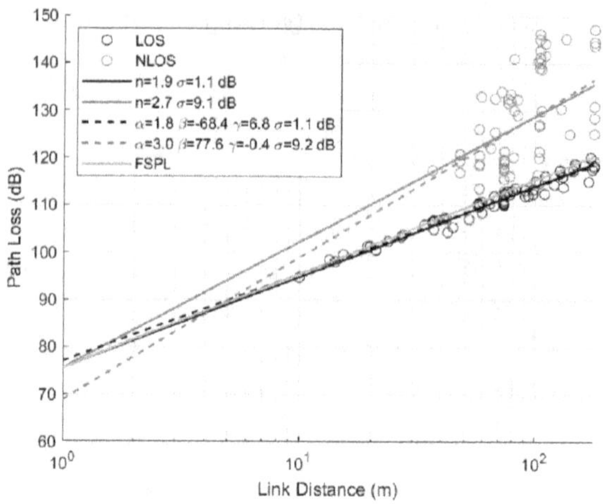

Figure 3.41. Path loss model for outdoor scenario.

Note that the outdoor scenario is an ensemble of path loss estimates from the three mentioned outdoor scenarios, since there are only limited NLoS links that can provide sensible fitting to the model when individual scenario is considered. It can be noticed that both the CI and ABG models provide equally good fit for LoS and NLoS links in both scenarios.

The mean μ and standard deviation σ of angular and delay spread values are listed in Table 3.5. The angular spread values are based on azimuth and zenith angle of departure (AoD/ZoD) and azimuth and zenith angle of arrival (AoA/ZoA) estimated from measurement-based ray launcher presented in [104]. Similar to the path loss modelling, the spread statistics for the outdoor case include the spread values from the three outdoor scenarios.

3.6.5.2 Stored channel model

The use of measured channel responses for PHY design and evaluation on a computer has been a well-recognized approach, e.g., [105], as it allows repeatable tests and comparison between different PHY schemes. The fact that measured channel responses serve as the ground-truth of any simulation-based channel modelling also justifies the use for realistic evaluation of any radio systems. There are, however, also challenges of using measured channels for PHY studies, i.e., (a) it is not straightforward to apply the measured channel responses to simulations that assume different hardware requirement than the measurement, e.g., signal dynamic range, antennas/arrays, system bandwidth, and moving speed of a mobile; and (b) a sufficient amount of measurements, e.g., for Monte-Carlo simulations and evaluation of packet error rates, may not be available due to limited capability of channel sounding. These challenges are overcome, e.g., (i) by under-sampling or interpolating the measured channels in space, bandwidth and time; and (ii) by making a general

Table 3.5. Angular and delay spread statistics in indoor and outdoor scenarios.

		AoD [deg]		ZoD [deg]		AoA [deg]		ZoA [deg]		Delay Spread [ns]	
Scenario		μ	σ	μ	σ	μ	σ	μ	σ	μ	σ
Indoor	**LoS**	25	9	6	5	14	8	3	2	14.6	4.4
	NLoS	38	13	8	5	22	11	4	2	26.3	11.8
Suburban	**LoS**	10	7	2	1	9	10	1	1	25.7	24.2
	NLoS	9	11	3	3	5	4	1	2	15.1	14.8
Residential	**LoS**	13	9	2	2	11	10	1	1	24.9	19.4
	NLoS	20	18	3	3	6	6	1	2	26.9	37.9
City Centre	**LoS**	18	9	4	6	13	4	2	1	21.3	9
	NLoS	24	17	4	3	14	9	2	2	25.4	19.9
Outdoor	**LoS**	12	9	2	3	10	9	1	1	24.7	20.6
	NLoS	21	18	3	3	9	8	2	2	25.6	31.0

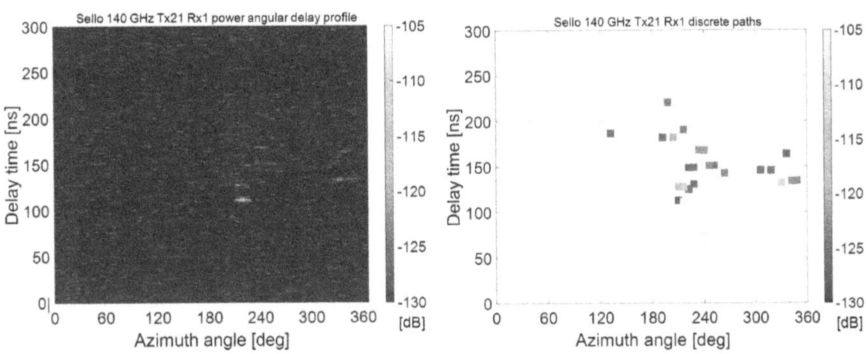

Figure 3.42. (a) Band- and aperture-limited channel response from a measurement and (b) its band- and aperture-unlimited model as propagation paths. Data are from a shopping mall measurement at 140 GHz [107].

mathematical description that allows us to over-sample, extrapolate, or invent the measured reality, which is called a channel model. The approach (ii) is exemplified in [105], where band-/aperture-/time-sampling-limited measurements of channels are approximated by the band-/aperture-/time-sampling unlimited form as illustrated in Figure 3.42. The former, coming from calibrated measurements, is represented by power spectrum, while the latter is a discrete model of propagation paths represented by Dirac delta functions. It is possible to synthesize infinite amount of small-scale fading realizations from the propagation paths of channels by applying the uniform randomly distributed phases to each path before summing them up at the antenna.

The measured sub-THz multi-path channels in the discrete path format are available in [106] for an entrance hall and outdoor scenarios, such as suburban, residential, and city centre. Details of the measurement channel sounder and measurement sites are available in [86]. Even though the measurements are single-directional to cover only one link end to estimate multi-path angles, double-directional data were derived by exploiting the available detailed geometric database of the measurement environment and measured channel response using a tool called a measurement-based ray launcher [104]. These single-directional and double-directional multi-path data, in the form of discrete multi-paths, are published in [105].

Stored multi-path data cover several transmit and receive locations in different environments. Each link, i.e., transmit and receive location, is characterized by a band-/aperture-unlimited double-directional power angular delay profile (PADP)

$$P_q(\Omega^{\mathrm{rx}}, \Omega^{\mathrm{tx}}, \tau) = \sum_{l=1}^{L_q} P_{l,q}\delta(\Omega^{\mathrm{rx}} - \Omega^{\mathrm{rx}}_{l,q})\delta(\Omega^{\mathrm{tx}} - \Omega^{\mathrm{tx}}_{l,q})\delta(\tau - \tau_{l,q}) \tag{3.19}$$

where q is the link index, L_q is the number of paths, $\delta(\cdot)$ is the delta function, $P_{l,q}$, $\Omega^{\mathrm{rx}}_{l,q}$, $\Omega^{\mathrm{tx}}_{l,q}$, and $\tau_{l,q}$ are the power (squared magnitude of path gain), the direction of arrival, the direction of departure, and the propagation delay of the lth path, respectively. Arrival and departure directions contain both azimuth and elevation angles. Current data are measured with vertically polarized antennas; hence, we restrict the definitions here to a single-polarized case only, neglecting the polarization characteristics.

Receiver and transmitter antennas can be specified by complex radiation patterns $\mathbf{g}_{\mathrm{rx}}(\Omega^{\mathrm{rx}}) \in \mathbb{C}^{M\times 1}$ and $\mathbf{g}_{\mathrm{tx}}(\Omega^{\mathrm{tx}}_l) \in \mathbb{C}^{N\times 1}$, respectively, where M is the number of Rx antennas and N is the number of transmitter antennas. Now, the channel frequency response matrix is determined as

$$\mathbf{H}_q(f) = \sum_{l=1}^{L_q} \mathbf{g}_{\mathrm{rx}}(\Omega^{\mathrm{rx}}_{l,q})\sqrt{P_{l,q}}\, e^{-j2\pi f \tau_{l,q}} \mathbf{g}_{\mathrm{tx}}(\Omega^{\mathrm{tx}}_{l,q})^T \in \mathbb{C}^{M\times N}. \tag{3.20}$$

Random snapshots of frequency response matrices can be generated by introducing random initial phase $\theta_{l,q}$ for each multi-path, where phase terms are drawn from the uniform distribution in $[0, 2\pi]$. Moreover, the temporal dimension and time variability can be included by introducing small Doppler frequencies $\nu_{l,q}$ for each path. This models a small-scale virtual motion, where only phases of path component change over time, but other propagation parameters remain constant.

The resulting snapshot/time variant frequency response matrix is

$$\mathbf{H}_q(t,f) = \sum_{l=1}^{L_q} \mathbf{g}_{\mathrm{rx}}(\Omega_{l,q}^{\mathrm{rx}}) \sqrt{P_{l,q}}\, e^{j(\theta_{l,q}+2\pi \nu_{l,q} t)}\, e^{-j2\pi f \tau_{l,q}}\, \mathbf{g}_{\mathrm{tx}}(\Omega_{l,q}^{\mathrm{tx}})^T \in \mathbb{C}^{M\times N}. \tag{3.21}$$

3.6.5.3 The number of independent beams

Wireless communication over sub-THz radio frequencies demands high-gain antennas to compensate for the high propagation loss. This leads to very directive antenna patterns, which illuminate only sub-sets among all available propagation pathways. Communication systems operating at lower frequencies have extensively used spatial multiplexing and beamforming to optimally utilize all degrees of freedom provided by the propagation channel. Now, at sub-THz, partly due to the channel sparsity and mainly due to foreseen RF technology limitation, such flexible transmission schemes might not be possible. Due to aforementioned reasons, it is interesting to study how many independent beams of practical beamwidth the propagation channel support. Directional wideband propagation measurements mentioned in Section 3.6.5.2 are used for this study. One can rather easily estimate how many significant paths are present in a measurement location, but interpreting that to separable beams is not evident.

Three methods to assess the number of useful beam directions are introduced in [108]. The second method is based on measured single-directional PADPs $P_q(\Omega, \tau)$ and a synthetic beam pattern $G(\Omega)$ defined in [109]. Only vertically polarized antennas were used in the channel measurement, and hence only single polarization is considered in the following number of beams.

Measured PADPs were evaluated using 10° half power bandwidth (HPBW), 10 dB dynamic range, 2 GHz BW, and 0.5 correlation threshold. An example PADP, beam power, and identified beam directions are illustrated by blue circles, a red curve, and orange squares, respectively, in Figure 3.43. Identified beam numbers in 132 measured indoor Tx and Rx locations are shown in Figure 3.44. Figure 3.45 depicts empirical CDFs of beam numbers in 132 indoor and 157 outdoor locations using both 10 and 20 dB dynamic ranges. Median values of beam numbers are two in both environments using a dynamic range of 10 dB, and using a 20 dB range, they are five and three in indoor and outdoor environments, respectively.

3.6.5.4 Comparison with below-100 GHz bands

Comparing the radio performance between current cmWave or low mmWave wireless communication systems and future 6G systems operating at frequencies between 100 and 300 GHz require the knowledge of the propagation channel from

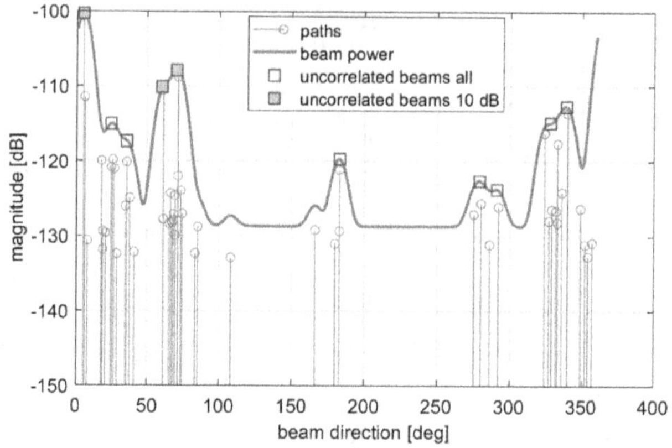

Figure 3.43. Measured path powers, the beam power, and independent beam azimuth directions in an example link.

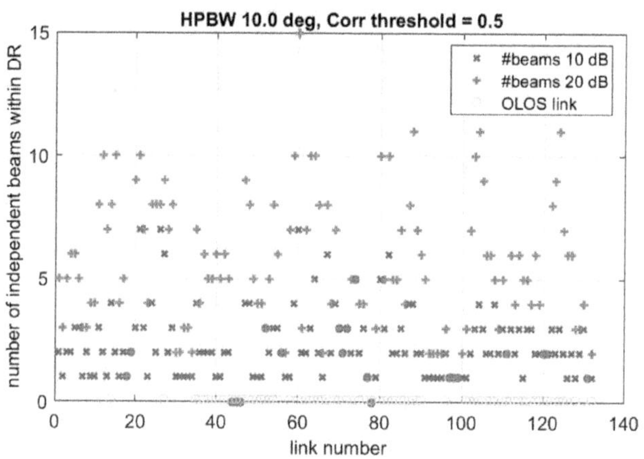

Figure 3.44. Number of independent beams in 132 indoor links.

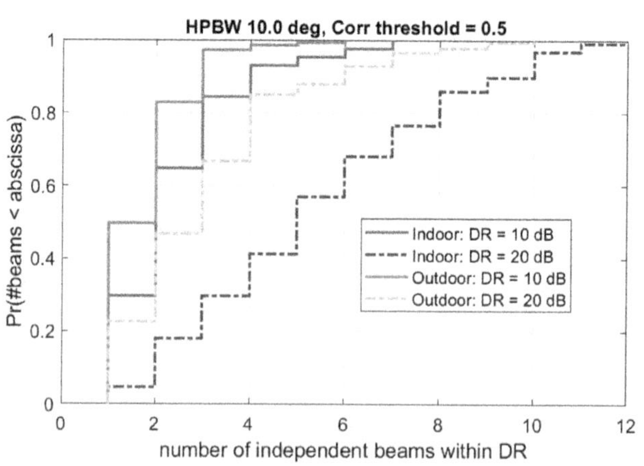

Figure 3.45. CDF of number of independent beams in 132 indoor and 157 outdoor links using either 10 or 20 dB dynamic range.

a few GHz to 300 GHz. The propagation channel frequency dependency may be analysed by a pure theoretical approach in free space but needs measurement in more complex environments. Performing wideband channel measurements in real environments over such a bandwidth of several hundred of GHz is technically and organizationally challenging; therefore, a more practical approach is to focus on the material characterization. Knowing the transmission and reflection losses, frequency dependency of usual building material is the first step to assess the potential differences between a propagation channel below 100 GHz and above 100 GHz. A propagation measurement campaign was performed using a vector network analyser (VNA) and frequency extenders to measure continuously from 2 to 260 GHz the reflection (R) and transmission (T) losses of common building material slabs at normal incidence. Figure 3.46 shows results from three typical materials compared with an International Telecommunication Union (ITU)-like model that keeps the ITU theoretical framework for a slab [110] but proposes new parameters fitted by measurement.

Homogeneous and flat surface materials, such as the plexiglass, respect the ITU-fitted model. The observed fading is due to the interference between the direct reflected (or transmitted) path and the multi-paths created inside the material by multiple reflection/transmission on the two air-material interfaces. The

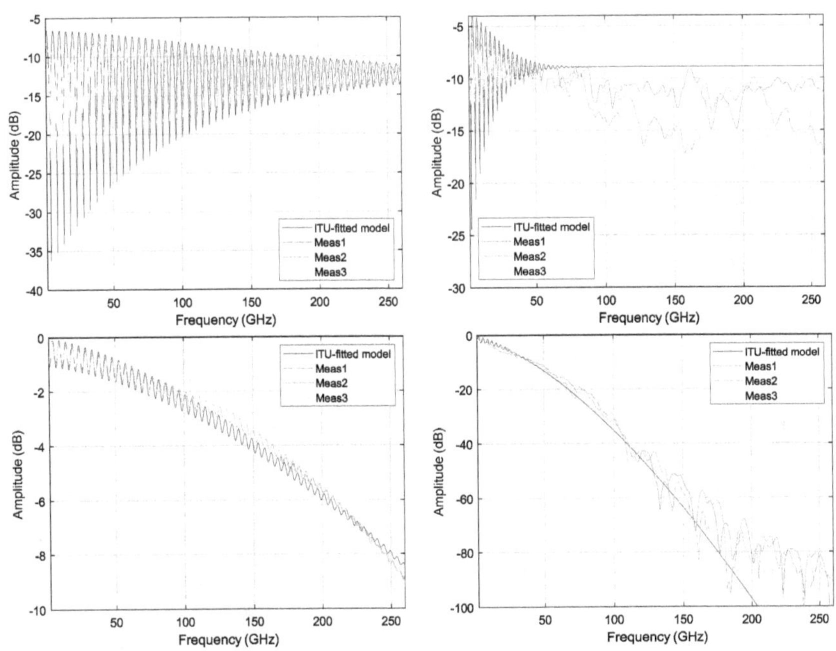

Figure 3.46. Material reflection (line a) and transmission gain (line b) compared with the ITU-fitted model fitted by measurement. The measurement is performed at three different points separated by 10 cm on the slab.

fading decreases with the frequency as multi-paths travelling inside the material are attenuated but the R loss average value is constant indicating a constant permittivity. Mortar represents non-homogeneous material with a rough surface. Above 100 GHz, R losses are impacted by the scattering and may be higher than 10 dB compared to a similar material with flat surface. T losses can be still calculated from the conductivity σ, expressed as $\sigma = cf^d$, f being the frequency and c and d being two parameters. But above 100 GHz, there are strong variations of up to 10 dB around the fitted model. For composite materials such as chipboard or glass wool, R losses can decrease with the frequency increase.

The ITU model is defined up to 100 GHz and needs to be improved for frequencies above 100 GHz from a propagation point of view, but could be used as it is in many simulations related to 6G sub-THz scenarios. Reflection loss errors due to rough surfaces may not be a concern in an office environment, shopping mall, airport, etc., as most of the materials are quite smooth. Transmission loss errors due to the material inhomogeneity may not be a concern as they are related to high transmission losses at high frequencies. Simulating a transmission loss of 20 dBs instead of 30 dBs may not significantly impact system simulation if we consider that a transmission loss higher than 20 dBs corresponds to a blockage. Dedicated multi-frequency measurement at cmWave and sub-THz frequencies in complex environments is required to check these assumptions.

3.7 Summary and Outlook

The current chapter provides an overview of selected RAN technologies with a high potential for future 6G networks. It is argued that for the limitless connectivity requirement of 6G, D-MIMO can provide expected macro-diversity (to exploit maximum diversity gain), design flexibility, and interference management, and the related challenges and opportunities are discussed. The main challenge for large-scale D-MIMO roll-out is arguably the cost of installation, as it requires fast and high-speed fronthaul connections. In addition, problems such as beam management aspects, practical approaches to non-coherent operation in higher bands, and transport solutions, e.g., wired/wireless, optical/electrical, and analogue/digital, satisfying the requirements need to be addressed. An optimized wired fronthaul can be realized through radio-stripe/RadioWeaves, while wireless fronthaul/backhaul can be obtained efficiently by IAB. The disaggregation is vital for creating scalable versions of D-MIMO architectures, and thus ORAN support can be viewed as an attractive feature facilitating the progress of the technology. Various ORAN-based deployment scenarios can be considered based on UE clustering and the engagement of O-DUs and O-RUs in serving the clusters. The definition of relevant deployment scenarios and new interfaces, such as inter-DU interface, is required.

RIS is a key technology for 6G due to its low-cost solution for controlling the propagation channel in favour of the communication link. It shows great potential in several use cases, for not only cellular scenarios, but in UAV and satellite system, and also in IoT scenarios. An architecture and associated three logical entities are introduced in this chapter to facilitate the automated controlling of the RIS, as a main requirement towards its widespread application. Multi-access connectivity is an important and necessary feature for beyond 5G and 6G technologies, for instance in private network scenarios. The 5G CN supports integration of non-3GPP access to provide secure connection for the UE accessing over a non-3GPP access network. Improvements on such framework could be envisioned for more flexible operation using an ORAN-based architecture to benefit the link's throughput, latency, and reliability. The large bandwidth available at mmWave and sub-THz bands introduces the opportunity to overcome the challenge of high-data rate delivery for some demanding 6G use cases. Hardware implementation and modelling the non-linear behaviour of the RF components are a challenge. The modelling of the system is essential for the computation of the link budget and the design of the waveform to mitigate the non-linearity in the system, as well as to estimate the performance of the system.

References

[1] J.G. Andrews, S. Buzzi, W. Choi, S.V. Hanly, A. Lozano, A.C. Soong, J.C. Zhang, "What will 5G be?" In *IEEE J. Sel. Area. Commun.*, vol. 32, no. 6, pp. 1065–1082, 2014.

[2] X. Lin, "An Overview of 5G Advanced Evolution in 3GPP Release 18," In *IEEE Communications Standard Magazine*, vol. 6, no. 3, September 2022.

[3] RP-213468, "Summary for RAN Rel-18 package," 3GPP RAN#94-e, December 2021. Accessed: April 6, 2023. [Online]. Available: http://www.3gpp.org/ftp/tsg_ran/TSG_RAN/TSGR_94e/Docs/RP-213469.zip.

[4] 5GPPP, "Beyond 5G/6G KPIs and Target Values," White Paper, June 2022. Accessed: April 6, 2023. [Online]. Available: https://5g-ppp.eu/wp-content/uploads/2022/06/white_paper_b5g-6g-kpis-camera-ready.pdf.

[5] Hexa-X, "D1.2 – Expanded 6G vision, use cases and societal values", Dec. 2021, Accessed April 6, 2023. [Online]. Available: https://ec.europa.eu/research/participants/documents/downloadPublic?documentIds=080166e5dc8b611b&appId=PPGMS.

[6] S. Chen, J. Zhang, J. Zhang, E. Bjornson, B. Ai, "A survey on user-centric cell-free massive MIMO systems," Digital Communications and Networks, 2022.

[7] S. Tripathi, N. V. Sabu, A. K. Gupta, H. S. Dhillon, "Millimeter-wave and Terahertz Spectrum for 6G Wireless," arXiv:2102.10267v1, February 2021. Accessed: April 6, 2023. [Online]. Available: https://arxiv.org/pdf/2102.10267.pdf.

[8] J. ITU, "Provisional final acts," in World Radiocommunication Conference 2019. ITU Publications, 2019.

[9] Ericsson, "Integrated access and backhaul – a new type of wireless backhaul in 5G," Ericsson Technology Review, June 23, 2020. Accessed: April 6, 2023. [Online]. Available: https://www.ericsson.com/en/reports-and-papers/ericsson-technology-review/articles/introducing-integrated-access-and-backhaul.

[10] A. Wolf, P. Schulz, M. Dörpinghaus, J. C. S. Santos Filho, and G. Fettweis, "How reliable and capable is multi-connectivity?," In *IEEE Trans. Commun.* 2018, 67, 1506–1520.

[11] T. Sylla, L. Mendiboure, S. Maaloul, H. Aniss, M. Chalouf, S. Delbruel, "Multi-Connectivity for 5G Networks and Beyond: A Survey," In *MPDI Sensors*, October 2022.

[12] M. Di Renzo, et al., "Smart radio environments empowered by reconfigurable intelligent surfaces: How it works, state of research, and the road ahead," In *IEEE J. Sel. Areas Commun.*, vol. 38, no. 11, pp. 2450–2525, Nov. 2020.

[13] Y. Liu, X. Liu, X. Mu, T. Hou, J. Xu, M. Di Renzo, and N. Al-Dhahir, "Reconfigurable Intelligent Surfaces: Principles and Opportunities," In *IEEE Communications Surveys and Tutorials*," vol. 23, no. 3, pp. 1546–1577, 2021.

[14] N. Rajatheva, et al., "White paper on broadband connectivity in 6G", arXiv preprint, April 2020. Accessed: April 6, 2023. [Online]. Available: https://arxiv.org/pdf/2004.14247.pdf.

[15] O. Tervo, T. Levanen, K. Pajukoski, J. Hulkkonen, P. Wainio, and M. Valkama, "5G New Radio Evolution Towards Sub-THz Communications," In *2020 2nd 6G Wireless Summit (6G SUMMIT)*, pp. 1–6, 2020. doi: 10.1109/6GSUMMIT49458.2020.9083807.

[16] E. Björnson and L. Sanguinetti, "Cell-Free versus Cellular Massive MIMO: What Processing is Needed for Cell-Free to Win?," In *2019 IEEE 20th International Workshop on Signal Processing Advances in Wireless Communications (SPAWC)*, pp. 1–5, 2019.

[17] H. Q. Ngo, A. Ashikhmin, H. Yang, E. G. Larsson, and T. L. Marzetta, "Cell-Free Massive MIMO Versus Small Cells," In *IEEE Transactions on Wireless Communications*, vol. 16, no. 3, pp. 1834–1850, 2017.

[18] T. L. Marzetta, "Noncooperative cellular wireless with unlimited numbers of BS antennas," In *IEEE Trans. Wireless Commun.*, vol. 9, no. 11, pp. 3590–3600, 2010.

[19] J. G. Andrews, S. Buzzi, W. Choi, S. V. Hanly, A. Lozano, A. C. K. Soong, and J. C. Zhang, "What will 5G be?", In. *IEEE J. Sel. Areas Commun.*, vol. 32, no. 6, pp. 1065–1082, 2014.

[20] V. Jungnickel, K. Manolakis, W. Zirwas, B. Panzner, V. Braun, M. Lossow, M. Sternad, R. Apelfröjd, and T. Svensson, "The role of small cells, coordinated multipoint, and massive MIMO in 5G," In *IEEE Commun. Mag.*, vol. 52, no. 5, pp. 44–51, 2014.

[21] E. Björnson, J. Hoydis, and L. Sanguinetti, "Massive MIMO has unlimited capacity," In *IEEE Trans. Wireless Commun.*, vol. 17, no. 1, pp. 574–590, 2018.

[22] D. Gesbert, S. Hanly, H. Huang, S. S. Shamai, O. Simeone, W. and Yu, "Multi-cell MIMO cooperative networks: A new look at interference," In *IEEE J. Sel. Areas Commun.*, vol. 28, no. 9, pp. 1380–1408, 2010.

[23] D. Lee, H. Seo, B. Clerckx, E. Hardouin, D. Mazzarese, S. Nagata, K. and Sayana, "Coordinated multipoint transmission and reception in LTE-advanced: Deployment scenarios and operational challenges," In *IEEE Commun. Mag.*, vol. 50, no. 2, pp. 148–155, 2012.

[24] G. Interdonato, E. Björnson, H. Q. Ngo, P. Frenger, and E. G. Larsson, "Ubiquitous cell-free massive MIMO Communications," In *EURASIP J. Wireless Commun. and Networking*, vol. 2019, no. 1, pp. 197–209, 2019a.

[25] *Future Technology Trends of Terrestrial Systems Towards 2030 and Beyond*, Report ITU-R M.251-0, November, 2022.

[26] G. Wikström, P. Persson, S. Parkvall, G. Mildh, E. Dahlman, B. Balakrishnan, P. Öhlen, E. Trojer, G. Rune, J. Arkko, Z. Turányi, D. Roeland, B. Sahlin, W. John, J. Halén, and H. Björkegren, "6G – Connecting a cyber-physical world", Ericsson White Paper, Feb. 2022. Accessed: April 6, 2023. [Online]. Available: https://www.ericsson.com/en/reports-and-papers/white-papers/a-research-outlook-towards-6g.

[27] *NR; Multi-connectivity; Overall description; Stage-2*, Technical Specification (TS) 37.340, v17.3.0, 3GPP, Jan. 2023.

[28] J. Zhang, S. Chen, Y. Lin, J. Zheng, B. Ai, and L. Hanzo, "Cell-free massive MIMO: A new next-generation paradigm," In *IEEE Access*, vol. 7, pp. 99878–99888, 2019.

[29] S. Buzzi, C. D'Andrea, A. Zappone, and C. D'Elia, "User-centric 5G cellular networks: Resource allocation and comparison with the cell-free massive MIMO approach," In *IEEE Trans. Wireless Commun.*, vol. 19, no. 2, pp. 1250–1264, 2019.

[30] E. Nayebi, A. Ashikhmin, T. L. Marzetta, H. Yang, and B. D. Rao, "Precoding and power optimization in cell-free massive MIMO systems," In *IEEE Trans. Wireless Commun.*, vol. 16, no. 7, pp. 4445–4459, 2017.

[31] G. J. Foschini, K. Karakayali, and R. A. Valenzuela, "Coordinating multiple antenna cellular networks to achieve enormous spectral efficiency," In *IEEE Proceedings Communications*, vol. 153, no. 4, pp. 548–555, 2006.

[32] X. Hong, Y. Jie, C.-X. Wang, J. Shi, and X. Ge, "Energy-spectral efficiency trade-off in virtual MIMO cellular systems," In *IEEE J. Sel. Areas Commun.*, vol. 31, no. 10, pp. 2128–2140, Oct. 2013.

[33] J. Zhang, E. Björnson, M. Matthaiou, D. W. K. Ng, H. Yang, and D. J. Love, "Prospective Multiple Antenna Technologies for Beyond 5G," In *IEEE Journal on Selected Areas in Communications*, vol. 38, no. 8, pp. 1637–1660, Aug. 2020.

[34] REINDEER "D2.1 Initial assessment of architectures and hardware resources for a RadioWeaves infrastructure," 2022. Accessed: April 6, 2023. [Online]. Available: https://ec.europa.eu/research/participants/documents /downloadPublic?documentIds=080166e5e7bcfeeb&appId=PPGMS.

[35] E. Emfors, "Radio Stripes: re-thinking mobile networks," 2019. Accessed: April 6, 2023. [Online]. Available: https://www.ericsson.com/en/blog/201 9/2/radio-stripes.

[36] L. Van der Perre, E. G. Larsson, F. Tufvesson, L. De Strycker, E. Björnson, and O. Edfors, "RadioWeaves for efficient connectivity: analysis and impact of constraints in actual deployments," Asilomar Conf., November, 2019.

[37] *NG-RAN; Architecture description*, Technical Specification, TS 38.401 v16.0.0, 3GPP, Jan. 2020,

[38] *ORAN Architecture Description*, Technical Specification (TS) ORAN.WG1.ORAN-Architecture-Description-v04.00, March 2021.

[39] G. Interdonato, P. Frenger, and E.G. Larsson, "Scalability Aspects of Cell-Free Massive MIMO," In *ICC 2019 IEEE International Conference on Communications (ICC)*, vol. 68, no. 7, pp. 1–6, 2019.

[40] V. Ranjbar, A. Girycki, M. A. Rahman, S. Pollin, M. Moonen, and E. Vinogradov, "Cell-Free mMIMO Support in the ORAN Architecture: A PHY Layer Perspective for 5G and Beyond Networks," In *IEEE Communications Standards Magazine*, vol. 6, no. 1, pp. 28–34, March 2022, doi: 10.1109/MCOMSTD.0001.2100067.

[41] *Draft Standard for Ethernet – Amendment: Physical Layer and Management Parameters for DTE Power via MDI over 4-Pair*, IEEE P802.3bt/D1.5, 30 Nov. 2015.

[42] U. K. Ganesan, E. Björnson, and E. G. Larsson, "RadioWeaves for extreme spatial multiplexing in indoor environments," In *2020 54th Asilomar Conference on Signals, Systems, and Computers*, 2020, pp. 1007–1011.

[43] REINDEER "D3.1 Analytical Performance Metrics and Physical-Layer Solutions," 2022. Accessed: April 6, 2023. [Online]. Available: https://ec.europa.eu/research/participants/documents/downloadPublic?documentIds=080166e5e7c62ba7&appId=PPGMS.

[44] I. Ahmed, H. Khammari, A. Shahid, A. Musa, K. S. Kim, E. De Poorter, and I. Moerman, "A Survey on Hybrid Beamforming Techniques in 5G: Architecture and System Model Perspectives," In *IEEE Comm. Surveys & Tutorials*, vol. 20, no. 4, 2018.

[45] *Study on channel model for frequency spectrum above 6 GHz*, Technical Report (TR) 38.900, v14.0.0, 3GPP July 2016.

[46] W. B. Abbas, F. Gomez-Cuba, and M. Zorzi, "Millimeter wave receiver efficiency: A comprehensive comparison of beamforming schemes with low resolution ADCs," In *IEEE Trans. Wireless Commun.*, vol. 16, no. 12, pp. 8131–8146, Dec. 2017.

[47] X. Gao, L. Dai, S. Han, C. L. I, and R. W. Heath, "Energy-Efficient Hybrid Analog and Digital Precoding for MmWave MIMO Systems with Large Antenna Arrays," In *IEEE J. Sel. Areas Commun.*, vol. 34, no. 4, pp. 998–1009, Apr. 2016.

[48] C. Fang, B. Makki, J. Li, and T. Svensson, "Hybrid Precoding in Cooperative Millimeter Wave Networks," In *IEEE Trans. Wireless Commun.*, vol. 20, no. 8, pp. 5373–5388, Aug. 2021.

[49] M. Polese, M. Giordani, T. Zugno, A. Roy, S. Goyal, D. Castor, and M. Zorzi, "Integrated access and backhaul in 5g mmwave networks: Potential and challenges," In *IEEE Communications Magazine*, vol. 58, no. 3, pp. 62–68, 2020.

[50] Techplayon, "5G self-backhaul-integrated access and backhaul," 2019. Accessed: April 6, 2023. [Online]. Available: https://www.techplayon.com/5g-self-backhaul-integrated-access-and-backhaul.

[51] M. N. Islam, S. Subramanian, and A. Sampath, "Integrated access backhaul in millimeter wave networks," In *2017 IEEE Wireless Communications and Networking Conference (WCNC)*, pp. 1–6, IEEE, 2017.

[52] A. Ghosh, A. Maeder, M. Baker, D. Chandramouli, "5G evolution: A view on 5G cellular technology beyond 3GPP release 15," In *IEEE Access* vol. 7,

pp. 127639–127651, 2019. Accessed: April 6, 2023. [Online]. Available: http://dx.doi.org/10.1109/ACCESS.2019.2939938.

[53] O. Teyeb, A. Muhammad, G. Mildh, E. Dahlman, F. Barac, and B. Makki, "Integrated access backhauled networks," In *2019 IEEE 90th Vehicular Technology Conference (VTC2019-Fall)*, pp. 1–5, 2019.

[54] M. Polese, M. Giordani, A. Roy, D. Castor, and M. Zorzi, "Distributed path selection strategies for integrated access and backhaul at mmwaves," In *2018 IEEE Global Communications Conference (GLOBECOM)*, pp. 1–7, 2018.

[55] 6G BRAINS, "D4.1 Design and Description of the Intelligent IAB and RmUE/mUE and human-centric control interfaces over Dynamic Ultra-dense D2D Cell Free Network," 2021. Accessed: April 6, 2023. [Online]. Available: https://ec.europa.eu/research/participants/documents/downloadPublic?documentIds=080166e5e63588d2&appId=PPGMS.

[56] C. Saha, M. Afshang, and H. S. Dhillon, "Integrated mmWave access and backhaul in 5G: Bandwidth partitioning and downlink analysis," In *Proc. IEEE Int. Conf. Commun. (ICC)*, pp. 1–6, May 2018.

[57] S. M. Azimi-Abarghouyi, B. Makki, M. Haenggi, M. Nasiri-Kenari, and T. Svensson, "Coverage analysis of finite cellular networks: A stochastic geometry approach," In *Proc. Iran Workshop Commun. Inf. Theory (IWCIT)*, pp. 1–5, Tehran, Iran, Apr. 2018.

[58] C. Madapatha, B. Makki, C. Fang, O. Teyeb, E. Dahlman, M. S. Alouini, and T. Svensson, "On Integrated Access and Backhaul Networks: Current Status and Potentials," In *IEEE Open Journal of the Communications Society*, vol. 1, pp. 1374–1389, 2020, doi: 10.1109/OJCOMS.2020.3022529.

[59] C. Madapatha, B. Makki, A. Muhammad, E. Dahlman, M. -S. Alouini, and T. Svensson, "On Topology Optimization and Routing in Integrated Access and Backhaul Networks: A Genetic Algorithm-Based Approach," In *IEEE Open Journal of the Communications Society*, vol. 2, pp. 2273–2291, 2021, doi: 10.1109/OJCOMS.2021.3114669.

[60] E. Björnson, H. Wymeersch, B. Matthiesen, P. Popovski, L. Sanguinetti, and E. de Carvalho, "Reconfigurable intelligent surfaces: A signal processing perspective with wireless applications," In *arXiv preprint,* 2021, arXiv:2102.00742.

[61] E. C. Strinati, G. C. Alexandropoulos, Vincenzo Sciancalepore, M. Renzo, H. Wymeersch, D. P. Huy, M. Crozzoli, R. D'Errico, E. Carvalho, P. Popovski, P. Lorenzo, L. Bastianelli, Mathieu Belouar, J. Mascolo, G. Gradoni, S. Phang, G. Lerosey, and B. Denis, "Wireless environment as a service enabled by reconfigurable intelligent surfaces: The RISE-6G perspective," In *2021 Joint European Conference on Networks and Communications & 6G Summit (EuCNC/6G Summit)*, 2021.

[62] E. C. Strinati, G. C. Alexandropoulos, H. Wymeersch, B. Denis, V. Sciancalepore, R. D'Errico, A. Clemente, D. T. Phan-Huy, E. De Carvalho, and P. Popovski, "Reconfigurable, Intelligent, and Sustainable Wireless Environments for 6G Smart Connectivity," in IEEE Communications Magazine, vol. 59, no. 10, pp. 99–105, October 2021, doi: 10.1109/MCOM.001.2100070.

[63] Y. Cao and T. Lv, "Intelligent reflecting surface enhanced resilient design for MEC offloading over millimeter wave links," 2019. arXiv:1912.06361.

[64] C. Pan, H. Ren, K. Wang, W. Xu, M. Elkashlan, A. Nallanathan, and L. Hanzo, "Multicell MIMO communications relying on intelligent reflecting surfaces," In *IEEE Trans. Wireless Commun.*, vol. 19, no. 8, pp. 5218–5233, Aug. 2020.

[65] S. Li, B. Duo, X. Yuan, Y. C. Liang, and M. Di Renzo, "Reconfigurable intelligent surface assisted UAV communication: Joint trajectory design and passive beamforming," In *IEEE Wireless Commun. Lett.*, vol. 9, no. 5, pp. 716–720, May 2020.

[66] D. Ma, M. Ding, and M. Hassan, "Enhancing cellular communications for UAVs via intelligent reflective surface," In *Proc. IEEE Wireless Commun. Netw. Conf. (WCNC)*, pp. 1–6, 2020.

[67] X. Mu, Y. Liu, L. Guo, J. Lin, and R. Schober, "Intelligent reflecting surface enhanced indoor robot path planning: A radio map-based approach," In *IEEE Trans. Wireless Commun.*, vol. 20, no. 7, July 4, 2021, doi: 10.1109/TWC.2021.3062089.

[68] RISE-6G "D2.5: RISE network architectures and deployment strategies analysis: first results," 2022. Accessed: April 6, 2023. [Online]. Available: https://ec.europa.eu/research/participants/documents/downloadPublic?documentIds=080166e5ef286723&appId=PPGMS.

[69] A. Albanese, F. Devoti, V. Sciancalepore, M. Di Renzo, and X. Costa-Pérez, "MARISA: A self-configuring metasurfaces absorption and reflection solution towards 6G," In *IEEE INFOCOM 2022-IEEE Conference on Computer Communications*, pp. 250–259, May 2022.

[70] I. Alamzadeh, G. C. Alexandropoulos, N. Shlezinger, and M. F. Imani, "A reconfigurable intelligent surface with integrated sensing capability," In. *Sci. Rep.* vol. 11, pp. 20737, 2021. Accessed: April 6, 2023. [Online]. Available: https://doi.org/10.1038/s41598-021-99722-x.

[71] G. Stratidakis, S. Droulias, and A. Alexiou, "An analytical framework for reconfigurable intelligent surfaces placement in a mobile user environment," In *Proceedings of the 19th ACM Conference on Embedded Networked Sensor Systems, ser. SenSys '21*. New York, NY, USA: Association for Computing

Machinery, pp. 623–627, 2021. Accessed: April 6, 2023. [Online]. Available: https://doi.org/10.1145/3485730.3494038.

[72] G. Stratidakis, S. Droulias, and A. Alexiou, "Optimal position and orientation study of Reconfigurable Intelligent Surfaces in a mobile user environment," In *IEEE Transactions on Antennas and Propagation*, vol. 70, no. 10, 2022, doi: 10.1109/TAP.2022.3208036.

[73] A.-A. A. Boulogeorgos, N. D. Chatzidiamantis, H. G. Sandalidis, A. Alexiou and M. D. Renzo, "Cascaded Composite Turbulence and Misalignment: Statistical Characterization and Applications to Reconfigurable Intelligent Surface-Empowered Wireless Systems," In *IEEE Transactions on Vehicular Technology*, vol. 71, no. 4, pp. 3821–3836, April 2022, doi: 10.1109/TVT.2021.3140084.

[74] A.-A. A. Boulogeorgos, N. D. Chatzidiamantis, H. G. Sandalidis, A. Alexiou, and M. D. Renzo, "Performance Analysis of Multi-Reconfigurable Intelligent Surface-Empowered THz Wireless Systems," In *IEEE International Conference on Communications (ICC)*, 2022.

[75] C. Liaskos, A. Tsioliaridou, A. Pitsillides, S. Ioannidis, and I. Akyildiz, "Using any surface to realize a new paradigm for wireless communications," In *Communications of the ACM*, vol. 61, no. 11, pp. 30–33, Oct. 2018.

[76] A. -A. A. Boulogeorgos, A. Alexiou, and M. D. Renzo, "Outage performance analysis of RIS-assisted UAV wireless systems under disorientation and misalignment," In *IEEE Transactions on Vehicular Technology*, 2022, doi: 10.1109/TVT.2022.3187050.

[77] 3GPP, "Study on Access Traffic Steering, Switch and Splitting Support in the 5G System Architecture," 3GPP Technical Report, TR 23.793, 2018.

[78] 5G-CLARITY "D3.2 Design Refinements and Initial Evaluation of the Coexistence, Multi-Connectivity, Resource Management and Positioning Frameworks," 2021. Accessed: April 6, 2023. [Online]. Available https://ec.europa.eu/research/participants/documents/downloadPublic?documentIds=080166e5e8fc7cff&appId=PPGMS.

[79] A. Ford, C. Raiciu, M. J. Handley, and O. Bonaventure, "TCP Extensions for Multipath Operation with Multiple Addresses," RFC 6824, Jan. 2013. Accessed: April 6, 2023. [Online]. Available: https://www.rfc-editor.org/info/rfc6824.

[80] C. Paasch, F. Duchêne, and G. Detal, "Multipath TCP – Linux Kernel implementation," 2023. Accessed: April 6, 2023. [Online]. Available: https://multipath-tcp.org/.

[81] 5G-CLARITY "D3.3 Complete Design and Final Evaluation of the Coexistence, Multi-Connectivity, Resource Management, and Positioning Frameworks," 2022. Accessed: April 6, 2023. [Online]. Available: https://ec.europ

a.eu/research/participants/documents/downloadPublic?documentIds=080166e5f3094463&appId=PPGMS.

[82] Virtualized multi-connectivity testbed, 5G-CLARITY Project. Accessed: April 6, 2023. [Online]. Available: https://github.com/jorgenavarroortiz/multitechnology_testbed_v0.

[83] A. Bourdoux, A. N. Barreto, B. van Liempd, C. de Lima, D. Dardari, D. Belot, E. S. Lohan, G. Seco-Granados, H. Sarieddeen, H. Wymeersch, J. Suutala, J. Saloranta, M. Guillaud, M. Isomursu, M. Valkama, M. R. K. Aziz, R. Berkvens, T. Sanguanpuak, T. Svensson, and Y. Miao, "6G white paper on localization and sensing." In *arXiv preprint* arXiv:2006.01779, 2020.

[84] M. E. Leinonen, M. Jokinen, N. Tervo, O. Kursu, and A. Pärssinen, "System EVM Characterization and Coverage Area Estimation of 5G Directive mmW Links," In *IEEE Trans. on Microw. Theory and Techn.*, vol. 67, no. 12, pp. 5282–5295, Dec. 2019.

[85] Hexa-X, "D2.1 – Towards Tbps Communications in 6G: Use cases and Gap Analysis", Dec. 2021, Accessed April 6, 2023. [Online]. Available: https://ec.europa.eu/research/participants/documents/downloadPublic?documentIds=080166e5df2529fd&appId=PPGMS.

[86] Hexa-X, "D2.2 – Initial radio models and analysis towards ultra-high data rate links in 6G", Dec. 2021, Accessed April 6, 2023. [Online]. Available: https://ec.europa.eu/research/participants/documents/downloadPublic?documentIds=080166e5e648d315&appId=PPGMS.

[87] A. Karakuzulu, A. Malignaggi, and D. Kissinger, "Low Insertion Loss D-band SPDT Switches Using Reverse and Forward Saturated SiGe HBTs," 2019 IEEE Radio and Wireless Symposium (RWS), 2019, pp. 1–3, doi: 10.1109/RWS.2019.8714362.

[88] X. Zhao, O. Glubokov, J. Campion, A. Gomez-Torrent, Aleksandr Krivovitca, U. Shah, and J. Oberhammer, "Silicon Micromachined D-Band Diplexer Using Releasable Filling Structure Technique," In *IEEE Transactions on Microwave Theory and Techniques*, vol. 68, no. 8, pp. 3448–3460, Aug. 2020, doi: 10.1109/TMTT.2020.3004585.

[89] H. Wang, et al., "Power Amplifiers Performance Survey 2000-Present," Georgia Tech. Accessed: April 6, 2023. [Online]. Available: https://gems.ece.gatech.edu/PA_survey.html.

[90] L. Belostotski, S. Jagtap, "Down With Noise: An Introduction to a Low-Noise Amplifier Survey," In *IEEE Solid-State Circuits Magazine*, vol. 12, no. 2, pp. 23–29, Spring 2020, doi: 10.1109/MSSC.2020.2987505.

[91] E. Dalman, S. Parkval, J. Skold, 5G NR, *The Next Generation Wireless Access Technology*, 2nd ed., Academic Press 2020, Chapter 26.

[92] B. Murmann, "ADC Survey," Accessed April 6, 2023. [Online]. Available: https://github.com/bmurmann/ADC-survey.

[93] P. Skrimponis, S. Dutta, M. Mezzavilla, S. Rangan, S. H. Mirfarshbafan, C. Studer, J. Buckwalter, and M. Rodwell, "Power Consumption Analysis for Mobile MmWave and Sub-THz Receivers," In *2020 2nd 6G Wireless Summit (6G SUMMIT)*, pp. 1–5, 2020. doi: 10.1109/6GSUMMIT49458.2020.9083793.

[94] A. Pärssinen, M. Alouini, M. Berg, T. Kuerner, P. Kyösti, M. E. Leinonen, M. Matinmikko-Blue, E. McCune, U. Pfeiffer, and P. Wambacq, (Eds.), "White Paper on RF Enabling 6G – Opportunities and Challenges from Technology to Spectrum," [White paper]. (6G Research Visions, No. 13). University of Oulu, 2020. Accessed: April 6, 2023. [Online]: Available: http://urn.fi/urn:isbn:9789526228419.

[95] P. Rodriguez-Vazquez, J. Grzyb, B. Heinemann, and U. R. Pfeiffer, "A QPSK 110-Gb/s Polarization-Diversity MIMO Wireless Link With a 220–255 GHz Tunable LO in a SiGe HBT Technology," In *IEEE Transactions on Microwave Theory and Techniques*, vol. 68, no. 9, pp. 3834–3851, Sept. 2020.

[96] A. V. Oppenheim and R. W. Schafer, *Discrete-Time Signal Processing*, Pearson Education Ltd., 3rd Ed., London, UK, p. 1056, 2013.

[97] D. Schreurs, M. O'Droma, A. A. Goacher, M. Gadringer et al., "Power Amplifier Behavioral Modeling," Cambridge University Press, p. 263, 2009.

[98] L. Belostotski and S. Jagtap, "Low-noise amplifier (LNA) performance survey," Univ. of Calgary, Canada, Jan. 2020. Accessed: April 6, 2023. [Online] Available: https://profiles.ucalgary.ca/sites/default/files/2022-08/lna_survey.xlsx.

[99] P. Rodríguez-Vázquez, M. E. Leinonen, J. Grzyb, N. Tervo, A. Parssinen, and U. R. Pfeiffer, "Signal-processing Challenges in Leveraging 100 Gb/s Wireless THz," In *2020 2nd 6G Wireless Summit (6G SUMMIT)*, pp. 1–5, 2020.

[100] *Study on supporting NR from 52.6 GHz to 71 GHz*, Technical Report (TR) 38.303, v17.0.0, 3GPP, March 2021.

[101] G. Fettweis, M. Dörpinghaus, S. Bender, L. Landau, P. Neuhaus, and M. Schlüter, "Zero Crossing Modulation for Communication with Temporally Oversampled 1-Bit Quantization," In *2019 53rd Asilomar Conference on Signals, Systems, and Computers*, pp. 207–214, 2019. doi: 10.1109/IEEECONF44664.2019.9048794.

[102] H. G. Myung, J. Lim, and D. J. Goodman, "Single carrier FDMA for uplink wireless transmission," In *IEEE Vehicular Technology Magazine*, vol. 1, no. 3, pp. 30–38, Sept. 2006, doi: 10.1109/MVT.2006.307304.

[103] M. P. Wylie-Green, E. Perrins, T. Svensson, "Introduction to CPM-SC-FDMA – A Novel Multiple-Access Power-Efficient Transmission Scheme." In *IEEE Transactions on Communications*, vol. 7, pp. 1904–1915, 2011.

[104] M. F. De Guzman, P. Koivumäki, and K. Haneda, "Double-directional Multipath Data at 140 GHz Derived from Measurement-based Ray-launcher," in Proc. 2022 IEEE 95th Vehicular Technology Conference (VTC2022-Spring), 2022, pp. 1–6.

[105] A. F. Molisch, M. Steinbauer, M. Toeltsch, E. Bonek, and R. S. Thoma, "Capacity of MIMO systems based on measured wireless channels," In *IEEE Journal on Selected Areas in Communications*, vol. 20, no. 3, pp. 561–569, April 2002.

[106] M. F. De Guzman and K. Haneda, "Double-directional multipath data at 140 GHz," available: 10.5281/zenodo.6653867.

[107] S. L. H. Nguyen, J. Järveläinen, A. Karttunen, K. Haneda, and J. Putkonen, "Comparing radio propagation channels between 28 and 140 GHz bands in a shopping mall," In *Proc. 12th European Conf. Ant. Prop. (EuCAP 2018)*, London, UK, pp. 1–5, Apr. 2018.

[108] P Kyösti, M. F. De Guzman, K Haneda, N Tervo, A Pärssinen, "How many beams does sub-THz channel support?," In *IEEE Antennas and Wireless Propagation Letters*, vol. 21, no. 1, pp. 74–78, 2021.

[109] *Study on channel model for frequencies from 0.5 to 100 GHz*, Technical Report (TR) 38.901 v14.1.1, 3GPP, Jul. 2017.

[110] *Effects of building materials and structures on radio wave propagation above about 100 MHz*, Recommendation ITU-R P.2040-1, ITU-R, Jul. 2015.

[111] A. U. Makarfi, K. M. Rabie, O. Kaiwartya, O. S. Badarneh, X. Li, and R. Kharel, "Reconfigurable intelligent surface enabled IoT networks in generalized fading channels," In *Proc. IEEE Int. Commun. Conf. (ICC)*, pp. 1–6, 2020.

[112] H. Q. Ngo, L. Tran, T. Q. Duong, M. Matthaiou, and E. G. Larsson, "On the total energy efficiency of cell-free massive MIMO," In *IEEE Trans. Green Commun. and Netw.*, vol. 2, no. 1, pp. 25–39, 2018.

[113] J. Liu, M. Sheng, L. Liu, and J. Li, "Network Densification in 5G: From the Short-Range Communications Perspective," In *IEEE Communications Magazine*, vol. 55, no. 12, pp. 96–102, December 2017.

[114] F. Sohrabi and W. Yu, "Hybrid Digital and Analog Beamforming Design for Large-Scale Antenna Arrays," In *IEEE Journal of Selected Topics in Signal Processing*, vol. 10, no. 3, pp. 501–513, April 2016, doi: 10.1109/JSTSP.2016.2520912.

DOI: 10.1561/9781638282396.ch4

Chapter 4

Towards Joint Communication and Sensing

By John Cosmas, et al.[1]

Localization of user equipment (UE) in mobile communication networks has been supported from the early stages of 3rd generation partnership project (3GPP). With 5th Generation (5G) and its target use cases, localization is increasingly gaining importance. Integrated sensing and localization in 6th Generation (6G) networks promise the introduction of more efficient networks and compelling applications to be developed.

Many use cases such as factories of future, healthcare, autonomous vehicles, energy, and urban environment require not only low latency, low jitter, and high availability data transmission applications that include closed-loop control on one side and ultra-high data rate communication applications that include video or large sensor data traffic on the other side, but also localization accuracy with a precision of up to 1 mm that would be an essential enabler for opening up a lot of new applications.

5G localization systems utilize received signal strength (RSS), time-of-arrival (ToA), and angle-of-arrival (AoA) technologies with sub-6 GHz, mmWave, and optical wireless communication (OWC) for estimating position of UE, whereas simultaneous localization and mapping (SLAM) systems could combine the utilization of communication technologies such as orthogonal frequency division

1. The full list of chapter authors is provided in the Contributing Authors section of the book.

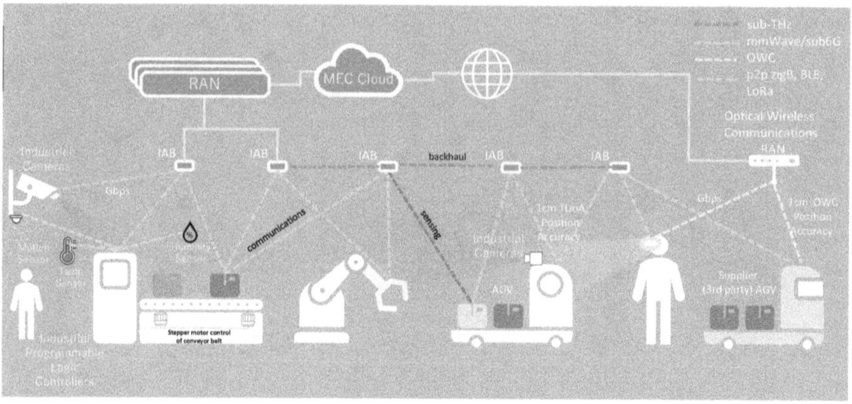

Figure 4.1. Expected sensing use cases in an Industry 4.0.

multiplexing (OFDM), orthogonal time frequency space (OTFS), OTFS-like, or frequency-modulated continuous wave (FMCW) modulation with sensing technologies such as terahertz (THz) beam technologies to sense point cloud of its environment by identifying landmarks for estimating position of UE. Essentially, the plurality of multiple-sensing technologies and multiple-sensing nodes will ensure the continued ability to obtain localization of UE despite the occurrence of any obstructions.

Figure 4.1 depicts expected sensing use cases in an Industry 4.0 setting, where, e.g., the localization accuracy requirement for automated guided vehicles (AGVs) and collaborative drones is 1 cm every 1 s, for augmented reality (AR) headsets it is 1 mm every 100 ms, for collaborating mobile robots (cobots) it is 1 mm with cycle time of motion control function of 1 ms, synchronicity of 10 ns, and for motion, temperature, humidity, etc., sensors is 10 cm every 1 to 10 s, whilst all require the provision of localization with 99.99% reliability. The motivation for using 6G mobile networks for localization as opposed to a dedicated radar/LIDAR system is the pervasiveness of mobile networks means that they can be applied universally and can employ the economies of scale to produce new and cheaper pervasive services that use high accuracy localization. The best LIDAR systems can obtain a precision from between 0.5 cm over 200 m (Ouster OS1) [1] to 2 cm over 400 m (Ouster OS2) [2] accuracy, which is an indication of what accuracy can be expected from a 6G communication and sensing system.

The sensing stratum uses the control plane for issuing reference signals for measuring RSS, ToA, and AoA for scheduling measurement frequency and accuracy for UE, the management plane for providing administrative data for computing UE position from distance measurements, and the vertical application Application Programming Interface (API) for accessing the UE position, see Figure 2.3 in Chapter 2.

In particular, highly advanced industrial environments will present significant challenges even for 5G specifications (up to Release 18 3GPP specifications), spanning congestion, interference, security and safety concerns, high power consumption, restricted propagation, poor location accuracy within the radio, and core backbone communication networks for the massive IoT use cases, especially inside buildings. 6G is preparing a new solution consisting of a combination of artificial intelligence (AI) methods with new communication technologies, potentially extending to OWC and THz to perform resource allocation over and beyond massive machine-type communications and to enhance performance with regard to capacity, reliability, latency, and localization accuracy. Examples of localization in industry are its use with video guides for facilitating maintenance of equipment and graphics superimposed on AR field of view of a factory or warehouse scene for the accurate location of electric, gas, and pneumatic facilities; mobile service robots in airports for physically guiding passengers through airport; and for AGV robots for carrying passenger luggage between baggage handling conveyor belt locations in airports [3].

6G and its visionary scenarios continue this trend and look at localization that is even more accurate and has even stricter latency requirements [4]. 6G SLAM will require multiple access technologies, such as sub-6 GHz, mmWave, sub-THz for RSS, Time Difference of Arrival (TDoA), and AoA localization, combined with sub-6 GHz sensing to produce a point cloud for producing a digital twin for updating the digital twin for obtaining location from environmental landmarks. Computational offloading such as on a multi-access edge computing (MEC) cloud is required for producing location from all these technologies using AI, for example, to recognize location from landmarks and artificial intelligence/machine learning (AI/ML), for example, to Kalman filtering to predict position and trajectory from measurements. This will provide novel and enhanced sensing and localization services leveraging on the positioning information of the users and their surroundings, for instance for merged reality or digital twins. For this, 6G will primarily act as a connector between the sensors and the users, e.g., with sensor fusion, where sensor data from different modalities are combined. However, 6G may also play a role in storing, aggregating, and analysing the sensor data before providing the mapping information to the user.

Another aspect of positioning and sensing in 6G is the possibility to incorporate the sensing data into the network operations. With an accurate map of the surroundings, the locations of the users and various stationary and moving obstacles, the user mobility, and optimal beam pattern can be predicted, improving the performance and resource utilization.

Finally, the research into 6G engenders a paradigm shift in the radio interface. For the past century, radio waves have been used for either communication or sensing, i.e., radar. However, as the radio communication begins using higher radio frequencies (>100 GHz), the potential sensing accuracies become viable. By repurposing the radio interface of the mobile network from only communication to joint communication and sensing, the ubiquity of the mobile devices and networks can provide a cost-efficient and widely spread sensing resource that can be used to enhance user services or network operations.

Section 4.1 presents the plurality of multiple sensing techniques to produce extremely accurate sensing information to a user to enhance services using, e.g., sensor fusion, i.e., 6G primarily acts to connect the sensors with user. Section 4.2 presents the plurality of multiple sensing nodes to enhance connectivity. Section 4.3 presents the repurposing of the radio interface to also act as a radar (i.e., Joint Communication and Sensing (JCAS)). All these technologies will ensure the continued localization of UE despite the occurrence of any obstructions.

4.1 Providing Extremely Accurate Sensing

4.1.1 Sub-1 cm Location Accuracy Using Sensor Fusion

Localization of UEs using ToA from sub-6 GHz wireless and RSS from optical wireless infrared techniques can be used to obtain an accuracy of less than 1 cm; however, it is highly dependent on the continuous direct line of sight access between the gNB access points (APs) and the UE. If there is no direct line of sight access to four or more gNB APs, then location ambiguity is introduced, and so other techniques can be used to maintain localization such as dead reckoning from inertial measurement unit (IMU), AoA from received radio signatures, iterative multi-lateration, or position from landmarks. The received radio signatures estimate AoA from more than two APs or AoA and distance from one AP, so that some forms of interim measures for obtaining location can continue to be made. Distance and speed of UE can be estimated using OTFS modulation from one AP due to its operation in the delay and Doppler domain as opposed to using OFDM, which operates in the time and frequency domains. If there is no direct line of sight access to any 6gNB APs, then position from landmarks calculates position from at least four landmarks identified from within point cloud data of a sensed environment obtained, for example, from a 6G sensing/LIDAR system and/or 360-degree image.

4.1.1.1 Distance measurement and localization using AI

3D localization using RSS or TDoA requires line-of-sight (LoS) time of flight (ToF) measurements to picosecond accuracy for estimating distance to mm accuracy from at least four accurately located 6gNB APs; otherwise, inaccuracies and ambiguities are obtained where more than one solution is produced. LoS propagation paths to each of these APs may not always be available as a result of incidental occlusion from moving objects, so a strategy of providing more, cheaper APs for providing alternative means for obtaining location and alternative methods using fewer APs are required such as AoA, which only requires two APs, since two lines intersect at or close to a point, or one AP and a distance, since a line and a distance define a point on the line.

In order to design such a system, a digital twin was obtained from 3D laser measurements in a real factory, and a ray tracing (RT) model was used to generate a rich data set of point cloud data, which was converted into a Siteviewer or Winprop CAD "digital twin" model, which uses a deterministic methodology to predict radio propagation of THz and mmWave frequencies based on the geometrical theory of propagation (GTP) that takes into account geometrical properties of the environment and propagation parameters such as ToF, direction of arrival (DoA), and direction of departure (DoD) [5].

The beam scheduler implements intelligent beam steering that, using an antenna array, can steer beams from 6gNB RUs to UE locations efficiently based on interpreted channel impulse response (CIR) knowledge of uplink isotropic transmissions from uplink UE. Then, it optimizes the beam direction using reinforcement leaning (RL) to maximize the SINR, and when the path between the transmitter and receiver is broken through an obstruction, it immediately finds and uses alternative secondary paths to steer the beam between the transmitter and receiver by avoiding the obstruction. The optimized beam direction from two BSs can also be used to obtain the location of UE locations.

The beamforming vector is obtained through a deep neural network (DNN) by using predicted digital twin isotropic transmissions from UE in a grid over the coverage area, and the predicted digital twin measured impulse response at each 6gNB RU receiver of four polarized transmissions and received power that acts as input information to train a DNN. Four polarizations were used to have more flexibility in terms of data generation for the training of AI, implying that the more polarizations the better for the AI training.

The beamform codebook index is the set of beamform vectors comprising of steering angles at the receiver of different alternative secondary paths to steer the beam between the transmitter and receiver by avoiding any obstruction. In the absence of a real environment, the beamforming power control and interference coordination are jointly carried out on a digital twin to enhance the performance of the 5G network. The network comprises of serving gNBs and one interfering gNB as shown in Figure 4.2. The deep reinforcement learning-neural network (DRL-NN) is modelled for the downlink scenario in which a gNB is serving a UE. The gNBs are a fixed, known distance apart, and different UEs are uniformly distributed in a particular service area. Also, the users are moving at a fixed speed.

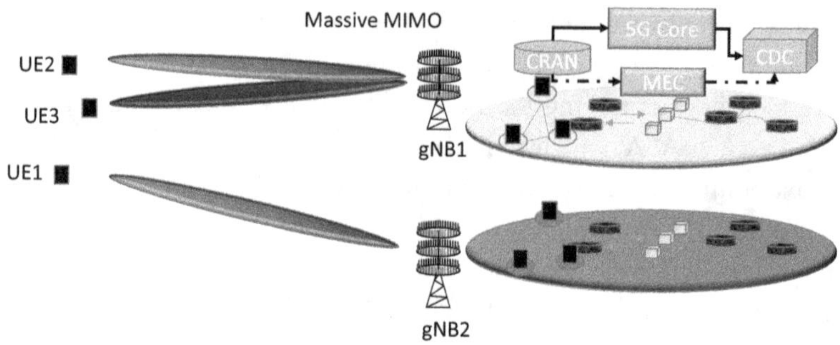

Figure 4.2. Radio beam control.

An UE is served by a maximum of one gNB. Hybrid beamforming for the downlink data transmission is employed to overcome the problem of high propagation loss.

RL is then to optimize the beam direction, whose state space, action space, reward, and learning algorithm are defined in [5]. Initial simulation is performed on MATLAB 2021a siteviewer for radio transmission (RT) using site viewer and simultaneous channel generation [6], but it could equally well have used Winprop, which models more accurate geometries and accommodates more surface properties [5]. The output of RT is fed for channel generation, which in turn produces input for the DNN for training. For RT and channel generation, different users' locations for fixed 6gNB RU positions are considered. Channel parameters vary with respect to varying user locations. Therefore, DNN is trained to different channel parameters, which in turn are dependent on user location. The DNN is trained on python AI module and is trained for efficient beamforming of the data towards the user. Post-training and during the working phase, the system performs beam scanning to obtain the channel parameters based on user locations. Beamforming weights are adjusted from channel parameters to form the beam pointing towards the current user location. In this way, based on the obtained channel parameters, the user locations could be estimated [5]. If a beamformer steers the main beam in a particular direction (θ, φ), then the directional accuracy is defined as $\pm\delta\theta$, $\pm\delta\varphi$ from the maximum signal-to-noise ratio (SNR) of the main lobe. The proposed deep learning integrated reinforcement learning (DLIRL) beamforming has an angle-of-departure (AoD) accuracy towards the UE location with a deviation of $\pm 2°$, whereas RL has a deviation of $\pm 3°$ and DNN's deviation is $\pm 5°$ [6].

4.1.1.2 Distance measurement using TDoA on mmWave Networks

The accurate 3D location of the 5G UE is calculated by the use of ToA measurements from at least four different locations and then exercising triangulation techniques as shown in Figure 4.3. In order to reach the required 1 cm accuracy, it

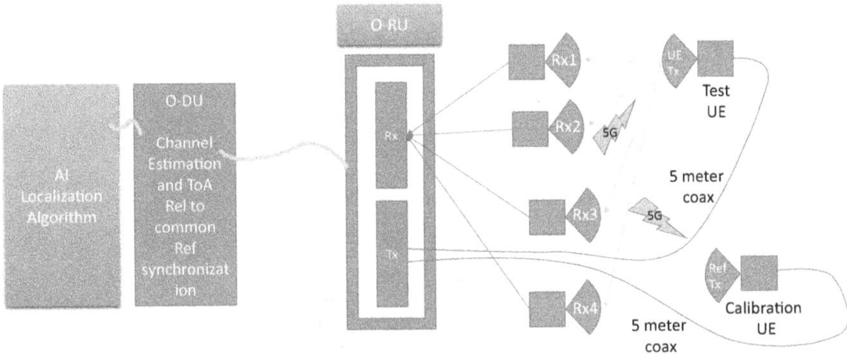

Figure 4.3. Setup for ultra-accurate UE distance measurement.

is required to measure the ToA at accuracy levels of ~33 picoseconds (calculated from 1 cm/speed of light).

The setup that is used to measure the test UE location consists of the O-RU and O-DU and the localization measurement algorithm application installed on an X86-based server. The O-RU is configured in a loop-back mode, which enables it to transmit 5G OFDM pseudo-random noise (PN) signals (Tx) at 3.5 GHz similar to the UE transmissions. These Tx signals are connected via coax to two patch antennas that simulate the test UE and the reference UE. The reference UE is a UE with a known location that is used to calibrate the accurate test UE location measurement. The test antenna is located on a scanner that can accurately move its position to new locations in space. The 5G signals transmitted over the air at 3.5 GHz by the test UE and the reference UE patch antennas are received by the O-RU receiver (Rx) via four different Rx antennas located in the four corners of the ceiling of the measurement room. The O-RU will continuously measure the ToA of the 5G signals coming from the test UE and the reference UE with a resolution of 0.3 ps (corresponding to 0.1 mm) and transfer the results to the O-DU that stores them on agreed memory sharing register. A localization algorithm reads the ToA registers and calculates the test UE location. Following the successful test UE location described herein, the test UE might be replaced by a commercial UE [7].

Some initial results, which are illustrated in Figure 4.4, show that although the measurement samples have been measured with a resolution of a timetic of 0.33 ps or 0.1 mm, the samples are approximately normally distributed about the mean of 69516 timetics or 6.9516 m. The 95% confidence interval falls within ±2 times the stdev = 833 × 2 = 1666 timetics = 16.66 cm about the mean, which can be considered as a measure of accuracy. It is assumed that the mean of the measurements represents the true distance that is being measured, which was not provided in these initial results. A Kalman filter is an effective method for

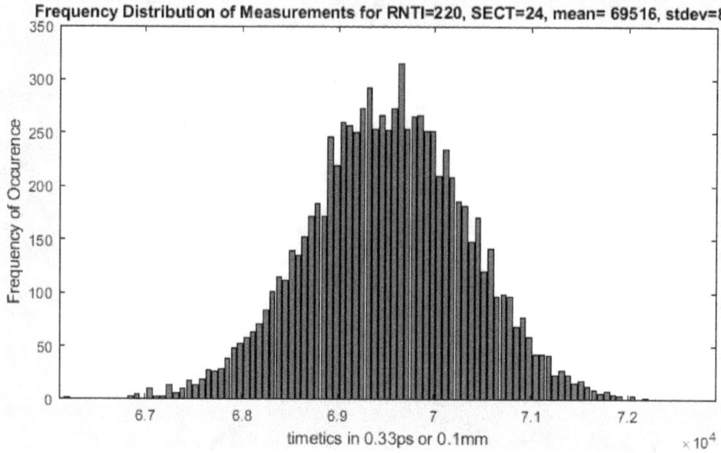

Figure 4.4. Frequency distribution of measurements for RNTI = 220, SECT = 24.

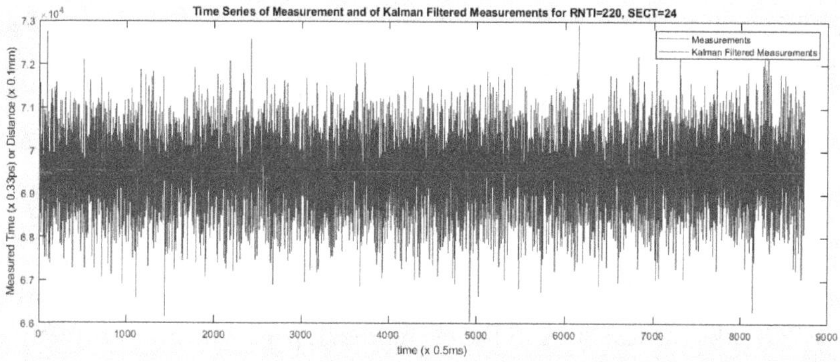

Figure 4.5. Time series of measurement and Kalman-filtered data for RNTI = 220, SECT = 24.

reducing normally distributed noise from measurement results, which when applied reduces the 95% confidence interval to ±2 times the stdev = 13.74 × 2 = 27.48 timetics = 2.748 mm about the mean, as shown in Figure 4.6, which is considered as a measure of accuracy of the filtered measurement data. The time series of the overall measurement results and Kalman-filtered measurement results are shown in Figure 4.5. Note that RNTI = 220 refers to the calibration UE, and Sect = 24 refers to the portion of spectrum being used for the measurements, as shown in Figure 4.5. These initial results seem to be sufficient to meet <1 cm accuracy but need further improvement to meet <1 mm accuracy.

4.1.1.3 Distance measurement using RSS on OWC networks

If obtaining location from 5G sub-6 GHz or mmWave communications is made difficult, for example, as a result of highly reflective metallic scatterers in the

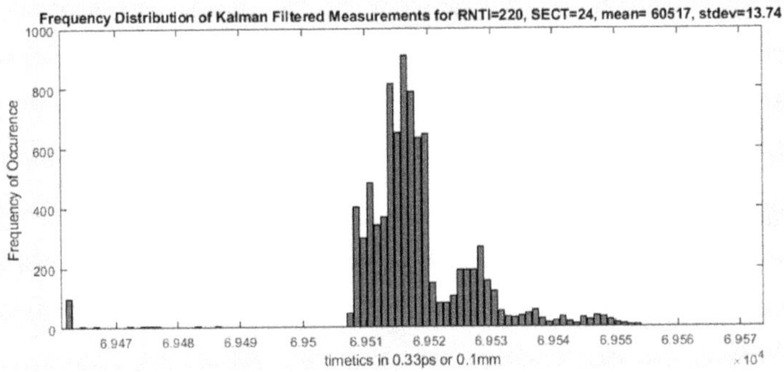

Figure 4.6. Frequency distribution of Kalman-filtered measurement for RNTI = 220, SECT = 24.

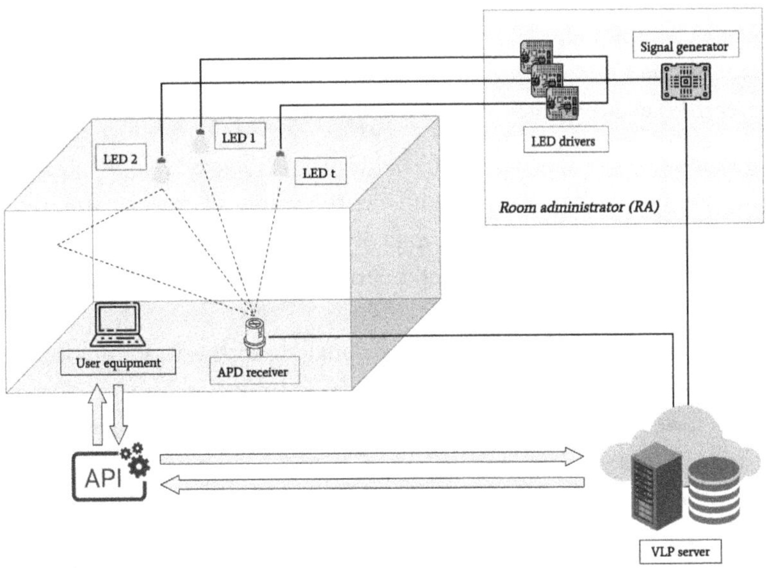

Figure 4.7. OWC position testbed.

coverage area, then RSS of an OWC communications system can be used as an alternate method for estimating position.

The OWC position testbed contains three main components for visible light positioning (VLP) LED or IR LED measurement campaign, as shown in Figure 4.7.

1. Location server: Including three parts: (1) LED control unit, (2) received signal collection unit, and (3) positioning unit. In an LED control unit, it is connected with an Arduino board, which has been set up to control LEDs in advance, including the LED switch settings and LED frequency settings. These settings can be used to control on/off of LEDs and change the frequencies of LEDs. When the VLP server receives setting change requests

from API, it will change settings of the LEDs according to the user's requests. In the received signal collection unit, it collects received signal data from APD receiver connected by a USB port, which can be downloaded by user through API. In the positioning unit, user can upload the received signal data file to execute positioning algorithm. Users can use the files obtained in the previous unit or upload them themselves. The positioning algorithm used in this server is traditional trilateration received signal strength indication (RSSI) algorithm. The output of the algorithm is the estimated coordinate of the receiver.

2. LED transmitter module: Including LED and its corresponding drivers. Every LED has to be driven by an LED driver in order to convert the power supply from 220 V into a suitable value and switch the frequency of LED.
3. Avalanche photodiode (APD) receiver: A highly sensitive semiconductor photodiode detector used to measure and collect received light signals. The received data are transmitted to the VLP server through the USB cable.

Some results in comparing with traditional RSSI-based VLP algorithm are shown in Figure 4.8. Two figures represent the values of ε when the orientation of the receiver was randomly in the range of $[-3^\circ, 3^\circ]$ and $[-5^\circ, 5^\circ]$, respectively. The results are very similar, which reflects that the algorithm has good immunity to changes in receiver angle.

Firstly, in Figure 4.8(a), when the traditional algorithm was employed and the orientation of the receiver θ was in $[-3^\circ, 3^\circ]$, values of ε were plotted using blue crosses for each measurement. They were mostly scattered between 0.25 and 0.35 m, and the average deviation is 0.31 m. It could be observed that even though the angle of the receiver θ was the same value, the value of ε is different for different measurements. In other words, not only did the estimate of the distance between the transmitter and the receiver always have a large deviation, but the deviation was also difficult to determine.

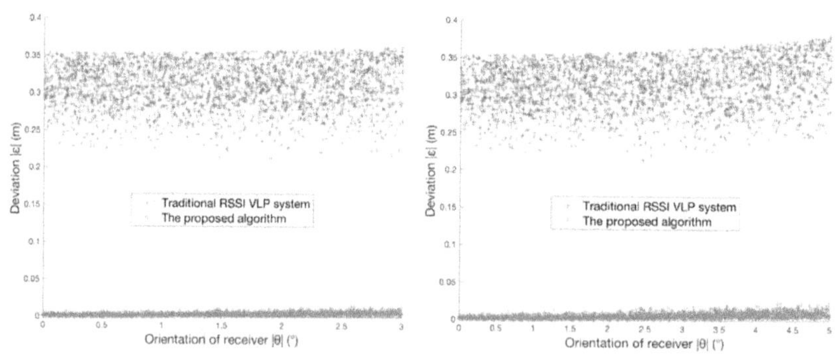

Figure 4.8. Deviation of the proposed algorithm and traditional RSS-based VLP algorithm with a variation of receiver angle (a) $\theta \in [-3^\circ, 3^\circ]$ and (b) receiver angle $\theta \in [-5^\circ, 5^\circ]$.

Afterwards, when the proposed algorithm was employed, the values of ε were scattered using red crosses. Compared with blue crosses, red crosses were much closer to the x-axis: this means that the value of ε was closer to 0. Indeed, the average of ε decreased to 3.5×10^{-3} m, and the maximum and minimum values of ε declined from 0.36 to 0.01 m, and from 0.2133 m to 1.33×10^{-5} m, respectively. Moreover, the distribution of red crosses was more compact: on the one hand, when the value of θ became the same, the values of ε calculated by different measurements were not very far from each other; on the other hand, the value of ε did not change greatly according to the change of θ.

Therefore, it could be inferred from the distribution of blue and red crosses that, in general, our proposal led to less deviation and a more concentrated distribution than traditional methods. As a result, estimated distances between transmitters and receivers would be closer to their real distances by using our proposal. Furthermore, it could be boldly predicted that it could lead to lower positioning errors (PEs).

This trend is also evident in Figure 4.8(b). When the traditional algorithm was employed and the orientation of the receiver θ was in $[-5^\circ, 5^\circ]$, blue crosses were mostly scattered between 0.25 and 0.4 m, and the average deviation was 0.3103 m. Compared to the situation of θ in $[-3^\circ, 3^\circ]$ in Figure 4.8(a), the average deviation has barely changed, but the distribution of their values was more spread out in Figure 4.8(b). It can be inferred that the deviation of the distance between the transmitter and the receiver caused by the random orientation of the receiver became larger and less predictable as the range of values of θ expands.

The CDF of positioning results is shown in Figure 4.9. 98.81% of the total results had the PE less than 10 cm, which means almost all of the receiver's positions could be accurately estimated.

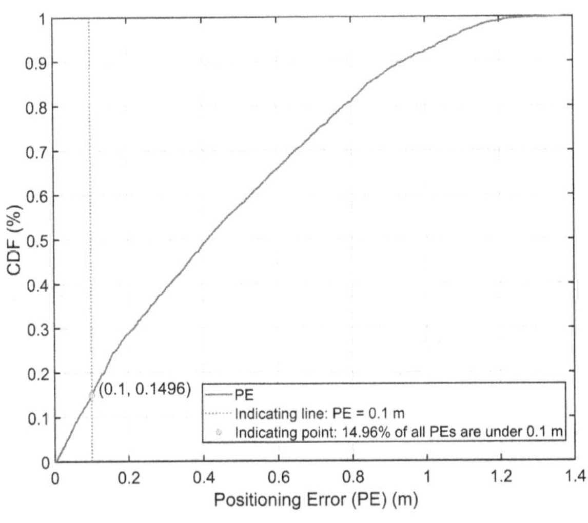

Figure 4.9. CDF of positioning results of the proposed algorithm and traditional RSS VLP algorithm in different receiver's orientation: using the proposal when $\theta \in [-5^\circ, 5^\circ]$.

4.1.1.4 Distance measurement using OTFS modulation

The 5G air interface and associated modulation have to support a number of diverse requirements and use cases (e.g., eMBB, high-speed use cases, mMTC, etc.), as detailed in many publications. The associated modulation waveform would then have to exhibit high performance in many diverse scenarios of high or low Doppler, delay spread, carrier frequency, etc. This is possible if the modulation scheme takes full advantage of the fading multipath nature of the channel and extracts the full diversity present in the channel in all dimensions of time, frequency, and space. Such a flexible waveform can serve as an integral part of a flexible air interface and associated core network. Therefore, researchers have recently introduced a novel modulation technique called OTFS [8], which may be useful for obtaining position of fast-moving UEs with high Doppler. It has been shown recently that OTFS arises as a well-suited modulation for the time and frequency selective fading channel. OTFS characterizes the Doppler-induced time-varying nature of the wireless channel and parameterizes it as a 2D impulse response in the delay-Doppler domain.

OTFS works in the delay Doppler domain rather than the time-frequency domain as shown in Figure 4.10. The delay Doppler domain representation of the channel converts the time-variant channel to the time-invariant channel, as shown in Figure 4.11. In addition to the OTFS diversity gains mentioned above, we have additional benefits of low reference signal overhead, enhanced channel state information (CSI) quality, and MIMO bit error rate (BER) performance of fast-moving UEs, as shown in Figure 4.13.

Because of the added advantages of OTFS over OFDM, the OTFS is an efficient way of estimating the localization via computation of ToA.

The above Figure 4.12 shows how a UAV can be efficient in locating the user in need via employing the OTFS for estimating an accurate ToA.

From the BER analysis for the OTFS and OFDM modulated signals shown in Figure 4.13, the comparative analysis of the OTFS and OFDM modulated signals

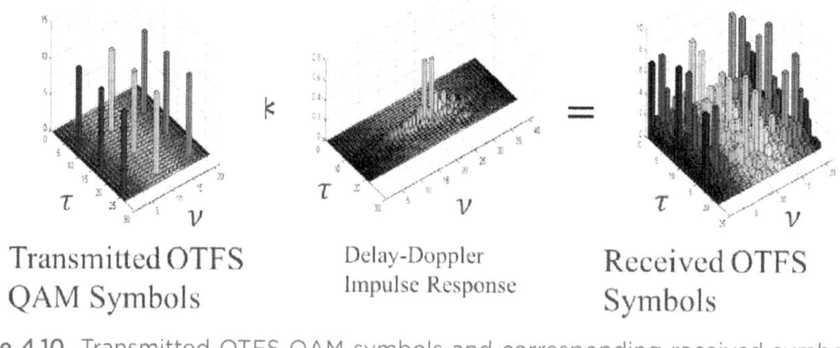

Figure 4.10. Transmitted OTFS QAM symbols and corresponding received symbols via delay Doppler channel.

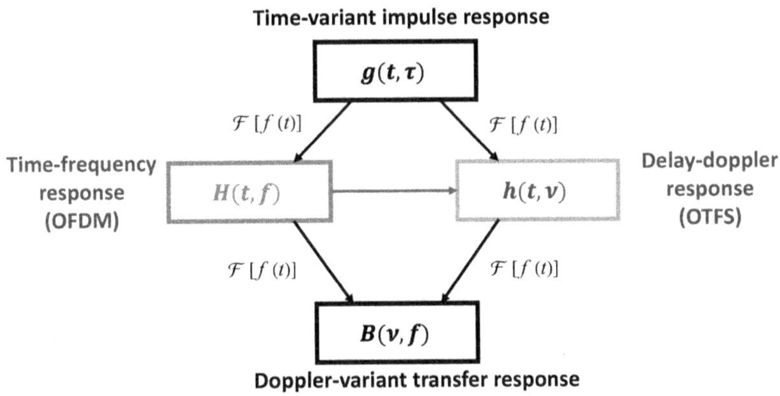

Figure 4.11. Different representations of linear time variant (LTV) wireless channels.

Figure 4.12. UAV with OTFS employed for the user localization.

for fast-moving UE (where the UE is moving at a constant speed of 28 km/h) can be visualized. The BER for OTFS-modulated signal is lower because its performance is not degraded even with the fast motion of the UE, and the bit error and bit loss are less as compared to the conventional OFDM. Thus, OTFS may be able to more reliably estimate the distance of UEs moving at a fast speed.

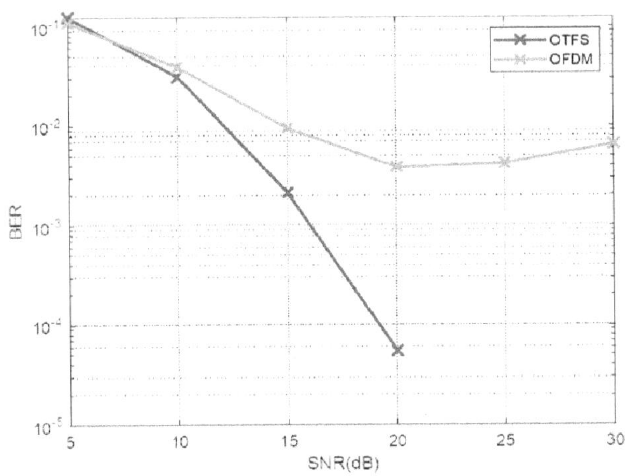

Figure 4.13. BER comparison of OTFS and OFDM when UE is moving at a constant speed of 28 km/h.

4.1.1.5 Distance and angle measurements using physically large arrays

To give an outlook towards the capabilities of large aperture arrays, a measurement-based analysis of beamforming and channel estimation is performed, targeted at positioning applications. The initial results are obtained using spherical-wave beamforming applied to data recorded using a (synthetic) large uniform rectangular array (URA), representing a large base station (BS) that is positioned along a wall in an indoor environment. The results are visually compared with super-resolution channel estimation results obtained when splitting the large URA into small sub-arrays. The channel measurements were recorded with a vector network analyser (VNA), with the synthetic arrays formed using mechanical positioners with a positioning accuracy below 1 mm.

Exploiting the full array aperture requires spherical-wave beamforming and perfect calibration of the full array, which can be difficult to achieve for thousands of array elements. The advantage of using the full available data is the superior resolution that can be achieved, allowing focusing power towards small spatial regions [9, 10], e.g., $\lambda = 2.16$ cm, for a carrier frequency of $f_c = 6.95$ GHz, is shown in Figure 4.14. The marginal spectra include modelled components that are computed from a geometric floorplan in combination with an image source model up to second order. In combination with the large aperture, a large signal bandwidth of BW = 3 GHz is used to show the optimum case regarding the achievable resolution. From the sub-figures, while direct paths and first-order reflections are well represented by the model, second-order reflections, which might no longer be visible over the full array, are not well associated with modelled components.

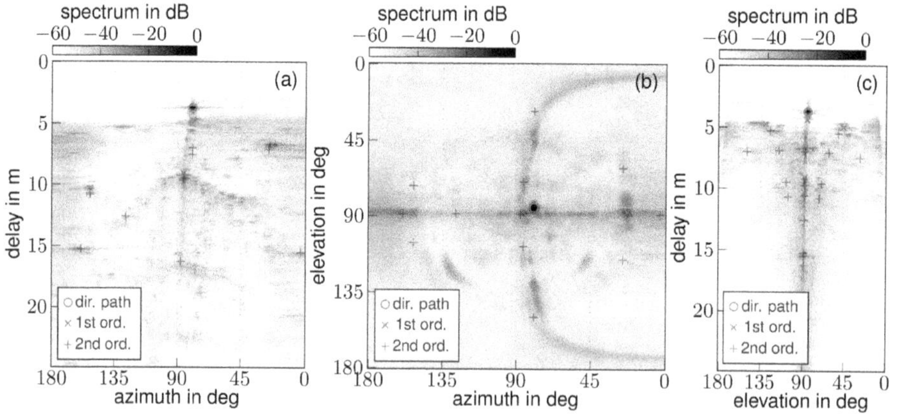

Figure 4.14. Spherical wave beamformer spectra for an exemplary LOS position in a medium-sized indoor environment. In the analysis, we use the full measurement bandwidth and all (synthetic) array elements. In the azimuth-elevation power spectrum shown in (b), 0 deg azimuth points towards a window in the environment and 0 deg elevation towards the ceiling.

Parametric approaches such as super-resolution channel estimation algorithms can achieve a high level of performance under the assumption that the channel is composed of discrete components that are parameterized in terms of, e.g., arrival times or angles. The separation into sub-arrays allows to make use of classic array processing assumptions, which are not feasible for application to the full array data, drastically improving the requirements for computational resources. Results obtained when applying a sparse Bayesian learning (SBL)-based channel estimation algorithm (described in [11]) to (4 × 4)-subarrays and B = 500 MHz of signal bandwidth are shown in Figure 4.15, as angle spectra (a), residual angle spectra after subtracting component estimates (b), and time-domain residual signals after subtracting the obtained component estimates (c). In Figure 4.15(a) and (d), the size of the markers represents the estimated amplitude, and the colour indicates whether a component was associated with a modelled component (cyan) or not (magenta). This association is performed by using an optimal sub-pattern assignment-based data association algorithm, which allows to associate estimated components to modelled components and thereby accounts for the varying visibility of multipath components (MPCs). While using only a fraction of the array elements and bandwidth, the obtained estimates still correspond well with the modelled MPCs. At the same time, the varying visibility can be accounted for, which will be an important

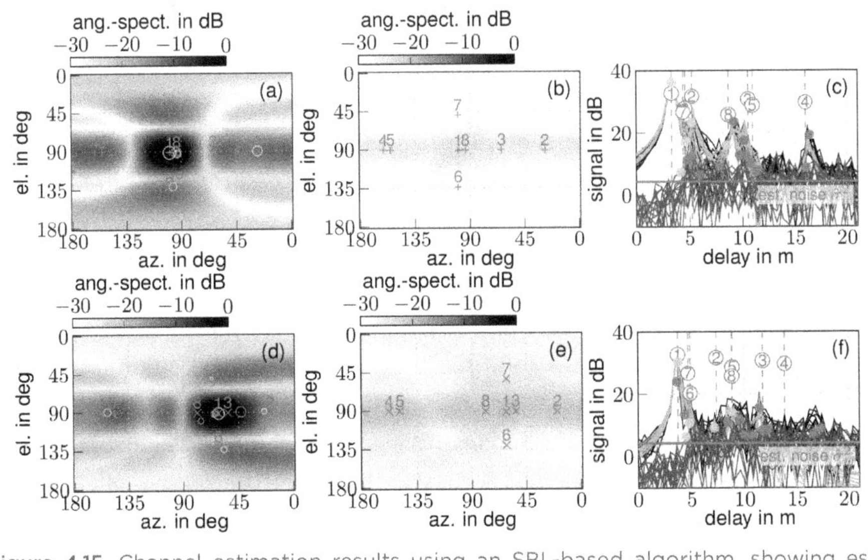

Figure 4.15. Channel estimation results using an SBL-based algorithm, showing estimated components over the angle-spectrum, residual angle-spectrum, and residual time-domain signals. Results are shown for (4 × 4) sub-arrays with 500 MHz bandwidth and for a subarray closest to the window (a, b, c) and farthest from the window (d, e, f). Component estimates are shown as magenta and cyan circles with the latter indicating association to a modelled environment feature (numbered 1–8), with 1 representing the direct path and 2 a window reflection.

aspect for multipath-based positioning and environment learning. Note the consistent changes that are visible in the modelled as well as estimated components, comparing the two subarray positions that are closest to the window (top row in Figure 4.15) and farthest from the window.

While analysis of the measurements performed has shown the effects of visibility as well as the spatial consistency of MPCs, without efficient fusion of sub-arrays, it will be difficult to achieve a similar performance compared to a large and fully coherent array. The next step will be the derivation of theoretical performance limits, e.g., in terms of the Cramer-Rao lower bound or similar, giving insight into system parameters such as signal bandwidth and the geometric distribution of sub-arrays. Of special interest will be the effect of imperfect phase or clock-synchronization between the subarrays, which is expected to be the main limiting factor to achieve the same performance as a fully coherent array.

4.1.1.6 Precise localization with the aid of synchronization signals and CIR

4.1.1.6.1 Introduction

Instead of adding an extra overload to the communication systems, one can rely on the mechanisms already in place to perform precise localization. In particular, approaches based on TDoA and ToA require the nodes not only to be synchronized with each other, but also to exchange localization-specific signals. However, the synchronization signals exchanged among the nodes for the purpose of synchronization as well as other sources of information such as CIR and AoA estimation, which are not primarily foreseen for localization purposes, can be utilized to precisely estimate the position of a UE.

In [12], the principles of network synchronization have been presented, which paves the way for an accurate mobile UE localization with the aid of synchronization signals, i.e., time stamps. In particular, a Bayesian recursive filtering (BRF)-based mobile unit (MU) joint synchronization and localization (sync&loc) approach is developed where Taylor expansion is utilized to linearize the non-linear relation between the measurements, i.e., time-stamp exchange and AoA, and the position parameters. While linearized BRF (L-BRF) can partially mitigate the destructive impact of non-linearities in the measurements, in addition to the covariance matrix underestimation, they are likely to diverge if a reliable estimate of the initial state is not available [13]. A promising approach, on the one hand, to avoid such shortcomings of L-BRF and, on the other hand, to boost the accuracy of position estimation, is estimating the prediction, measurement likelihood, and posterior distributions using particle Gaussian mixture (PGM) filters introduced in [14]. Specifically, in this approach, instead of a single Gaussian function, each distribution is approximated with a sum weighted of Gaussian functions or Gaussian mixtures [15]. Nevertheless, the problem that immediately arises when using PGM filters is dimensionality, rendering the approach computationally expensive for multi-variable estimations. To overcome this drawback, a hybrid parametric and particle-based approach was employed, which capitalizes on the linear relations in the measurements to reduce the dimensionality.

Here, based on [16], a DNN-assisted particle filter (PF)-based (DePF) joint sync&loc algorithm is proposed that draws on the CIR to estimate the AoA (using multiple signal classification (MUSIC) algorithm [17]) and to determine the link condition, i.e., LoS or NLoS, using a pre-trained DNN, thereby excluding the erroneous measurements to enable a more precise parameter estimation. It then estimates the joint probability distribution of MU's clock and position parameters using the PGM filter. The dimension of the PGM filter is then reduced by revealing and exploiting the existing linear sub-structures in the measurements, thereby tackling the dimensionality problem.

There are, however, several preliminaries for the PGM filter to return an accurate estimation of the MU's clock and position parameters. In addition to the timestamp exchange mechanism explained in [12], as mentioned above, AoA using the MUSIC algorithm and DNN-based NLoS identification are the prerequisites for the DePF algorithm. The former increases position estimation accuracy, while the latter prevents the PGM from diverging.

4.1.1.6.2 NLoS identification, AoA, and CIR

The capability to estimate CIR is highly ubiquitous among APs. Therefore, relying on the CIR to develop a localization algorithm appears to be a realistic approach. The AP-UE CIR is a rich source of information about the condition of the communication link, e.g., whether the channel is LoS or NLoS, and the location of the UE. More precisely, the former is crucial to know when estimating the latter, as is the accuracy of the distance or time, and AoA measurements significantly decline if conducted under NLoS conditions.

Figure 4.16 shows the architecture of the DNN deployed for NLoS identification. We have utilized such a network in [18] to identify the link condition in indoor environments. The input layer has one channel fed with N samples, i.e., the magnitude of the CIR. The number of hidden layers and neurons in each hidden layer is set to l_H and n_H, respectively. The rationale to rely on when selecting these numbers is that, according to [19], any classifier function can be realized by two hidden layers, i.e., currently there is no theoretical reason to use more than two. However, the lack of evidence does not imply that the DNNs with more hidden layers do not improve the classification accuracy. It rather suggests that the number of required hidden layers does not follow a well-established logic and is

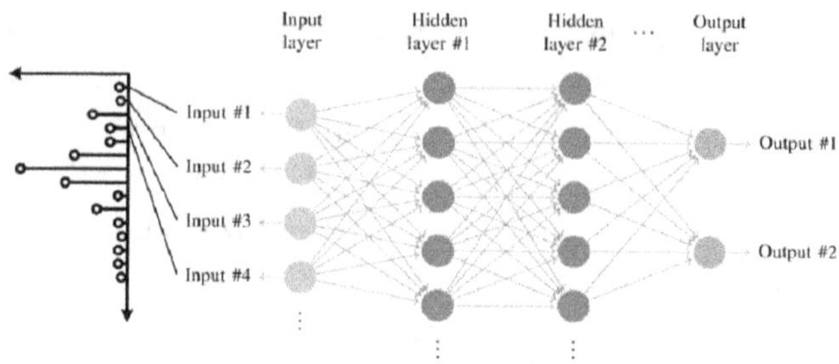

Figure 4.16. The DNN employed for NLoS identification. It has $l_H = 2$ hidden layers with n_H neurons and two output neurons [16].

mainly determined by a trial-and-error process. Therefore, for the algorithm proposed in this work, we empirically determine the number of hidden layers that delivers the best performance. Furthermore, as a rule of thumb, the number of neurons is suggested to be between the number of inputs and that of the outputs to prevent under/overfitting. Let the output probability vector of the DNN be $[1 - \hat{p}_{nlos}, \hat{p}_{nlos}]$, where $\hat{p}_{nlos}$ denotes the probability of the CIR being corresponded to an NLoS link. For the NLoS identifier, we seek to train the DNN such that the output probability vector is as close as possible to the [1,0]/[0,1] for the LoS/NLoS CIRs. In other words, from the optimization point of view, we aim to design a loss function whose output is small when the DNN returns the correct vector and is large otherwise. It turns out that the function that possesses the above-mentioned property is the logarithmic function [20]. The loss function is known in the literature as the binary cross-entropy loss function. The goal of training is then to adjust the weights of the neurons such that the binary cross-entropy loss function is minimized. Finally, when the trained DNN is employed in the context of joint synchronization and localization algorithm, the decision on the link condition is fed into the algorithm using a binary parameter, which is set to one when $\hat{p}^i_{nlos} > 0.5$ and zero otherwise. The CIR fed into the DNN to identify the link condition can be treated as an input signal to the state-of-the-art AoA algorithms such as MUSIC or Estimation of Signal Parameters via Rational Invariance Techniques (ESPRIT) to obtain the AoA. We do not mention the details of these algorithms here, as rich literature on these algorithms is available online. In what follows, we elaborate on how PGM filters fuse the above-mentioned pieces of information to reach an estimation of clock and position parameters.

4.1.1.6.3 Particle Gaussian mixture filter

The data obtained from the synchronization signals, i.e., time-stamps, the channel condition obtained by means of DNN, and the AoA estimated using the CIR can be fused using PGM filters to estimate the location and clock parameters of an UE.

The idea underpinning PGM filters is to approximate the posterior PDF by the sum of weighted Gaussian density functions (GDFs) [15] as shown in Figure 4.17.

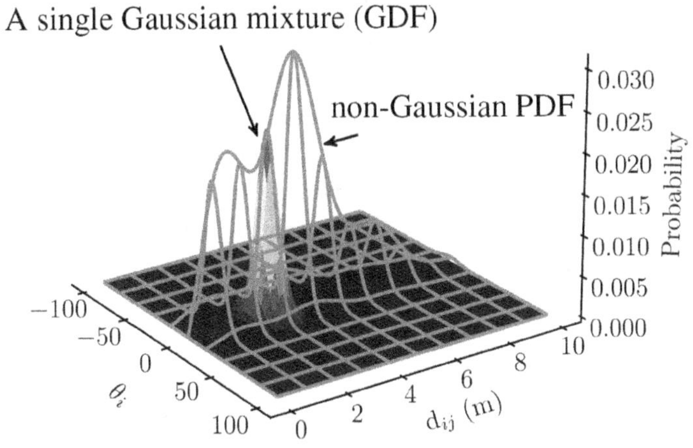

Figure 4.17. An example distribution of the clock and position parameters. θ_i denotes the clock parameters and d_{ij} represents the position parameters [16].

Considering the time-stamp mechanism introduced in [12], we can conclude that the clock parameters, on the one hand, are linearly dependent on the time-stamps and, on the other hand, do not depend on the position parameters. This suggests that, although the likelihood of the measurements is not Gaussian distributed in general, it is indeed Gaussian across the clock parameters. By capitalizing on the linear Gaussian substructures in the model, the state dimensions are kept low.

Consequently, the GDFs can be employed only across the position parameters, transforming the structure of the posterior distribution into the multiplication of a single GDF across the clock parameters and the sum of multiple weighted GDF across the position parameters (visualized in Figure 4.17). Such a structure not only lays the ground for the hybrid parametric and particle-based implementation of BRF-based joint sync&loc estimation but also dramatically reduces the computational burden.

4.1.1.6.4 Takeaways

As mentioned in the previous subsections, synchronization signals, i.e., time stamps and CIRs can be utilized to perform precise localization. In particular, we can employ PGM filters in a hybrid parametric and particle manner to track the clock parameters of a UE and estimate its position. Such a technique does not require an exchange of any localization-specific signal and relies only on the pre-existing signal exchange mechanisms.

4.1.1.7 MEC cloud database and server for localization and mapping data fusion

SLAM achieves the purpose of simultaneous positioning and map construction based on combination of self-perception LIDAR-like sensing data combined with localization data. These data are necessarily captured by sensors and collated in a database on which processing can be performed for the purposes of different applications, and in particular location estimation applications, an example of which is now presented below. RSS OWC distance measurements from UEs, TDoA mmWave distance measurements from 6G access nodes, and AoA direction measurements from 6G access are recorded on MEC location database VNF and processed by a location server VNF to produce a data fusion estimate of the location of UEs that is stored on the Location Database VNF for access by network and user applications. Data fusion combines estimates of position from mmWave OWC RSS, TdoA, and AoA position measurements using the Kalman filter machine learning algorithm to obtain a more accurate position estimate that compensates for the different sampling times of the different distance and direction measurements.

The initial structure of the location database (LD) consists of:

1. The antenna table, shown by Figure 4.18 (Table 1 within the figure), stores all antenna/LED emitter coordinates and, in this instance, the relevant calibration parameters required, for example, the VLC Halo-Lens Compensation approach [21, 22]. Currently, VLP relies strongly on the assumed Lambertian properties of light sources. In practise, not all lights are Lambertian. To facilitate and benefit from the widespread deployment of VLC technology in numerous environments, measurements from non-Lambertian sources are analysed, and a novel calibration VLC halo-lens compensation technique was developed that enables high-accuracy positioning in a wider setting [21, 22].

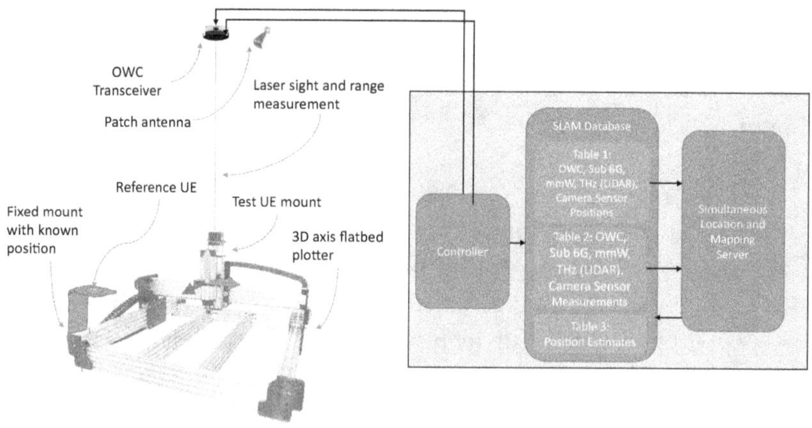

Figure 4.18. SLAM database and server system.

2. The measurement table, shown in Figure 4.18 (Table 2), stores all the latest measured location parameters obtained by the RAN. Here, we illustrate the OWC-RSS, sub-6GHz-TdoA, mmWave AoA, and eventual THz (LIDAR) and camera components. Each dataset (row) pertains to a singular UE with an appointed ID and timestamp of the measurement acquisition instant.
3. The estimates table, shown in Figure 4.18 (Table 3), is where the processed position estimates are stored. For data fusion applied later in the location server (LS), the states to be stored in the estimates table must be consistent with the states of the target applied to the location server. This enables the previous UE state estimates to be used within the Kalman filter as priory data. Asynchronous sampling of range-based measurements is known to have negative impacts on localization performance. Various asynchronous sampling localization techniques (ASLT) exist to mitigate these effects. The exact suitability of such solutions is not evident, due to their additional processes, subsequent complexity, and increased costs. Extensive simulations were conducted to demonstrate the effectiveness of ASLT under variable sampling latencies, sensor measurement noise, and target trajectories. These draw attention to the computational trade-off and lead to the development of a novel solution achieving optimal localization performance with a significant energy reduction of over 50% [22].

4.2 Enhancing Connectivity

4.2.1 Positioning and Position-aided Communication in Distributed Access Architectures

Envisioned interactive applications foresee that physical and virtual worlds will get blended, as introduced in Chapter 2. This requires for wireless connectivity support providing "real-time" and "real-space" functionality; the differences should be unnoticeable to the human and the potential machines and applications. Distributed access architectures are key candidates to provide this. For example, RadioWeaves technology is being developed to that end, presenting a platform of interconnected communication and computation resources [9].

Position information in these networks can play a double role:

1. It is directly required to enable novel applications in professional and care environments, entertainment, including gaming, and personal and home spaces. Clear examples include tracking and location-based services, as well as robotic applications in industrial environments and extended reality (XR) applications. The analysis of diverse use cases indicates that an accuracy of 1–0.1 m is required for many applications [23].

2. Position information can support wireless communication and power transfer functionality. For the latter, location information enables beamforming solutions, which can greatly benefit the efficiency of the transfer [24]. Furthermore, in the pursuit of ultra-reliable communication, position information can be exploited to anticipate bad connections and support "break before make" decisions.

Different device classes [23] have been categorized for nodes that will need to be positioned in the same environment. The selection of position-related measurements will strongly depend on the capabilities of these devices. In particular, the positioning of low-power devices without any battery or with only limited energy storage relies on wireless charging, which poses stringent limits.

Distributed architectures hosting a very large number of antennas offer hyper-diversity that can be exploited well to extract both accurate and precise position information. Accuracy is most often related when specifying requirements and algorithmic progress are reported. However, precision, which is a measure for variations on the accuracy and also quantifies bad outliers, is also an essential performance measure, in particular when reliability is important. A distributed deployment is of particular interest in this regard, as it can support the "zero-outage."

The distributed architectures also enable location-aware proactive redundancy for guaranteed ultrafast exchange of critical data. Cell-free access is being deployed to simultaneously sustain links to multiple arrays [25]. It is expected that distributed compute-connectivity infrastructures will be an essential part of the anticipated heterogeneous 6G networks, e.g., in the context of Industry 4.0 and in smart home and care environments [23]. A sudden bad connection to one array will always be covered in advance by another array without a latency penalty. The locations of devices were tracked, and information was used to provide ultra-robustness on critical links. Learning of the environment will support an optimal allocation of array resources to prevent retransmissions and link outages.

Adequate positioning techniques that can offer both good accuracy and good precision in distributed architecture open an opportunity to address the requirements of many novel applications mentioned as drivers for 6G [23]. Techniques leveraging on distributed resources will extract information from both LoS links and multipath reflections. Therefore, it is essential to characterize and model the propagation environment in distributed architectures hosting a very large number of antennas. Novel experimental campaign-based channel modelling work is being conducted. Algorithmic development exploits the ultra-wide aperture and near-field properties (in the sense of being within the Rayleigh distance) of the signals. Furthermore, the hypothesis that location and environmental awareness allow for a large-scale predictability of the channel state information and thus an optimization of the efficient exploitation of the available hyper-diversity is confirmed in the first instance in the context of initial access [24].

4.2.2 Sub-6GHz, mmWave, and sub-THz RT Model and its Verification from Measurements and Applications Within Digital Twin of Factory for 6G

4.2.2.1 Construction of digital network twins in factory environments for access optimizations

6G envisions the integration of novel spectrum bands to support the development of ubiquitous smart wireless communications in industrial scenarios. Future industrial tasks and services rely on the simultaneous utilization of sub-6 GHz, mm-waves, sub-THz, and OWC [26]. The free blocks of spectrum available at THz and OWC enable the implementation of high-data-rate wireless links with enhanced capacity, latency, and with an unprecedented level of accuracy and resolution in sensing applications. Therefore, reliable channel models are of exceptional importance for the design, performance evaluation, standardization, and deployment of future 6G networks.

However, the development and parametrization of a single model covering such a wide spectrum of frequency bands and applications is challenging in multiple aspects. Localization and imaging applications require a precise correspondence between the geometrical properties of the propagated paths and the locations of users and scatterers. Moreover, testing heterogeneous localization methods, based on the combination of methods in different bands, requires models with consistency not only in the spatial domain, but also in the frequency domain: the scatterers must be in the same position for the different simulated bands [27, 28]. Therefore, models with deterministic components, such as RT, are more appropriate for these applications, since they allow simultaneous simulations at different frequencies with a precise geometrical representation of the environment in the propagation parameters of the paths. Nonetheless, the comparability of the results with reality depends on the complexity of the RT map/model. Thus, obtaining an accurate RT model with a high level of detail and a large number of objects is of crucial importance.

Therefore, a precise RT model was obtained from point cloud data from extensive 3D laser scans in an arbitrary industrial scenario, in which multi-band RF measurements were also conducted simultaneously for propagation characterization [29] and calibration and verification of the model at sub-6 GHz and mmWave [30]. In addition, simultaneous sub-6 GHz, mmWave, sub-THz, and OWC measurements will be conducted for verification purposes in the future. The ultimate goal of this model is to perform realistic analysis on different algorithms for sensing, localization, multi-band, and sensor-aided beamforming based on the different properties of this digital twin of the environment [31].

4.2.2.2 Channel modelling for heterogeneous networks and joint communications and sensing applications

From point cloud to RT simulations

The methodology to generate the RT model is summarized in Figure 4.19.

Two different scanners were used according to the size of the object/environment: the Leica BLK 360 for the macro scenario and the hand-held scanner Artec Leo for the machines and other details. The macro scenario was scanned in multiple positions, as shown in Figure 4.20, and the data were later combined using special tags that were placed in the environment.

The CAD model was obtained by reconstructing the surfaces out of the points. Once all the points from the different scans are merged, the surfaces are reconstructed with basic shapes out of polygons that can be interpreted by the RT tool. Figure 4.21 shows the level of detail on the CAD model. Finally, material-dependent electromagnetic properties are assigned to each surface to have an accurate calculation of the RF components at different frequencies with the RT tool.

Model validation with RF measurements

Simultaneously with the point-cloud scans, ultrawideband multiband RF measurements at sub-6 GHz, 30 GHz, and 60 GHz were conducted in the same area. These measurements have been used, as shown in Figure 4.22(a)–(c), to validate the RT model in the time-delay and angular domains. In the latter case, (c) shows the measured power at the different scanning angles (with 30° half power bandwidth (HPBW) horn antennas) at the TX side and the direction of the simulated paths,

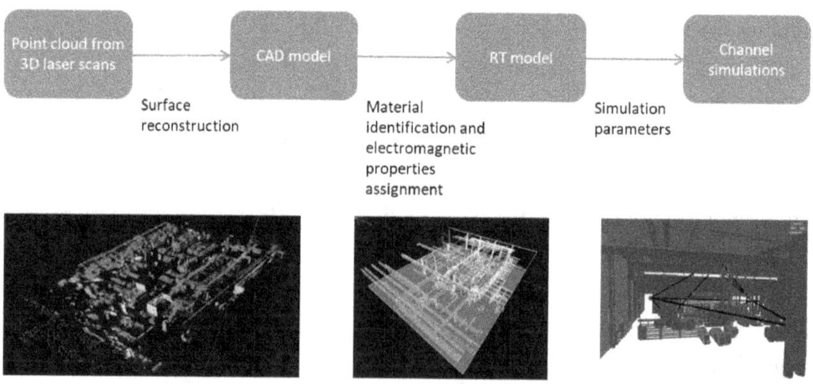

Figure 4.19. Processing methodology from point-cloud scans to RT models and simulations.

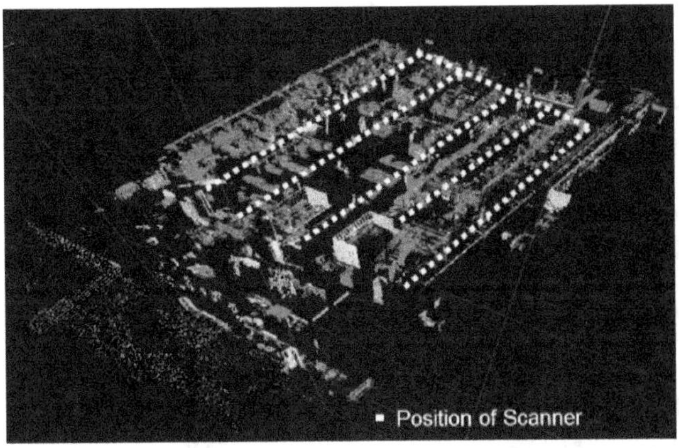

Figure 4.20. Point-cloud scanning positions.

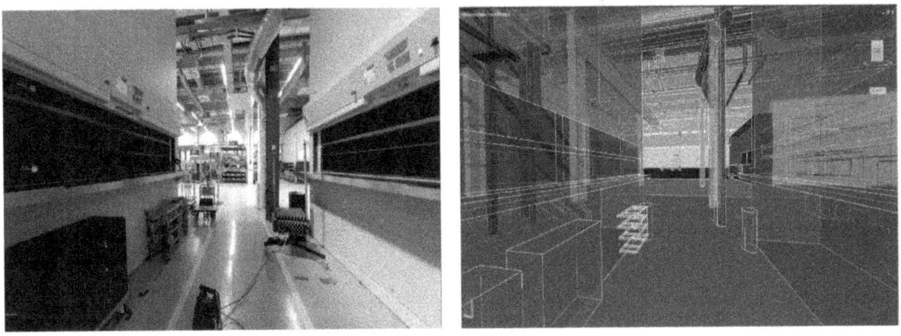

Figure 4.21. Picture of one of the corridors in the scenario and CAD model after surface reconstruction.

showing a high correlation between the density of the simulated path and the measured power.

After the RT model was validated, the simulations can be used to assist the interpretation of the measurement results by identifying scatterers and propagation mechanisms, as shown Figure 4.22(d).

A digital twin is a digital representation of a real object or process. In this case, the digital twin corresponds to the physical environment of a particular industrial scenario of interest. This model is used with RT to obtain different CIRs or RF parameters such as received power, delay spread, etc. that can be used in different tasks such as facilitating localization or the beam-steering process at high frequencies.

4.2.2.3 Applications of digital twins

However, one of the key requirements for digital twin components is their real-time representation or an acceptable update rate capability in order to tackle changes in high-dynamic environments in a factory.

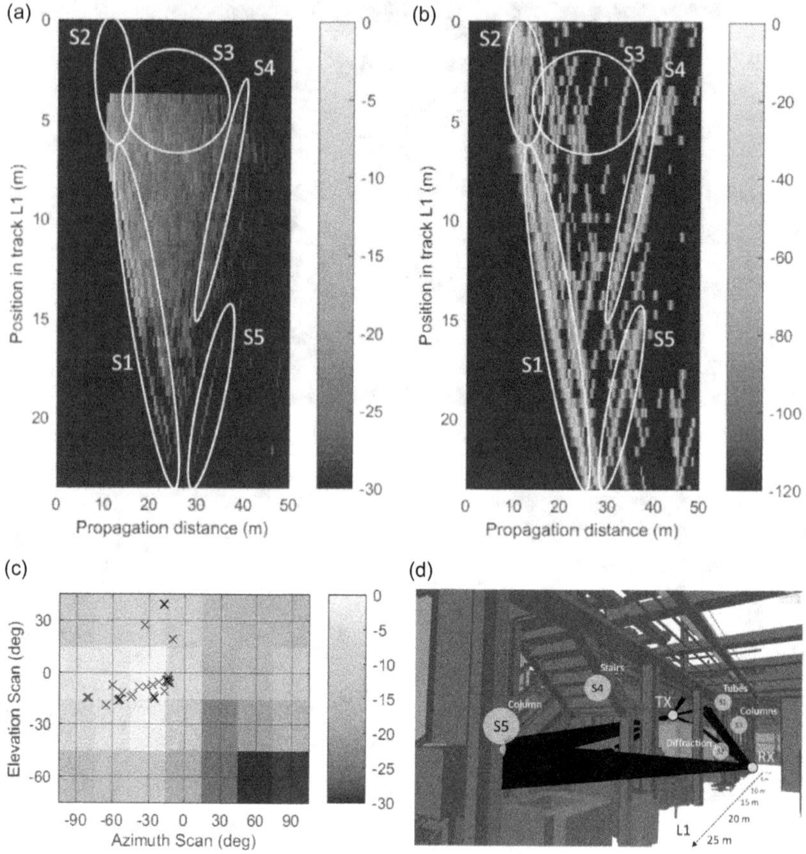

Figure 4.22. Power delay profiles of different RX positions over a line from (a) RF measurements at 6.75 GHz and (b) RT simulations. (c) Measured power azimuth/elevation profile and direction of the simulated paths in the RT model. (d) Identification of scatterers and propagation mechanisms.

An example of the latter case is that, given the location of the UE, RT can be used to determine the visibility condition or to estimate the pointing direction from the BS of a beam-former at mmWave or sub-THz, minimizing the training overhead.

4.2.3 Enhanced Connectivity with Channel Knowledge Map

Sensing to enhance communication services

A channel knowledge map (CKM) is a site-specific database containing transceiver locations and channel-related information that can be utilized to improve environmental awareness and improve CSI acquisitions. CKM is, therefore, critical to achieving high capacity, ultra-low latency, and ultra-massive connectivity in 6G networks. CKM can provide vital parameters of the wireless channels without requiring channel training. It is also possible to reduce training overhead in large MIMO systems by advancing localization and environmental awareness. The authors in [32] examined environment-aware beamforming for RIS-aided communication enabled by the CKM, which does not require online training. In the simulation, CKM in active/passive beamforming led to significant rate improvements over training-based beamforming, and it also proved to be robust against UE location errors.

Together with CSI, dynamic blockage information from sensing services can also be useful for enhancing the communication performance, especially at high frequencies where the signal could be affected by the poor propagation environment with high path loss. To realize context information-assisted communications, the BS could potentially predict the position of UE and blockers with advanced sensors and localization techniques, and this information could be useful for dynamic blockage avoidance. Moreover, channel variance issues could be mitigated by using the recently proposed concept of predictor antenna (PA) [33]. PA system refers to a setup with two groups of antennas deploying on the top of a vehicle, where the front antennas (called PAs) sense and report back the CSI to the BS. Then, the receive antenna(s) (RA(s)) following behind the PAs could use the CSI from PAs when they reached the same positions as the PAs. In this way, the quality of CSI is improved, leading to better system performance. In [34], focusing on highway scenario, the PA concept was incorporated into a large-scale cooperative PA (LSCPA) setup with cooperative communications among BSs and utilize the information provided by different vehicles to avoid not only temporal blockages but also the CSI outdating. Results summarized in [32] indicate that the E2E throughput for a given time slot of the network can be improved by 32% using context information in the form of coarse localization and trajectory information, whereas with highly accurate localization and trajectory information, reliable CSI can be obtained with the PA concept, resulting in a 335% gain.

4.3 Joint Communication and Sensing

4.3.1 Introduction

JCAS is are already one of the main differentiators of the 6G vision with respect to 5G communication systems. While sensing includes positioning, it will also encompass novel functionalities that are not present in 5G, which in turn may lead to new services. Supporting these services will have architectural implications. This section will discuss these novel types of sensing, the services they may enable, and the implications for the 6G architecture. In addition, recent developments towards the practical implementation of joint communication and sensing will be detailed.

4.3.2 Sensing as a Service

The term sensing is often reduced from its broader definition (detection of events, measuring changes in the environment of physical properties) to mean radar-like sensing. However, sensing can also cover channel estimation, radio frequency sensing, spectroscopy, weather monitoring, and any downstream processes that rely on sensing data.

4.3.2.1 Technical concept

Radar-type sensing

Radar-type sensing can be classified into three categories: monostatic sensing, bistatic sensing, and positioning as sensing, with illustrations shown in Figure 4.23.

Monostatic sensing is an extensively studied topic in JCAS, where the transmitter and the receiver are co-located (e.g., radar, as shown in Figure 4.23(a)) and where the transmitted signal is reflected by surrounding objects or targets and then processed at the receiver. If the transmitter and the receiver are located at different places, it is called bistatic sensing, as shown in Figure 4.23(b). In these two scenarios, the receiver can estimate the position of the incidence point that reflects the signal (e.g., buildings, cars, bicycles, and pedestrians) and map the surrounding environment (mainly passive objects). In contrast, the concept of positioning as sensing refers to estimating the position of a connected device directly (i.e., UE) using signals from multiple anchors (shown in Figure 4.23(c)), with examples such as the global positioning system (GPS), ultra-wideband (UWB) positioning, and 5G positioning. The mentioned scenarios can also be combined. For example, when a UE with an unknown position is involved in bistatic sensing (e.g., one BS and one UE instead of two BSs), this becomes a SLAM.

Sensing involves several processing steps, some of which can be optimized in a closed-loop tracking scenario to progressively improve the sensing performance over time: (i) signal design, (ii) signal detection and acquisition, (iii) channel parameter estimation, and (iv) position estimation. The task of signal design could be

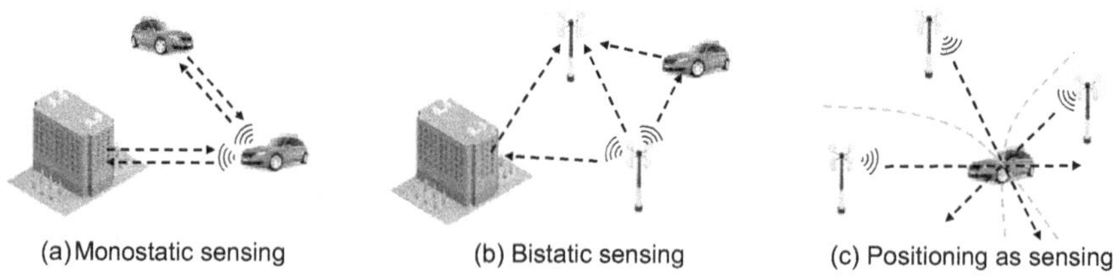

Figure 4.23. Illustration of three categories of sensing. (a) Sensing the environment with a vehicular radar. (b) Sensing the environment with two BSs. (c) Positioning of a vehicle with three BSs.

performed in time, frequency, and space to enhance sensing performance based on a priori knowledge of the environment obtained through position estimates in previous measurement epochs [35]. The designed signal is transmitted over the radio channel and acquired at the receiver, which then performs synchronization, phase noise tracking, and filtering. The sampled signal is applied to a parametric channel estimation routine, which returns estimates of ToA, AoD, AoA, and Doppler of the LoS path and possibly MPCs [36]. The channel parameters are provided for the localization, mapping, or tracking routine. This routine aims to solve the inverse problem of inferring the state of the UE (i.e., location and orientation) or the environment from the estimated channel parameters. In monostatic sensing, the transmitted data could be known at the receiver due to being co-located on the same hardware, which enables sensing using communication payload data. However, in bistatic sensing and positioning, pilot (reference) signals are needed, and there is a trade-off between communication and sensing resources.

In general, the sensing estimation accuracy depends on a variety of factors, which can be summarized as follows: (i) resolution, (ii) SNR, (iii) sensing scenario, and (iv) model mismatch. To estimate the parameters of a signal path during the channel estimation, the path should be resolvable (separable) in at least one domain among delay, AoD, AoA, or Doppler. However, even under sufficient resolution, the received signal may be weak and limit the performance, which can be determined by the transmit power, path loss, the reflection coefficient of the object, etc. In addition, the sensing scenario (deployment of transmitters and receivers, operating frequency, etc., which have implications on blocking, scattering, or molecular absorption characteristics) and the mobility of objects and users (the coherent processing duration of the waveforms will be limited) play a vital role. Finally, unmodelled effects due to radio hardware impairments or propagation effects will lead to reduced localization and sensing performance, which can have severe impacts in applications with stringent accuracy requirements.

Consequently, offline sensing system design (e.g., hardware selection and deployment), online optimization (e.g., signal design), and signal processing algorithms (taking into account the model mismatch or not) constitute essential ingredients to guarantee an acceptable level of performance for a sensing system.

Non-radar type sensing

Typical sensing methods for using communication infrastructure try to mirror a radar. In these methods, focus is on detecting the target and estimating its parameters such as velocity, range, etc. However, there are sensing use cases where the focus is not on range and Doppler but on other aspects, such as sensing weather, human activity, local environment, etc. Figure 4.24 illustrates the communication infrastructure for non-radar type of sensing.

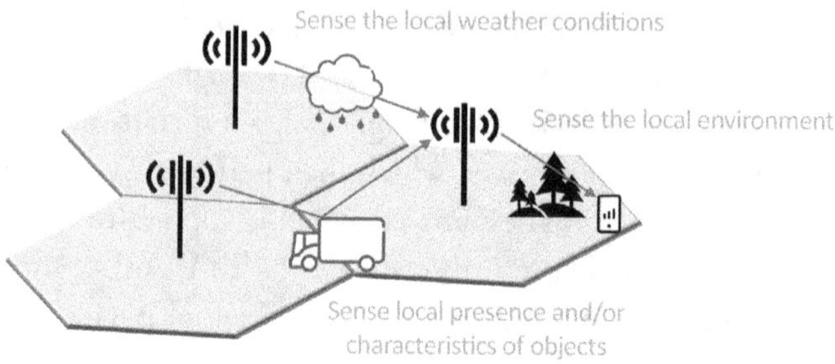

Figure 4.24. Radar and non-radar type of sensing using mobile networks.

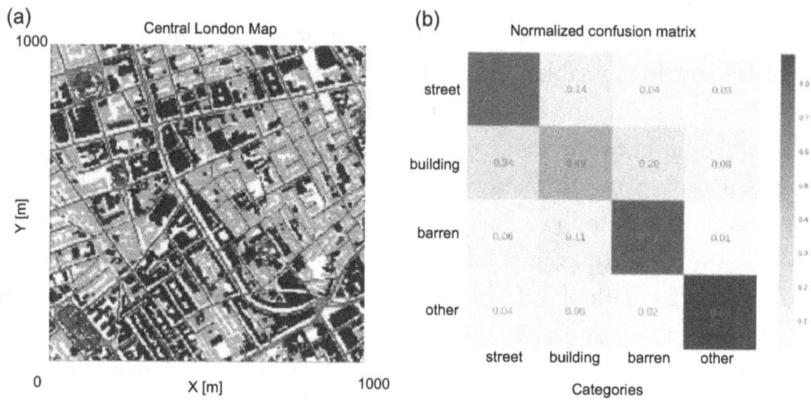

Figure 4.25. (a) Central London map with different landscapes shown in different colours. (b) Classification accuracy denoted via confusion matrix for landscape sensing algorithm discussed in [37].

4.3.2.2 Use of sensing information in services

Landscape sensing and contextual sensing (incl. weather monitoring)

Landscape sensing methods identify the macro-environment around the UE, such as forests, streets, buildings, water body, etc. It can aid in several scenarios in RAN automation such as tailoring the signal to the UE from the BS. The landscape sensing methods proposed in [37] detect the UE landscape using measurements from the central London metropolitan area (refer to Figure 4.25(a)). Figure 4.25(b) shows the classification accuracy achieved using the methods proposed in [37]; here, landscape categories such as those that denote "street," "building," "barren landscapes," and "other" categories can be applied.

Contextual sensing is another type of non-radar type of sensing that deals with identifying the target context and aiding in the higher-level tasks. For example, a fall detection in an assisted living scenario can trigger an emergency call to the hospital. In [32], the authors proposed active sensing where target-UE uses its communication capability to share its inertial sensor values with the edge-server in the

BS. Using these inertial sensor values, an AI agent on the edge server will identify the human activity.

Sensing to optimize factory environments

The digitalization within factories is progressing, but the integration of the digital and physical worlds still faces many challenges. Localization, sensing, and integrated communication will play an important role, and next-generation mobile communication systems can support the Industry 4.0 vision in a great manner. Sensing the environment and creating a digital twin of objects and layouts of the building help digitalize the factory environment. Localization of mobile units such as AGVs or assets (respectively products) within and outside of the factory allows for seamless tracking and analytics of business flows. Errors in production processes can be detected early, and inefficient flows can be optimized. The evaluation of the current location and communication requirements of mobile units is examined and optimized bidirectionally. Mobile units can take routes that may be a little longer but more efficient in regard to communication requirements (throughput, latency, etc.). Next-generation mobile networks with integrated sensing capabilities will be even more valuable if sensor fusion with existing localization and sensing technologies (e.g., camera, LIDAR, or UWB-based) is able by offering open interfaces.

4.3.2.3 Architectural implications of sensing as a service

Signal resource allocation

To implement sensing on the OFDM transceivers, the sensing resources need to be periodically allocated, as shown in Figure 4.26. Typical sensing (radar-type) use cases can be mapped to the requirements on the range resolution, maximum unambiguous range, velocity resolution, and unambiguous velocity. These requirements put constraints on the OFDM waveform, specifically on periodicity, symbol duration, and frame time, as illustrated in Figure 4.26. In [38], the sensing overhead for a traffic sensing use case was computed, considering a range resolution of 0.5 m, a maximum unambiguous range of 100 m, a velocity resolution of 0.5 m/s, and unambiguous velocity in the range of [−20 to 20] m/s, for realizing traffic sensing use case, and showing that this will result in an overhead of 2.7% for a mmWave deployment with numerology, centre frequency, subcarrier spacing, and bandwidth of 28 GHz, 120 kHz, and 300 MHz, respectively.

Network, spectrum, and hardware requirements

To sense target in mono-static way, the transceiver must simultaneously transmit and receive signals; this creates a full duplex requirement on the communication infrastructure. Such a full-duplex-capable transceiver requires a high level of self-interference cancellation, which is very challenging. This can be overcome by using

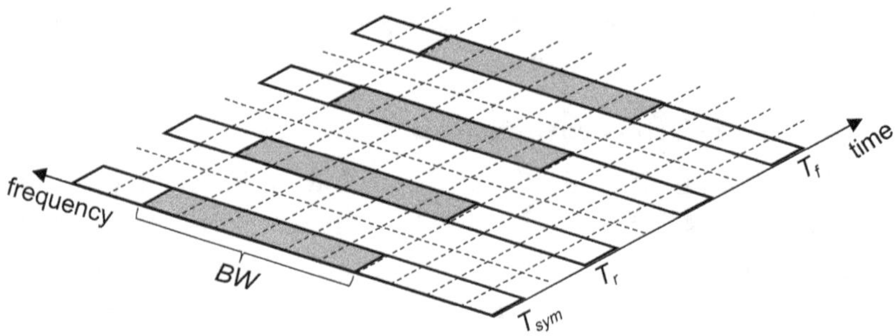

Figure 4.26. Sensing overhead in OFDM frame.

bi-static or multi-static sensing; however, these methods require a high level of synchronization between the co-operating BSs, which is very challenging to achieve.

Though large bandwidth and high frequency of operation are beneficial from a sensing perspective as they can enable very high range and angle resolutions, they come with a cost of very limited coverage and expensive hardware. In contrast, the lower-frequency spectrum provides much wider coverage. Depending on the accuracy and coverage requirements for the use case spectrum, these need to be carefully chosen.

Services in the 6G ecosystem

Sensing the environment will be a completely new feature, and the capabilities are gaining more relevance as localization becomes more accurate and reliable. Especially indoors, where no GPS is available, the next-generation mobile networks can add high impact on certain applications. But with these developments, the service offering from pure communication service is broadened, and the stakeholders and users' groups might grow. The complexity of offering multiple services will be challenging. The configuration and even the usage of these services should be flexible. In some scenarios, all capabilities will be needed at the same time, which might even require an integrated approach. The next-generation mobile networks should be as flexible as possible in each phase of the lifecycle (planning, analysis, design, development, testing, implementation, and maintenance). Especially in indoor environments, the requirements can change over time, and it might be necessary to include new hardware, even from third-party vendors to enhance the service quality. Also, accessing sensing data in a standardized manner at various processing stages (from raw to finalized contextual data) is beneficial to allow an ecosystem to flourish and enable, for example, a truly seamless sensor fusion between different sensing and localization technologies. Privacy might become more sensitive for two reasons. First, as localization information becomes more accurate, misusage can damage or harm people and institutions much more severely, and secondly, sensing people

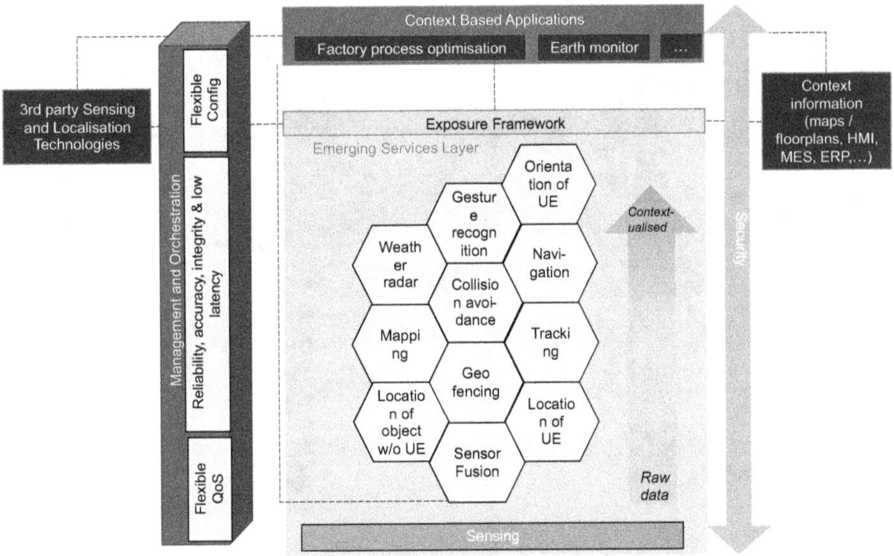

Figure 4.27. Location-based services ecosystem from sensing perspective.

(radar-like) and detecting the presence of humans pose a new category of privacy concerns, especially if, with pattern analysis, the identification of this detected person can be deduced. If critical use cases are built on top of sensing and localization data, the sensing data itself must be secured and verified by next-generation mobile network mechanisms.

New services may emerge based on sensing and localization capabilities, or existing services may integrate into the next generation of mobile network capabilities. A very straight-forward synergy between services is to enhance communication based on the current location. The farther away a service resides in Figure 4.27 from pure and raw sensing data, the more context information and classification of data are required by the service.

4.3.3 Joint Communication and Sensing in Practice

When developing new technologies, there is always a big step to go from a theoretical idea to showing that it works in practise. In Figure 4.28, the results from some initial JCAS tests are shown. In this example, a bi-static setup operating at 69 GHz is used. The signal transmitted is a 400 MHz-wide OFDM signal with 960 kHz subcarrier spacing. In the setup, there is a reflective wall and a person (wearing a tin foil hat to enhance reflections) that should be detected. Both the transmitter and receiver consist of a 1 × 16-element phase array, where the transmitter uses 50 different beams and the receiver uses 56 different beams, both in the range ±45°. In the bottom left plot, the RSS is plotted as a function of the beam indices of the

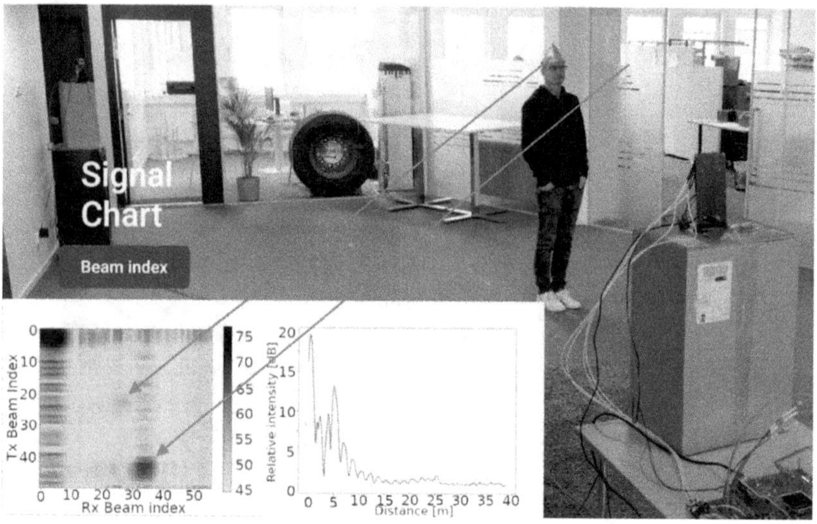

Figure 4.28. Initial demonstrations of JCAS using a bistatic setup at 69 GHz.

transmitter and receiver. At indices 0,0, the strongest response is found, and that corresponds to the LoS channel. In addition, there are two more blobs indicated by the red arrows, one originating from reflections at the wall and the other from the person in the picture. In the right plot, the distance to the targets is calculated based on the channel estimation. The distance indicated is relative to the LoS distance, meaning that the 0 m peak is due to the LoS channel, the peak at 5 m is caused by the wall, and the small peak at 2 m is caused by the person. For a more detailed description of the demonstration and more pictures, see [32].

4.4 Conclusions

The global impact of the existing global positioning system (GPS) has transformed the lives of many people with many new applications that use outdoor location information, which has impacted people's lives in many different ways, such as: (1) find your path, (2) avoiding traffic, (3) tracking your phone or child's phone, (4) finding the nearest place such as restaurant, (5) finding places such as schools & colleges, (6) track stolen phone, (7) preventing car theft, (8) tracking for law enforcement, (9) discover unknown places, and (10) find nearby places. It is expected that indoor and outdoor location systems with an accuracy of millimetres, which cannot be performed by GPS due to the building obscuring the direct line of site to GPS satellites, will similarly transform the lives of many people by inspiring many new indoor and outdoor applications, which will have a global impact on people's lives in many different new ways in many different verticals. Furthermore the introduction 6G JCAS capabilities will allow the spatial sensing of the environment since by measuring the delay of the return echo in the line-of-sight path between the transmitter and object, the distance to the object can be calculated and therefore its position and velocity. This is very useful information for not only appraising the LoS/NLoS radio transmission in which mMIMO dart-like beams propagate but also sensing object for collision avoidance, proximity sensing and location from landmark sensing.

References

[1] *OS1 Ultra-Wide View High-Resolution Imaging Lidar*, Ouster Specification OS1 rev. 7 v.3.0, 2023. Accessed: April 6, 2023. [Online]. Available: https://data.ouster.io/downloads/datasheets/datasheet-rev7-v3p0-os1.pdf.

[2] *OS2 Ultra-Wide View High-Resolution Imaging Lidar*, Ouster Specification OS2, rev 6, v2.4.x, 2022. Accessed: April 6, 2023. [Online]. Available: https://data.ouster.io/downloads/datasheets/datasheet-rev06-v2p4-os2.pdf.

[3] 6G BRAINS, "D2.1 Definition and Description of the 6G BRAINS Primary Use Cases and Derivation of User Requirements," July 2021. Accessed: April 6, 2023. [Online]. Available: https://ec.europa.eu/research/participants/documents/downloadPublic?documentIds=080166e5f5607765&appId=PPGMS.

[4] Hexa-X, "D3.1 Localisation and sensing use cases and gap analysis," Dec. 2021. Accessed: April 6, 2023. [Online]. Available: https://ec.europa.eu/research/participants/documents/downloadPublic?documentIds=080166e5e648d02a&appId=PPGMS.

[5] 6G BRAINS, "D3.1 3D Laser measurement of one factory at Bosch with 3D cloud scanner and 3D hand scanner," Sep. 2021. Accessed: April 6, 2023. [Online]. Available: https://ec.europa.eu/research/participants/documents/downloadPublic?documentIds=080166e5e2a9ad6f&appId=PPGMS.

[6] G. Eappen, J. Cosmas, T. Shankar, A. Rajesh, R. Nilavalan, and J. Thomas "Deep learning integrated reinforcement learning for adaptive beamforming in B5G networks," In *IET Communications*, Sept. 2022. Accessed: April 6, 2023. [Online]. Available: https://doi.org/10.1049/cmu2.12501.

[7] 6G BRAINS, "D6.1 Technical Specification of the 3D Location Architecture," Jan. 2022. Accessed: April 6, 2023. [Online]. Available: https://ec.europa.eu/research/participants/documents/downloadPublic?documentIds=080166e5f561136a&appId=PPGMS.

[8] S. S Das and R. Prasad, *Orthogonal Time Frequency Space Modulation OTFS a waveform for 6G*, New York, NY, USA, River Publishers, September 1, 2022,

eBook ISBN9781003339021, Accessed: April 6, 2023. [Online]. Available: https://doi.org/10.1201/9781003339021.

[9] L. Van der Perre, E. G. Larsson, F. Tufvesson, L. D. Strycker, E. Björnson and O. Edfors, "RadioWeaves for efficient connectivity: analysis and impact of constraints in actual deployments," In *2019 53rd Asilomar Conference on Signals, Systems, and Computers*, pp. 15–22, Pacific Grove, CA, USA, 2019, doi: 10.1109/IEEECONF44664.2019.9048825.

[10] T. Wilding, S. Grebien, E. Leitinger, U. Mühlmann and K. Witrisal, "Single-Anchor, Multipath-Assisted Indoor Positioning with Aliased Antenna Arrays (Invited Paper)," In *2018 52nd Asilomar Conference on Signals, Systems, and Computers*, pp. 525–531, Pacific Grove, CA, USA, 2018, doi: 10.1109/ACSSC.2018.8645163.

[11] T. L. Hansen, M. A. Badiu, B. H. Fleury and B. D. Rao, "A sparse Bayesian learning algorithm with dictionary parameter estimation," In *2014 IEEE 8th Sensor Array and Multichannel Signal Processing Workshop (SAM)*, pp. 385–388, A Coruna, Spain, 2014, doi: 10.1109/SAM.2014.6882422.

[12] 5G-CLARITY, "D2.3 Primary System Architecture Evaluation," July 2021. Accessed: April 6, 2023. [Online]. Available: https://ec.europa.eu/research/participants/documents/downloadPublic?documentIds=080166e5f30947a2&appId=PPGMS.

[13] A. S. Stordal, H. A. Karlsen, G. Naevdal, H. J. Skaug, and B. Valles, "Bridging the ensemble kalman filter and particle filters: the adaptive Gaussian mixture filter," In *Computational Geosciences*, pp. 293–305, 2011.

[14] D. Alspach and H. Sorenson, "Nonlinear Bayesian estimation using Gaussian sum approximations," In *IEEE transactions on automatic control*, vol. 17, pp. 439–448, 1972.

[15] F. Gustafsson, "Particle filter theory and practice with positioning applications," In *IEEE Aerospace and Electronic Systems Magazine*, vol. 25, pp. 53–82, 2010.

[16] M. Goodarzi, V. Sark, N. Maletic, J. Gutiérrez, G. Caire, and E. Grass, "DNN-assisted Particle-based Bayesian Joint Synchronization and Localization," In *IEEE Transactions on Communications*, vol. 70, no. 8, July 2022.

[17] R. O. Schmidt, "Multiple emitter location and signal parameter estimation," In *IEEE Trans. Antennas Propag.*, vol. AP-34, no. 3, pp. 276–280, Mar. 1986.

[18] 5G-CLARITY, "D4.2 Validation of 5G-CLARITY SDN/NFV Platform, Interface Design with 5G Service Platform, and Initial Evaluation of ML Algorithms," July 2021. Accessed: April 6, 2023. [Online]. Available: https://ec.europa.eu/research/participants/documents/downloadPublic?documentIds=080166e5f309449b&appId=PPGMS.

[19] J. Heaton, *Introduction to neural networks with Java 2nd edition.*, Heaton Research Inc. Oct. 1, 2008.

[20] S. Boyd and L. Vandenberghe, *Convex optimization*, Cambridge university press, 2004.

[21] B. Meunier, J. Cosmas, K. Ali, N. Jawad, G. Eappen, W. Li, and H. Zhang "Visible Light Positioning with Lens Compensation for Non-Lambertian Emission," In *IEEE Transactions on Broadcasting*, vol. 69, no. 1, pp. 289–302, March 2023.

[22] B. Meunier "Wireless Indoor Localisation within the 5G Internet of Radio Light," Ph. D. Thesis, Brunel University, London, March 2022.

[23] REINDEER, "D1.1 Use case-driven specifications and technical requirements and initial channel model," Sep. 2021. Accessed: April 6, 2023. [Online]. Available: https://ec.europa.eu/research/participants/documents/downloadPublic?documentIds=080166e5e2ac056c&appId=PPGMS.

[24] B. Deutschmann, T. Wilding, E.G. Larsson, and K. Witrisal, "Location-based Initial Access for Wireless Power Transfer with Physically Large Arrays," In *2022 IEEE International Conference on Communications Workshops (ICC Workshops)*, 16–20 May 2022.

[25] REINDEER, "D2.1 Initial assessment of architectures and hardware resources for a RadioWeaves infrastructure," Jan. 2022. Accessed: April 6, 2023. [Online]. Available: https://ec.europa.eu/research/participants/documents/downloadPublic?documentIds=080166e5e7bcfeeb&appId=PPGMS.

[26] European Commission, "Bring Reinforcement-learning Into Radio Light Network for Massive Connections (6G BRAINS)," CORDIS, 2023. Accessed: April 6, 2023. [Online]. Available: https://cordis.europa.eu/project/id/101017226.

[27] D. Dupleich, R. Müller, M. Landmann, E. Shinwasusin, K. Saito, J. Takada, J. Luo, R. Thomä, and G. del Galdo, "Multi-Band Propagation and Radio Channel Characterization in Street Canyon Scenarios for 5G and Beyond," In *IEEE Access*, vol. 7, pp. 160385–160396, 2019.

[28] D. Dupleich, R. Müller, M. Landmann, J. Luo, G. Del Galdo and R. S. Thomä, "Multi-band Characterization of Propagation in Industry Scenarios," In *14th European Conference on Antennas and Propagation (EuCAP)*, Copenhagen, Denmark, 2020.

[29] D. Dupleich, N. Han, A. Ebert, R. Müller, S. Ludwig, A. Artemenko, J. Eichinger, T. Geiss, G. Del Galdo, and R. Thomä, "From Sub-6 GHz to mm-Wave: Simultaneous Multi-band characterization of Propagation from Measurements in Industry Scenarios," In *16th European Conference on Antennas and Propagation (EuCAP)*, Madrid, Spain, 2022.

[30] H. Niu, D. Dupleich, Y. Völker-Schöneberg, A. Ebert, R. Müller, J. Eichinger, A. Artemenko, G. Del Galdo, and R. Thomä, "From 3D Point Cloud Data to Ray-tracing Multi-band Simulations in Industrial Scenario," In *2022 IEEE 95th Vehicular Technology Conference: (VTC2022-Spring)*, Helsinki, 19–22 June 2022.

[31] 6G BRAINS, "D6.2 3D Location Simulation Models and Lab Prototypes," Jan. 2023.

[32] Hexa-X, "D3.2 Initial models and measurements for localisation and sensing," Oct. 2022.

[33] H. Guo, B. Makki, D. T. Phan-Huy, E. Dahlman, M.S. Alouini, and T. Svensson, "Predictor antenna: A technique to boost the performance of moving relays," In *IEEE Communications Magazine*, vol. 59, no. 7, pp. 80–86, 2021.

[34] H. Guo, B. Makki, M. Alouini, and T. Svensson, "High-rate uninterrupted internet-of-vehicle communications in highways: Dynamic blockage avoidance and CSIT acquisition," In *IEEE Communications Magazine*, vol. 60, no. 7, pp. 44–50, July 2022.

[35] A. Kakkavas, H. Wymeersch, G. Seco-Granados, M. H. Castañeda García, R. A. Stirling-Gallacher, and J. A. Nossek. "Power Allocation and Parameter Estimation for Multipath-based 5G Positioning," In *IEEE Transactions on Wireless Communications*, vol. 20, no. 11, pp. 7302–7316, Nov. 2021.

[36] H. Chen, H. Sarieddeen, T. Ballal, H. Wymeersch, M. S. Alouini, and T. Y. Al-Naffouri, "A Tutorial on Terahertz-Band Localization for 6G Communication Systems," In *IEEE Communications Surveys & Tutorials*, vol. 24, no. 3, pp. 1780–1815, May. 2022.

[37] V. Yajnanarayana, D. Huang, D. Shrestha, Y. Geng, A. Behravan, and E. Dahlman, "AI Based Landscape Sensing Using Radio Signals," In *2021 IEEE 32nd Annual International Symposium on Personal, Indoor and Mobile Radio Communications (PIMRC)*, pp. 1–5, 2021, doi: 10.1109/PIMRC50174.2021.9569617.

[38] A. Behravan, R. Baldemair, S. Parkvall, E. Dahlman, V. Yajnanarayana, H. Björkegren, and D. Shresta, "Introducing sensing into future wireless communication systems," In *2022 2nd IEEE International Symposium on Joint Communications & Sensing (JC&S)*, pp. 1–5, 2022, doi: 10.1109/JCS54387.2022.9743513.

DOI: 10.1561/9781638282396.ch5

Chapter 5

Towards Natively Intelligent Networks

By Marco Gramaglia, Xi Li, Ginés García-Aviles, et al.[1]

Over the past few years, network automation has become an increasingly important topic in both research and industry. With the growing complexity of modern networks and the need for efficient network management, artificial intelligence-based algorithms have gained attention as a promising solution. Standardization efforts have also been underway to support analytics derived from network data. As we look towards the future of mobile networks, it is expected that the integration of network intelligence (NI) will be a crucial component of the next-generation 6th Generation (6G) mobile network. While the foundational pillars of 6G have been defined by recent research and standardization works (e.g., 3GPP Release 18), it will be necessary for the upcoming waves of mobile network architecture to natively integrate NI solutions, making lifecycle management of NI an integral part of network services. This will require a novel architectural framework that supports the integration of NI into the network services, as well as the development of novel algorithms that can leverage NI to optimize network performance and enable new services and applications. In this context, this chapter proposes a comprehensive framework for the integration of NI in the 6G mobile network, which includes the key components of NI functions, management and orchestration (MANO), and infrastructure. We also introduce novel algorithms that can leverage NI to optimize

1. The full list of chapter authors is provided in the Contributing Authors section of the book.

network performance and enable new services and applications. Overall, the goal is to provide a roadmap for the integration of NI into the 6G mobile network, with the aim of creating a more efficient, intelligent, and flexible network architecture that can meet the demands of the next generation of mobile services.

This chapter is structured as follows: the overall set of enablers needed for the transition towards intelligent networks is discussed in Section 5.1. Then, Sections 5.2 and 5.3 discuss the overall network intelligent architectural framework that has to deal with the specificities of the application of NI to future 6G mobile network. In Section 5.4, we then draw some guidelines for the design of such network intelligence function (NIF) that empower the vision presented in this chapter. Finally, from Section 5.5 to Section 5.12, we discuss different NI solutions that will address from different perspectives the challenges posed by the management and operation of a 6G network to eventually maximize the impact envisioned in Chapter 1.

5.1 Enablers for an Intelligent Network

Top-down network architecture design process typically starts by identifying a targeted service portfolio, given specific environmental conditions and business goals. These characteristics are captured in a set of representative use cases that are then used to derive the needed network functionality. This approach is, however, limited in terms of how well expected advancements in the application technology (e.g., computing, sensing, and actuation) and how end-user needs can be envisioned. Ideally, a network architecture would be adaptable beyond its original design goals to meet the unforeseeable changes in the service portfolio and related business goals. The network should hence allow its automatic reconfiguration with minimal manual operations in response to new and changing demands. Self-adaption of the network functionality is addressed by NIs that embed and extend artificial intelligence/machine learning (AI/ML) functionality beyond orchestration and network management into network functions and services. Intelligent Networks functions reside across the common service platform, the network functions (NFs) domain, and the management, see Figure 5.1.

It consists of AI/ML-enabled closed-loop control of NFs with the necessary supporting enabler frameworks to automate and modify network operations based on given policies and goals. The first prerequisite for NIs is a common end-to-end (E2E) analytics framework that collects and provides accurate and timely data for inference and training the AI/ML functionality. The E2E analytics framework is

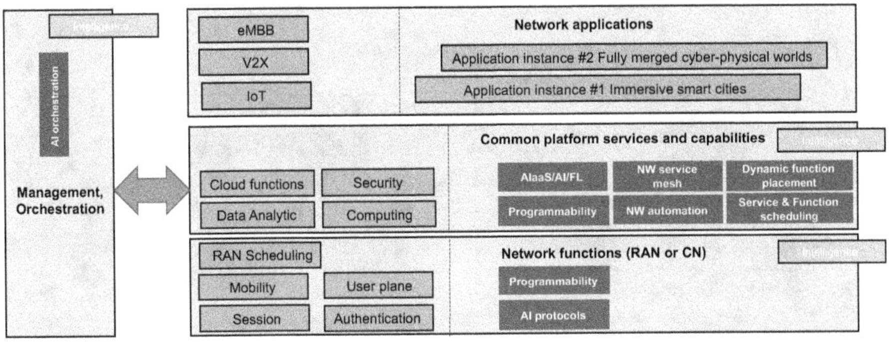

Figure 5.1. 6G enablers for the NIs in dark blue boxes in the context of the different 6G architecture domains.

responsible for exchanging knowledge across various network layers and domains, enabling AI agents and ML models training and operation. The AI-model's training depends on the concrete location where the model will be applied and the type of data that is accessible to it in a given configuration. As 6G architectures may span across a cloud continuum from the central cloud to the distributed edge and user equipment (UE), two deployment solutions are identified: first, a fully distributed edge-based solution, where all AI functions are instantiated at the edge nodes as cloud native applications; and second, a hybrid solution, where computationally expensive AI functions are trained and possibly executed in the central cloud using wider and richer data sets than would be available in a given edge cloud. As 6G UEs are assumed to consume intelligent services from the network, they can also train their own AI models collaboratively in a privacy-preserving way according to the federated learning (FL) paradigm [41]. To facilitate this, it is required to provide specific means to discover and join learning federations of UEs, allowing them to train their own AI-models collaboratively while maintaining privacy. Trained AI/ML models are stored and maintained in an AI repository, from which specific ones are installed into resource-controlling AI agents. AI agents are distributed across the network to control and adjust the resources of the network, which they are responsible for (see Figure 5.2).

Specifically, the first step is to identify the set of AI functions that are required to support this closed-loop network automation approach, possibly allowing hierarchies of closed loops, and considering their virtualization, packaging, and orchestration as in-network cloud-native functions that can be controlled and supervised by the MANO layers. To effectively address this approach, four AI functions have been identified as fundamental: the AI model repository function, the AI training function, the AI monitoring function, and the AI agent. Table 5.1 provides a short description for each of them.

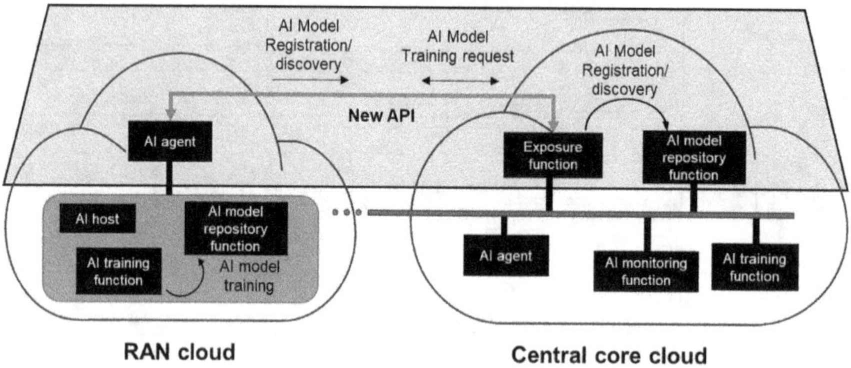

Figure 5.2. Architecture enablers for supporting AIaaS.

Table 5.1. AI functions identified as enablers for supporting AIaaS.

AI Function	Description
AI model repository function	It is the function that provides a catalogue of the available AI/ML trained models (including their metadata for capability specification), which are either already deployed or ready to be deployed within new or existing instances of AI agents. Multiple versions of the same model can be stored in the AI model repository function, e.g., related to subsequent training sessions over different datasets.
AI training function	It is the function that performs the training of AI/ML algorithms (including any required data pre-processing) and produces executable models that can be integrated in the AI agents. The AI training function is triggered either autonomously by the AI monitoring function when a performance degradation is detected, or by the management and orchestration layer whenever a new model has to be generated.
AI monitoring function	It is the function that takes care to evaluate the performance of the AI/ML models and consequently provide the trigger for training and re-training operations in the AI training function. This translates into evaluating the runtime accuracy of deployed models, as well as their performance in terms of action impact. This includes methods for identifying any potential conflict (direct or indirect) in the model inferences.
AI agent	It is the function that uses the trained AI/ML models (one or more) to perform the inference process (including any required data pre-processing functionality). According to the specific model(s) it executes, the AI agent requires specific data to be ingested. The outputs of the AI agents are then used to drive the actions and the behaviour of other functions, including 5th Generation (5G) NFs and application functions (AFs), as well as management and orchestration functions.

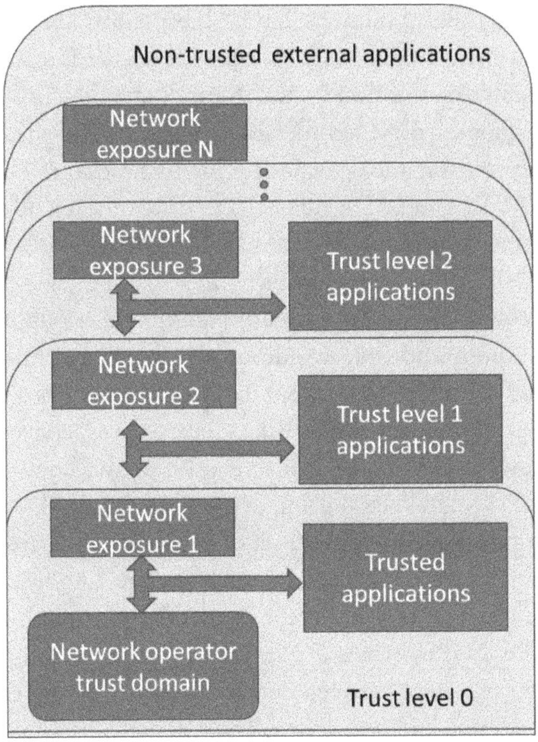

Figure 5.3. Network exposures for each trust level.

Collectively, the whole set of AI functions described in Table 5.1 form the AI as a Service (AIaaS) framework whose main purpose is to offer services to be consumed by any trusted NF, or even third-party applications that are external to the network through network exposure APIs (Figure 5.3). The consumer of the AIaaS can submit requests for inferencing decisions to AI agents for a particular action. The elements of the proposed AIaaS framework must comply with regulatory aspects of data governance that are accomplished by management of multiple trust levels between network domains to ensure data privacy requirements within each security domain. The proposed framework may be able to reduce the AI overhead, since operation of a given AI agent in one domain can be extended to different privacy domains, provided that the privacy requirements are addressed within each domain.

In addition, the underlying infrastructure layer needs to be adaptable to varying NF workloads, new functionality, and dynamic placement of NFs across the multi-cloud continuum. Dynamic instantiation of the AI agents and programmable NFs leverages AI-driven orchestration with the capability to select different processing points for network functions across the multi-cloud continuum. This is done by dynamic function placement (DFP), which introduces a two-level hierarchical orchestration solution where domain internal dynamicity is not fully exposed externally but still network functions can be executed in a multi-domain, multi-cloud environment on-demand basis. A top-level orchestrator interacts with the domain-level orchestrator and decides which candidate domains for NFs are to be created or moved. Final deployment details and selection of the domain's internal processing point are left for the domain-specific orchestrator. The newly instantiated or reconfigured NFs and AI agents need to discover each other, and they need to communicate efficiently via a network service mesh (NSM).

In order to satisfy communication efficiency requirements, NSM must support the most common communication patterns like publish-subscribe and request-response communication. Additionally, both synchronous and asynchronous communication must be supported. NSM provides a registry (or similar) for (de-)registering services that are then exposed via service discovery, which is used to find out what services are available in NSM and how.

Networking policies also play an important role in NSM, and they must be enforced in a distributed manner, i.e., a set of networking policies distributed over the related domains is dealt with in the same manner. Additionally, OAM and AI-related knowledge may have different connectivity requirements between the related domains and their internals to harmonize management operations.

NSM-managed tunnelling between domains needs to support different types of communication patterns, and typically this is indirect communication via a communication intermediary. So tunnelled data represents *NF:intermediary* communication and could be unidirectional or bidirectional depending on how routing is arranged in the domains.

One important thing to consider is the support for traffic prioritization inside the tunnel, because it might influence how inter-domain tunnelling should be configured. For instance, one straightforward option would be to avoid mixing different quality of service (QoS) classes in the same tunnel and instead have QoS class-specific tunnels. Maintaining multiple tunnels per domain boundary should not restrict scaling too much and technically be feasible, i.e., not so much state information should be maintained per tunnel endpoint.

5.1.1 A Two-level DFP in Multi-domain Landscape

In this section, further details are provided on the two-level design for DFP, which is shortly introduced in Section 5.1. Additionally, the impacts of adding multi-domain support to orchestration and service discovery, which are closely related to DFP, are also discussed.

The proposed design has the decision-making on two levels:

- Top-level decision-making, during which the best candidate domains are searched for.

- Domain-specific decision-making, where the network function deployment will be finalized according to the domain internal resource conditions and preferences.

The decision-making structure that follows is closely related to the logic of two-level hierarchical orchestration, in which multi-domain scope is handled first, followed by domain-specific scope. The same multi-level support shall be considered in other supporting functionalities, such as monitoring and management (e.g., NF scaling), which are part of DFP. Monitoring is naturally done within a domain, and then the information is exposed externally. One critical requirement of the design is the ability to expose information from domain(s). The top-level part is dependent on the information that is exposed by individual domains. It is crucial that the validity of the information is adequately scalable. If the exposed information is too dynamic ("short-lived"), a distributed system may end up wasting time and resources trying to keep dynamic information synchronized. Aside from the information's validity and lifetime, the amount of exposed information can also be a limiting factor. Therefore, domains may need to pre-process the information and expose it as information aggregates. However, in some cases, raw information might also need to be exposed.

NF scaling is a common operation for legacy orchestration, and it is used as an example for defining the impacts of the two-level design for decision-making and operation execution. This is the reason why multi-domain and intra-domain scaling should be considered separately:

- Multi-domain: scaling over multiple domains requires technical key performance indicators (KPIs) (e.g., utilization metrics representing the current resource conditions and load estimations for the near future from each domain) to make service offering-related decisions (e.g., requesting increased service capacity or relocating service offerings) from one domain to another.
- Intra-domain: traditional NF scaling up/down or in/out based on the local needs and/or external triggers such as requests from the top-level entity.

The duality of the decision-making closely corresponds to the timing aspects of the related control loops. For instance, any control loop running on top of multiple domains (top level) presumably has longer execution delays than those running inside a domain, which is a crucial piece of information since the potential execution delay of the control loop(s) determines how fast such a system can react to changes. This reaction sensitivity also relates to the dynamic environment in which the control loop is operating, and the information based on which the control loop is executed is part of this environment.

Similarly, to the exposed domain information, it is important that the decision-making information is not too dynamic or short-lived. This implies that longer control loops, such as those used in multi-domain scope, are better suited for wide-scope predictive decision-making. In addition, faster control loops, which are used inside domains or even inside logical or physical nodes, are better for reacting to rapid changes under dynamic conditions. In this design, the top-level domain does predictive operations with a longer lifetime, and each underlying domain operates on that basis. Therefore, from a decision-making perspective, domains are able to react faster to local changes and make short-term decisions and operations until longer-term decisions are propagated from the top level. It is important to have some form of feedback channel between the levels. For instance, domains can communicate their internal actions and decisions back to the top level, where the long-term solution(s) can be adjusted accordingly, i.e., predictions have to be updated according to what has happened at the lower (domain) level. Additionally, domains can have a certain degree of autonomy by having instructions and thresholds on when to report back to the upper (top level) level. It is essential that the used service discovery solution is efficient enough and scales well. To ensure proper scaling, service discovery can also implement a two-step design: (i) resolve the destination domain, and (ii) resolve the communication endpoint (NF) inside the domain. In multi-domain environment and especially when DFP is effectively in use, it is important that the establishment of inter-NF communication session could be optimized. According to this optimization, service discovery can be a domain internal thing without requiring longer loops between multiple domains, including some centralized functionalities.

For multi-domain environments, the usage of communication intermediaries (e.g., service communication proxy (SCP) in 5G environment) can help us in making scalable inter-NF communication by using indirect communication, where an intermediary hides a set of NFs and that for such NFs the service discovery results in the communication address of the corresponding intermediary. This ensures the scaling properties of service discovery and makes the data in (top level) service registry static since all the dynamics of NF instances are masked by communication intermediaries. Typical legacy schema for masking NFs has been grouping them based on NF type behind an intermediary and at the same time enable load balancing opportunity per NF type. However, in a more heterogeneous multi-domain environment where NF instances can be found in the core, (far) edge, or even behind the radio interface, the question of whether this is the only option arises, implying that new schemes based on which NFs can be grouped together behind intermediaries will likely emerge. It is well known that direct inter-NF communication, especially between domains, potentially limits how effectively load balancing can be done for NFs.

Context-related communication (e.g., context transfer or context sharing) could follow the normal inter-NF communication patterns, e.g., indirect communication via intermediary. This may be the only available option for inter-domain cases, but for the cases where the target NF instance is in the local domain, direct *NF:NF* communication might be preferred. In this way, context-related communication does not need to share communication medium with inter-NF communication and could receive higher priority, respectively, because NF context(s) must be in place before any inter-NF communication would be meaningful. There are no technical requirements to support traffic prioritization based on its type, but using separate communication mediums could further help in isolation and ensure separate scaling as well.

Regarding shared data between old and new NF instances, it could be categorized as (i) generic and (ii) instance-specific. The former is functional or practical as such between instances, but the latter has to be updated by a new instance, which receives the data provided by a context transfer functionality.

Shared data might have different real-time (RT) requirements based on how it is distributed and kept synchronized. For near-real-time (NRT) data, any changes must be distributed immediately to ensure that data updates are propagated fast enough. Respectively, for non-NRT (NNRT) data, there is no such urgency to make distributed data changes immediately; thus, the system can collect multiple updates and then distribute them as an aggregation. There are strict timing constraints in the interaction between the DFP and the service registry that determine whether the results of the DFP operations that impact service availability and resourcing are updated in time. It should be noted that updates with indirect communication via an intermediary, which is not directly visible in the service registry, can potentially be faster than the updates in the domain-wide service registry for direct communication updates. Naturally, this requires that DFPs be interfaced for distributed operations. DFP needs to support coordinated operations across domain borders to support such advanced use cases. Last, two topics are introduced for a new type of NF instance-related operation in DFP that derives especially from multi-domain support. Namely, they are NF relocation and NF offloading. While these two operations and the already-mentioned NF scaling are quite self-explanatory, it is often hard to differentiate them from each other rationally. Their existence can be justified in a multi-domain environment for this purpose, and they are defined as follows.

Scaling is called a "legacy" operation for NFs within this context. It is primarily driven by NF consumer demand and refers to scaling existing NF producer(s) capacity (up, down, in, or out) to ensure a specific NF's "optimal" service level. KPIs for "optimal" can be obvious in many ways. It should be noted that multi-domain implies new requirements for NF scaling as well. For instance, how to support distributed multi-domain transactions for NF scaling by coupling together domains' atomic transactions is a valid research question.

"NF Relocation" is a permanent operation that is used to change the physical location of NF instance(s) in the network topology. As soon as relocation and possible context transfers are done, the original instance is deleted. For example, a management entity might ask for an optimization to request NF relocation. The need for relocation can also be derived from the 6G service model, where it has been envisioned that end-user-specific (far) edge services that may have strict latency requirements are required to follow the end-user in the network topology.

NF offloading is the distribution of existing NF workloads to new network topology locations with newly created NF instance(s). One example of an offloading decision could be sudden changes in capacity offerings, i.e., load balancing. Offloading differs from relocation in a way it does not involve the deletion of the original NF instance(s). In other words, offloading can be considered a temporary expansion of NF capacity for offloaded workloads, which can be withdrawn once the offloaded tasks are finished. One typical scenario is NF offloading from the core domain to the (far) edge domain to meet the latency requirements.

5.1.2 AI Workload Placement Perspective

This section presents an intelligent network enabler for effectively managing and orchestrating computing resources. Computation tasks may vary, e.g., in relation to the source of the data to be processed. One critical computation tasks' category is related to AI and ML workloads. Specifically, (i) balancing the computation and communication load, (ii) moving computation, and (iii) efficient placement of computation further challenges are introduced. Physical nodes (e.g., user equipment, edge/cloud servers) that undertake the execution of AI workloads may encounter trust level problems, traffic load-related issues, or challenges/requirements related to their energy consumption.

The availability of the data, along with potential restrictions on its sharing among the various network segments/nodes due to privacy reasons, is one of the crucial factors that must be taken into account. In order to provide maximum privacy, but also network transmission-related communication overhead and, as a result scalability, the placement of AI workloads must consider the vicinity of the data source, e.g., in terms of network hops/available bandwidth for transferring the data, as well as the data volume and frequency based on which this must be transmitted. In addition to the aforementioned factors, one should also examine the special features/characteristics of AI workloads, namely different input data models, behaviour models, and knowledge sharing needs among nodes of the network, as well as considerations related to the trust level of the nodes that undertake computation tasks.

A management methodology focusing on AI workloads for Beyond 5G (B5G)/6G architectures could target a three-fold strategy, i.e., to minimize the energy consumption of the overall network towards sustainability, minimize the processing and transmission delay in relation to the data volumes that are to be transmitted towards the processing entities, and maximize the overall trust level of the system by prioritizing nodes/AI agents with high trustworthiness indexes [1].

5.2 NI Native Architecture Empowered by AI

In this section, an architectural model that stems from and integrates with current standards (e.g., O-RAN, 3GPP, and ETSI) and realizes the vision of native support for E2E NI coordination is described. The use of NI or AI to represent AI functions applied to network problems is alternated in this chapter. To achieve the specified result, the proposed architectural framework builds upon the initial guidelines provided by [2]. This previous study covered a subset of the current requirements from a higher-level perspective and focused on the relationship among NI degrees of freedom, input-output relationships, and RT constraints subject to infrastructure observability and controllability. In the following, the proposed architectural framework is discussed in detail, as depicted in Figure 5.4.

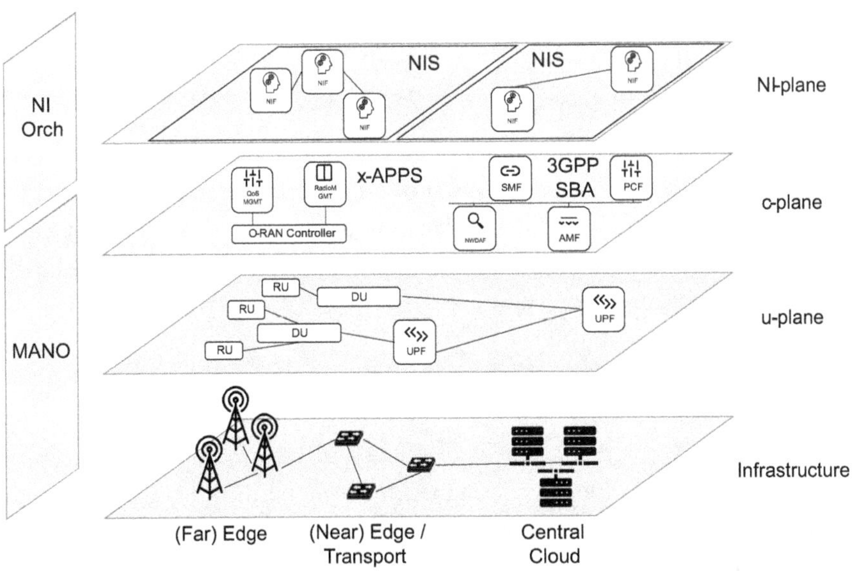

Figure 5.4. The framework for the NI integration in the overall architecture.

The network's softwarization and data-driven trends have led us to decompose the structure into four complementary layers. Along with the infrastructure layer, the control plane, and the user plane, which are the three essential building blocks of a software network like the 5G network, an additional layer is envisioned: the network intelligence plane (NIP). This NIP is introduced owing to the current trends in the industry that integrate the functions related to NI [3] in the network.

The MANO of this compound network is performed by two modules: the MANO, as traditionally done in 5G networks, which handles the typical lifecycle management of the network and network slices, and a new sibling element, the network intelligence orchestrator (NIO), which takes care of all the operations related to the management of the intelligence of the network. These operations include:

- The *selection* of the NIFs that come together to build a network intelligence service (NIS) to pursue one of the KPIs envisioned by 6G networks, such as energy efficiency.
- The *monitoring* of such functions, including the monitoring of their KPIs (e.g., their accuracy) and of the specific actions that may be taken to optimize them (e.g., meta-parameter change, re-training, or model changes).
- The specific *training* procedures in case of learning models.
- The *interaction* with the MANO to handle service and resource orchestration.

For MANO, all the definitions and functional components from ETSI can be reused, omitting these to avoid clutter. Instead, in the next sections, the internals of the novel NI orchestration, specifying interfaces and procedures, are described.

5.2.1 Detailed Architecture

In this section, an overall representation of architectural models is provided as well as the list of relevant definitions from standards. The newly proposed components (e.g., NIF, NIS, and NIO) and the explanation of how they fit such existing architectural models (e.g., NIF mapping to xApp/rApp in O-RAN and NWDAF in 3GPP) are presented in the next.

5.2.2 Taxonomy

As discussed in [3], the management of NI in a softwarized network can be performed in a similar manner as the management of network services is designed for 5G networks. This allows to reuse well-known concepts, adapting them to the context of NI. The high-level interactions are depicted in Figure 5.5, highlighting how the interactions take place.

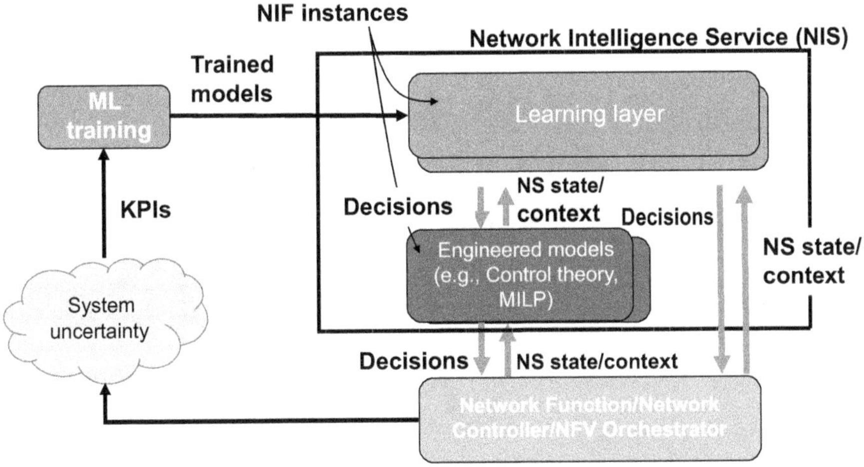

Figure 5.5. The taxonomy of the NI.

Thus, analogously to the information model specified for network management by, for instance, 3GPP, the concepts of *NI Service* (cf. Network Service, leveraging a slice such as enhanced mobile broadband (eMBB) or ultra-reliable low latency communications (URLLC)) and *NI Function* (like any function specified by, e.g., 3GPP or O-RAN) are defined as follows:

NIF: Functional block within a network (slice) that implements a decision-making functionality to be deployed in a controller, NFV orchestrator, or network function and has well-defined interfaces and behaviour.

NIS: Composition of NIFs that has a specific target, usually related to a specific set of targeted KPIs. Table 5.2 shows examples of NISs derived from NI functionalities developed in [3].

There is a one-to-many relationship between NIS and NIFs, as the former could be provided by one or more instances of the latter. Consequently, network operators or service providers can, for example, request specific sustainability and reliability services targeting one or more KPIs. The NI orchestration will take care of providing such a service by composing specific instances of NIFs.

NIFs themselves could be of different kinds. They could be learning models based on, e.g., deep neural networks (DNNs) or engineered models, or they could be built upon specific optimization algorithms such as those based on control theory or mixed-integer linear programming (MILP). These NIFs have two main interactions with the underlying layers (c-plane, u-plane, or infrastructure, proxied by a NFV orchestrator): NIFs inject decisions and receive information about the network slice state and the context of such a state. A NIS is hence a coordinated

Table 5.2. Examples of NI services derived by NI functionalities [3].

NI Service	KPIs	NIFs
Reliable virtualized RAN	Reliability	• Reliable distributed unit (DU) for RAN virtualization • Orchestration of radio and computing resources in vRANs
Sustainable network operation	Virtual network function (VNF) energy savings Compute resource savings OPEX savings	• Cloud acceleration for virtualized RAN • Compute aware scheduling analytics • AI-enhanced edge orchestration • Data-driven resource orchestration • Multi-timescale network slice reservation
Network capacity management	Wireless capacity increase	• Reconfigurable intelligent surfaces control
Edge orchestration	OPEX savings	• Network service auto-scaling • Capacity forecasting

effort of one or more NIFs that could be arranged hierarchically. For example, a NIS could be composed of a learning-type NIF sending decisions to an engineered model NIF, which in turn acts on the underlying infrastructure.

In order to manage the interaction between different NIFs, a further level of detail is needed, which decomposes each NIF into atomic elements that perform a specific operation. That is, besides the specific requirements associated with the algorithms, a mechanism to create a common framework to map the most common features of NI algorithms is needed, and subsequently integrate them into the overall architecture, and design the necessary interfaces that algorithms use to interact with their environment.

For the purpose of creating a common framework and integrating NI algorithms into the overall architecture, a methodology that is already utilized by the MAPE-K (*Monitor-Analyse-Plan-Execute* over a shared Knowledge) feedback loop is proposed. The MAPE-K feedback loop is considered one of the most influential reference control models for autonomous and self-adaptive systems [4]. This nomenclature adopted to label NI requirements allows classifying the algorithms that run at NI instances in a unified manner, based on how they interact with the other elements of the network.

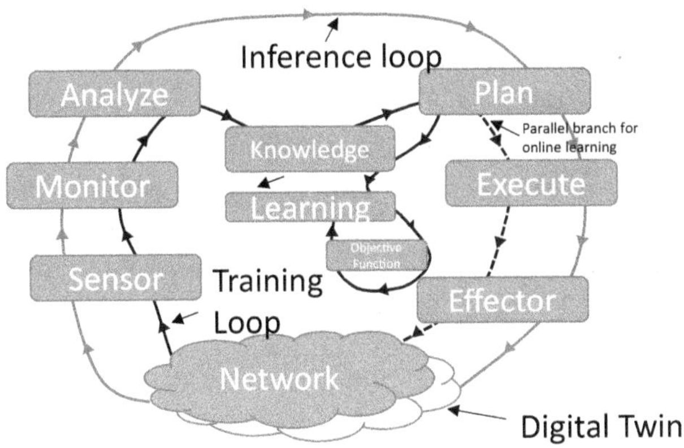

Figure 5.6. The N-MAPE-K framework.

It is worth noting that the original MAPE-K framework has limitations in the target context of mobile network functionalities supported by NI. Therefore, changes to the legacy MAPE-K are proposed to consider the specificities of the network environment, as depicted in Figure 5.6. In the figure, the different training/control loops that may be implemented by a NIF are depicted: the inference loop, the training loop, and the training loop with a branch for online learning. The model emerging from this adaptation is coined *Network MAPE-K* (N-MAPE-K). Specifically, the MAPE-K is extended along two dimensions:

- The purpose of the NIF, whether the knowledge is being trained or used in inference for the operation of the network, follows the MLOps paradigm.
- The nature of the NIF algorithm distinguishes between online learning and pre-trained/engineered models.

For the latter, the knowledge module shall be integrated with a training definition, containing all the attributes of the algorithm and thus specifying aspects such as the input data shape, batches, and most importantly, the used loss function (which could be dynamically adjusted) and the state/action representation, depending on whether the NIF algorithm belongs to the family of supervised learning or to the one of online learning. Additionally, the effector and the sensors can also be redirected to a digital twin element if needed by the specific NI instance, in order to disentangle the learning-loop process from the real operation of the network.

Hence, each NIF can be further split into NIF components (NIF-C) as follows:

- The Sensors specify all the probes that are needed to gather the input data and the kind of input data to be gathered. In principle, the APIs are specified at this level.
- The Monitor block specifies how the NIF interacts with the sensors.

- The Analyse block includes any pre-processing, summary, or preparation of the data, such as averages, autoencoders, and clustering algorithms.
- The Plan implies the specific NI algorithm that is implemented, for instance a neural network (NN) fulfilling categorization tasks.
- The Execute part specifies how the algorithm is going to interact with the system and how to possibly change configuration parameters.
- Finally, the Effector includes specific configuration parameters updated in the network function, again specifying the API.

With this framework, already presented in [5], NI algorithms are represented in a unified way and hence perform all the NI lifecycle management as discussed next.

5.3 NI Orchestrator

The structure of the proposed NIO is detailed next. The NIO is the element in charge of managing and orchestrating the NIS, NIF, and NIF-C, the elements that build the NI. The NIO structure is mutated from the layered structure of the ETSI NFV MANO framework, tailoring the components to the specificities of NI.

5.3.1 NIO Internals

The network intelligence orchestration framework, depicted in Figure 5.7, is structured over three levels, as in the ETSI NFV MANO framework. In the following,

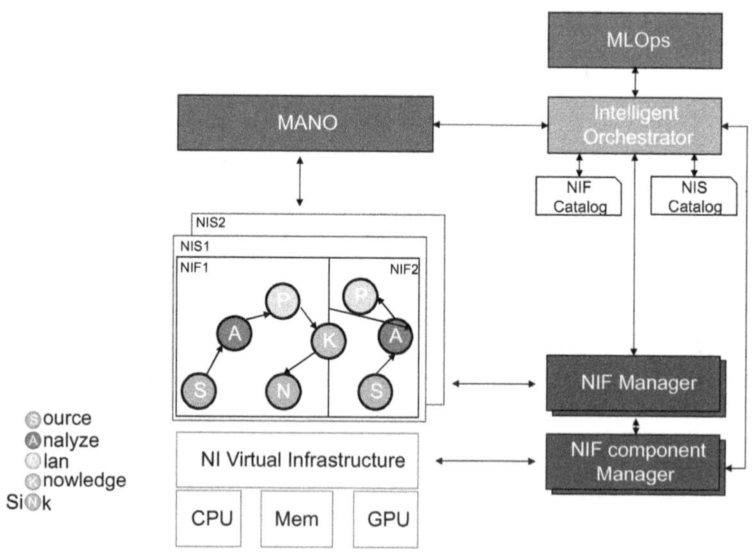

Figure 5.7. The NI orchestration framework.

references to lifecycle management refer to onboarding, instantiation, termination, scaling, and state retrieval.

NIF component manager: The NIF component manager is in charge of handling the lifecycle of the NIF-Cs, independently of their kind (i.e., independently of whether they are Source/Analyse/Plan/Knowledge/Sink) and their connection towards the infrastructure. For instance, in the case of Sources, the internet protocol (IP) addresses of the different data producers shall be provided, while in the case of Sinks, the specific configuration of application programming interface (API) endpoints has to be configured. This will have specific instantiations according to where this interaction shall take place. For instance, if the NIF is executed from the core, then Sinks and Sources shall integrate with the network registry function (NRF) and the network exposure function (NEF), properly synchronizing with the network data analytics function (NWDAF), whose analytics are captured as a set of Analyse, Plan, and Knowledge boxes. Similar considerations also apply for other network domains, such as the radio access network (RAN), where this framework can be fully integrated into the O-RAN x-Apps ecosystem.

NIF manager: The NIF manager, instead, has a global view of the set of NIF-C that composes every NIF: besides the lifecycle management of the NIF, this module is in charge of monitoring the health of the intelligence functions. This includes typical diagnostic information (e.g., constantly checking the KPIs yielded by the NIFs, like the accuracy) if the NIF is being used in inference or it is an online learning solution, or other metrics such as the loss and the training loops if the NIF is currently being trained (like the one presented in Section 5.2). The NIF manager is responsible for setting the meta-parameters of the models (through the interaction with the NIF-C manager) and reporting the health status of the NIF to the upmost module in the hierarchy, the Intelligent Orchestrator.

This module is in charge of the lifecycle management of the NIS, by properly coordinating the NIFs that build each of them. This includes the possibility of sharing NIF-C among different NIFs (e.g., two NIFs that require the same input) and also the arbitration policies in the case of two NIFs that share the same sink, that is, the configuration APIs.

Intelligent orchestrator: The intelligent orchestrator manages the connection towards the network MANO to gather important information, such as the expected network KPIs for the managed slice and service, as well as the information of the underlying network infrastructure. The intelligent orchestrator has catalogues of already-onboarded NIS and NIFs. In particular, NIFs may need to be (re-)trained to cope with changing or different conditions or on a periodic basis. In this case, the network orchestration interfaces with an external platform to build ML pipelines and perform such operations.

5.3.2 NI Distributed and Scalable MANO Framework for Massive Number of Network Slices

Sustainable, scalable, and energy-efficient massive slicing MANO in 6G systems will rely on distributed AI-native architectures, where both the analytics and the decisions will be performed close to the monitoring points via federated and multi-agent learning strategies. This will improve both scalability and reaction times, paving the way for zero-touch network service management (ZSM).

For 6G evolution, the proposed distributed and scalable MANO framework investigates zero-touch MANO in the support of network slicing at massive scales for B5G and 6G networks focusing on MANO scalability [6]. It proposes a novel autonomic MANO framework, heavily leveraging the distribution of operations together with state-of-the-art data-driven AI-based mechanisms. As shown in Figure 5.8, the proposed architecture splits the centralized management system into several management sub-systems, distributing both the intelligence as well as the decision-making across various components. Each technological domain may have one or several distributed management elements. Hierarchical, distributed, scalable, and AI-based management of a massive number of network slices across domains towards zero-touch management is achieved by distributing the management functions over all entities in charge of the life cycle management (LCM) and runtime management of network slices. This ensures the delegation of service-level management functions to be onboarded within the network slice.

The proposed distributed and scalable MANO framework has been designed for AI-driven MANO of massive numbers of network slices. It supports operations of fault management, self-healing, self-configuration, performance optimization

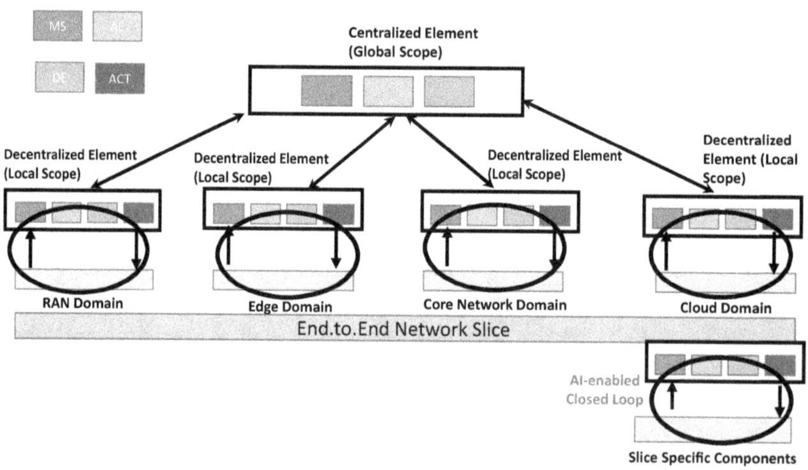

Figure 5.8. Proposed distributed and scalable MANO concept for 6G evolution – a high level outline.

(including energy-saving), and security operations. The proposed framework follows the so-called MAPE paradigm and is in line with key ETSI ZSM and ENI requirements and, hence, placed within the architecture introduced in Section 5.2. The AI-driven, multiple MAPE loops are used for level-specific, control loop-based optimization. Distributed closed control loops assist the LCM entities with state-of-the-art AI-based and data-driven mechanisms.

The proposed distributed and scalable MANO framework is based on four key component types: the monitoring system (MS), analytics engines (AE), decision engines (DE), and actuator (ACT) that present different parts of a control loop pipeline instantiation that can be used according to the target scope of analysis and decision with the goal of minimizing the raw data exchange to allow a fast decision analysis and decision. Each such loop is implemented using a pipeline composed of MS, AE, DE, and ACT. Typically, AEs and DEs are AI-driven. These four components present different levels of instantiation that can be used according to the target scope of analysis and decision, where the aim is to minimize the raw data exchange and allow a fast local analysis and decision. Their descriptions are as follows.

The MS is an entity responsible for gathering a set of different metrics from the systems that the DE is controlling. The MS can be common for multiple control loops.

- An AE performs time-series predictions as well as feature space regressions, clustering, and classification to extract insights from the measurements collected by the MS.
- A DE decides the LCM actions that need to be applied to face the issues detected by the AE. It may also control the MS measurement granularity or the AE prediction parameters.
- The ACT converts high-level (intent) reconfiguration commands obtained from the DE sublayer into a set of atomic reconfiguration commands. Hence, DEs do not have to deal with details of reconfiguration.

5.3.2.1 Distributed and scalable MANO architecture

For 6G evolution, the proposed distributed and scalable MANO architecture, its corresponding components, and interfaces are shown in Figure 5.9.

- Business layer: The layer consists of the business entities that operate the framework, provide slice management services to slice tenants, or own a slice (slice tenants).
- MANO layer: The layer consists of the core functions of the framework responsible for MANO of slices, slice LCM, and exposure of management interfaces to specific business entities.

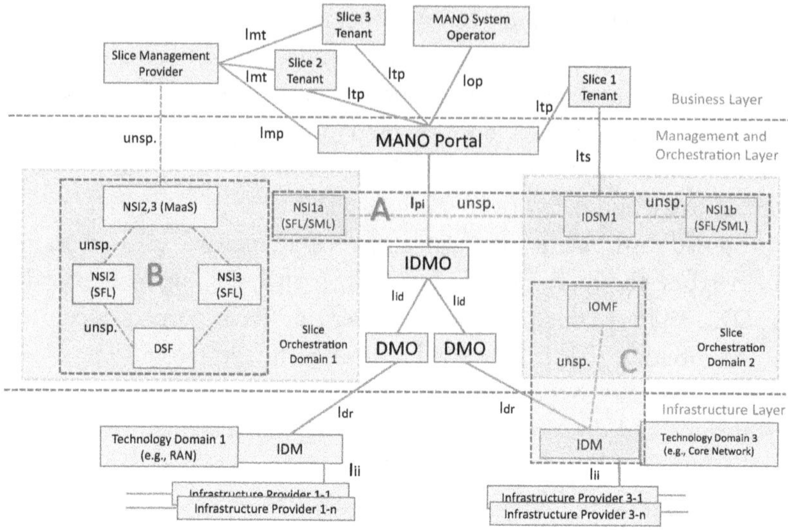

Figure 5.9. Architecture, components and interfaces of the proposed architecture distributed, and scalable MANO framework for 6G evolution.

- Infrastructure layer: This layer consists of the infrastructure, infrastructure providers, and functions enabling communication with the MANO layer and enabling optimization of the usage of infrastructural resources.

In the following, detailed descriptions are provided for each static and dynamic component of the framework belonging to the management and orchestration and infrastructure layers, with a focus on lifecycle management. The description of business entities and their interactions is presented from a high-level perspective. The components of management and orchestration and infrastructure layers are as follows:

- The MANO portal is used by slice tenants, slice management providers, and infrastructure providers to request operations regarding slice LCM, i.e., slice deployment, slice modification, and slice termination [6]. It also exposes the capabilities offered by the proposed distributed and scalable MANO framework (available slice templates, etc.) and partakes in negotiations related to the business dimension of the contract. The portal is also used to pass all the accounting and billing-related information.

Inter-Domain Manager and Orchestrator (IDMO) is equivalent to the 3GPP network slice management function (NSMF) [7] and exposes the northbound API (NBI) for the OSS/BSS or consumer service management function (CSMF). IDMO oversees the LCM of E2E network slices. It has full-scope slice MANO decision capabilities and takes global actions for network-wide, cross-slice, and cross-domain optimizations. The tenant or the slice owner interacts with the

OSS/BSS or CSMF to define the network slice to deploy using an already-generated blueprint to generate a network slice template (NST) that includes attributes and meta-data on the network slice (e.g., the start date and end date, slice owner, type of slice, etc.) and information on each sub-slice composing the network slice. For instance, in the case of computing resource (i.e., cloud or edge) domain, the NST may include information such as the number of CPUs, memory, and virtualization technology (i.e., virtual machine (VM) or containers) to be used. For the RAN domain, resources may be related to the functional split type [8], the MAC scheduler algorithm, the number of radio resource blocks (RB), and others. Finally, for the transport domain, resources may include the type of link (bandwidth, latency), number of virtual local area networks (VLANs), front-haul link capacity, virtual private network (VPN) links, and QoS. Each technological domain's needed resources are enclosed in the NST in the form of a technological domain-specific descriptor. For instance, for the NFVI domain, the resources are described using a network service descriptor (NSD) that includes the VNF(s) list and their descriptors.

Domain manager and orchestrator (DMO) is responsible for the orchestration and management of each of the slice orchestration domain (SOD) slices. Each technological domain is managed and orchestrated by its own entity, DMO, which is equivalent in 3GPP to the network sub-slice management function (NSSMF) [7]. Depending on the technological domain, a NSSMF may correspond to NFVO for cloud/edge, RANO for RAN, and software defined networking (SDN) controller for the case of the transport network.

Domain shared functions (DSFs) are a set of shared functions (VNFs) that can be implemented as physical network function (PNF)/VNF or CNF and can be reused by SFLs of multiple slices. The approach provides a reduced footprint of the deployed slice.

The inter-domain slice manager (IDSM) is a part of the slice template (a set of VNFs), and in some cases, it can be generated automatically by the IDMO (if IDMO is responsible for slice template split between multiple SODs). When IDSM is in use, it provides the slice tenant with a management interface. The IDSM is also responsible for the calculation of slice-related KPIs.

In the proposed distributed and scalable MANO framework slice structure, two separate layers can be distinguished – the slice management part called the slice management layer (SML) and the slice main part called the slice functional layer (SFL). The SML is an implementation of the ISM concept, having in mind the AI-based MAPE management. The SML is, therefore, split into MS sublayer (MS-S), analytic engines sublayer (AE-S), DE sublayer (DE-S), and actuating functions sublayer (ACT-S). The SFL contains a set of virtual functions that form the network slice to be deployed. The SFL part is composed of virtual functions that are dedicated solely to a slice (they are included in the slice template).

The infrastructure domain manager (IDM) provides the overall management of the infrastructure. Its interface to DMO allows for the allocation of resources (NFVI agent), the exchange of information related to the energy consumption of resources, and the exchange of the information related to the cost of resources that can be used by IDMO for resource brokering. The framework enables programmable infrastructure management. The DMO orchestrator can dynamically deploy management functions that cooperate with IDM to achieve infrastructure management. The IDM has an interface to the infrastructure provider, who can use the MANO portal asking for the deployment of additional infrastructure management functions, called IOMFs (see below). The functions are orchestrated in a similar way to slices, and LCM requests are sent by the infrastructure provider to the MANO Portal.

The infrastructure orchestrated management functions (IOMF) are specific functions that support infrastructure management. They can be orchestrated by the IDMO upon request of an infrastructure provider via the MANO portal.

Option A in Figure 5.9 concerns the deployment of a self-managed multi-domain slice. Such a type of slice requires a special component that is responsible for E2E slice management; it is worth noting that this component is implemented as a part of the E2E-slice template, not as a part of DMO. Option B shows the deployment of slices that use the PaaS approach, i.e., shared functions that implement the management as a service (MaaS) paradigm. Another set of shared functions, DSF (see further), is in this option exploited by the functional part of the slice. Option C shows infrastructure management-oriented and orchestrated DMO functions that are created in a similar way to slices, on the request of the infrastructure provider. The components that are deployed within each option are described in detail in the forthcoming sections. More details on the overall concept can be found in [9] and [6].

5.3.2.2 Monitoring system of distributed and scalable MANO framework for 6G networks

The MS of the distributed and scalable MANO framework (MANO-MS) implements the crucial concept of observing the network within the proposed architecture and gathering vital information that will feed the NI agents operating the network. In particular, MANO-MS is a distributed MS, developed within a container infrastructure system that relies on Kubernetes. Figure 5.10 illustrates the architecture of MANO-MS, with the central component being the sampling loop, implemented using sampling functions. The manager component is responsible for the life-cycle management of the sampling functions and provides access to the gathered telemetry data via *q* and *db* interfaces. The communication between these

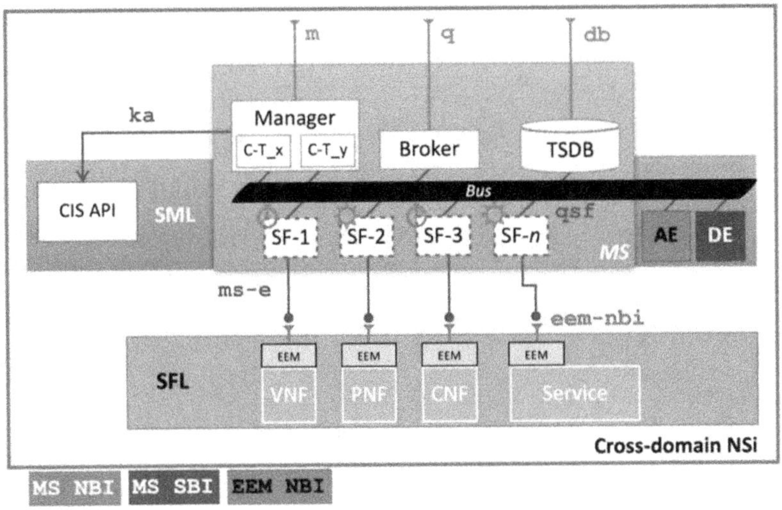

Figure 5.10. MANO-MS: Monitoring system of the proposed distributed and scalable MANO framework.

components is facilitated by the service bus of MANO-MS, which is implemented through Apache Kafka.

The platform is controlled through the *m* interface, as depicted in Figure 5.10. During deployment, the bus topics are configured, and the user then requests a sampling loop by submitting a configuration file. After validation, the manager generates the corresponding deployment instructions, which are passed on to Kubernetes (K8s) for deployment. The deployed sampling function, implemented as a CNF, enters the operational stage, and periodically samples the telemetry data as configured, publishing it to the bus.

The MANO-MS platform is versatile and capable of monitoring a wide range of telemetry data, as long as an embedded element monitoring is available (e.g., metrics from different network components as depicted in Figure 5.11).

5.3.3 Enabling SDN Control with NI

In previous sections, it has become clearer that NI can be provided at multiple network domains and through multiple orchestration layers. In this section, an architecture for enabling an SDN controller with intelligence is provided, as are several use cases.

Artificial intelligence, and more specifically machine learning, has been demonstrated as a feasible tool to support network traffic analysis in multiple scenarios. To this end, the introduction of ML in transport networks has been previously studied [10], but typically is focused on non-RT scenarios and with the usage of dedicated tools.

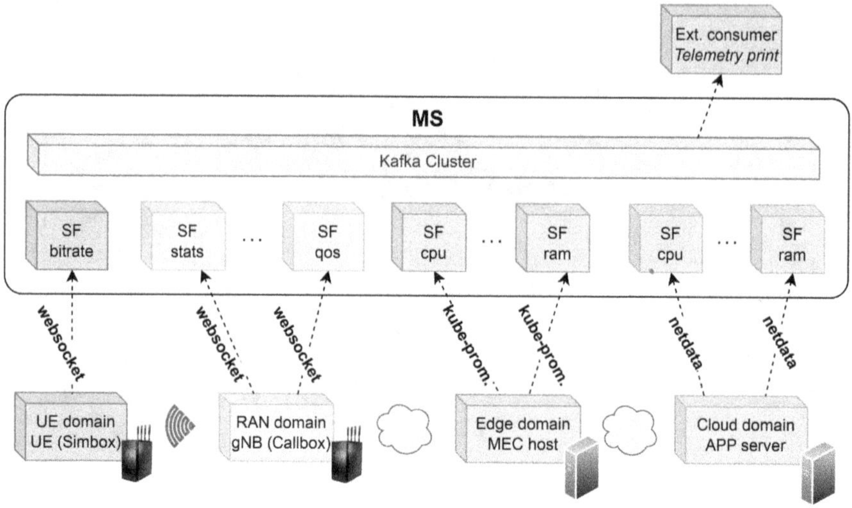

Figure 5.11. Multi-domain slice for 5G end-to-end monitoring.

Network programmability has been brought to transport network due to the introduction of SDN, which proposes the decoupling of control and data planes of underlying network equipment. To this end, it offers improved network programmability of the equipment and introduces the concept of SDN controller. The SDN controller is responsible for the control and management of the network equipment, allowing a logically centralized control and management.

TeraFlow project has proposed a new transport SDN architecture that enables an open environment for network applications and devices using full standard interfaces with container-based services, which are deployed as micro-services and managed on elastic infrastructure through agile DevOps processes and continuous delivery workflows [11].

Figure 5.12 shows the architecture for TeraFlowSDN (TFS), which is a cloud-native SDN controller composed of multiple container-based services using novel virtualization techniques, which are deployed as micro-services and managed on elastic infrastructure through agile DevOps processes and continuous delivery workflows. These micro-services structure an application as a collection of interconnected and related services using a common integration fabric. In a micro-services architecture, services are simple and detailed, and the protocols are lightweight. This architecture eases the introduction of novel services within the SDN controller that are able to provide NI. In this section, two functionalities are noted.

Figure 5.13 shows the first proposed functionality, which relates to traffic forecasting. SDN controller is able to monitor current network status and obtain timestamped traffic matrices on top of a pan-European core network (using Geant as an example [12]). Then, a traffic forecasting application (forecaster) can introduce

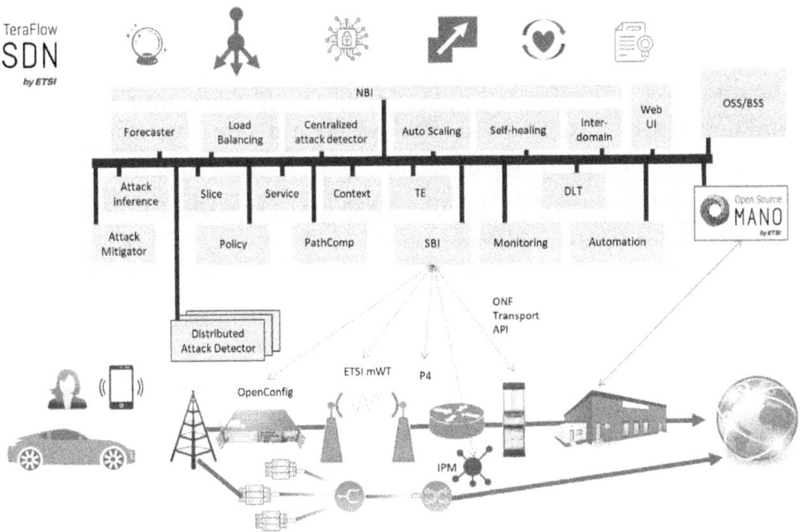

Figure 5.12. Enabling NI in the cloud-native SDN controller.

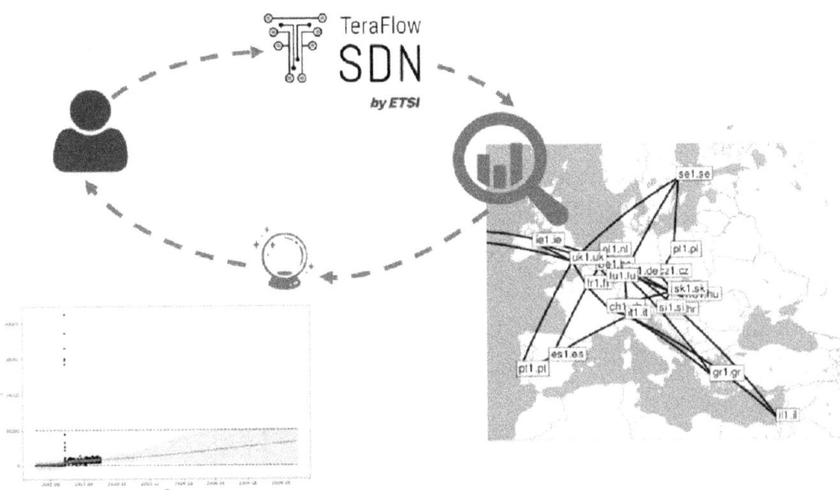

Figure 5.13. Traffic forecasting with SDN controller. Source of sample network: [12].

ML-based algorithms to predict the network status in the future. This information can be provided to the OSS/BSS in order to trigger necessary link updates, as well as provided back to the SDN controller in order to limit resources allocated to specific network links.

Forecaster is a novel TFS component that is able to perform proactive SDN traffic prediction (i.e., forecasts) by means of ML algorithms. For example, it is able to collect RT KPIs, such as link occupancy, and use ML algorithms to forecast where and when a problem (e.g., unavailable link resource) is likely to occur so as to reroute traffic before it happens.

The interfaces provided by the Forecaster component allow to do a complete network forecast by introducing a topology identifier, or obtain specific forecast of a link, by introducing a link identifier. A forecast is a data structure that contains an array of timestamped predictions for a specified period of time for a specific link. Another provided interface allows to check if a new requested connectivity service will have resources available in the future. To this end, a link resource availability threshold is configured, and the forecast component provides a *ForecastPrediction* with the decision.

Multiple traffic forecasting libraries can be introduced in order to provide the component with multiple engines. One of the aforementioned possibilities is the Prophet library, which already includes a seasonal model for data forecasts. Another possibility could be the introduction of AutoML. AutoML does not require training, and thus, no previous modelling of the data is required.

Another features that TeraFlowSDN includes is the introduction of NI to face different threats on different planes. Management and control planes expose capacities that could be impacted by multiple vector attacks: from insiders (e.g., unauthorized configurations), to exposed interfaces (northbound, southbound, or east-west bound), to SDN applications. At the data plane, the TeraFlowSDN controller is exposed to both classical and advanced network attacks (e.g., DDoS, malware, traffic manipulation, physical-layer intrusions, etc.).

Moreover, the inclusion of ML components as applications of the SDN controller also adds a new threat surface that can be utilized by the so-called adversarial techniques that try to fool ML components by introducing small perturbations in the input that cannot be perceived by humans.

Both the massive amounts of information flowing through the network infrastructure and the very short latencies in threat detection impose limitations on current intrusion detection systems that perform ML-based network management. To cope with this problem, TeraFlowSDN includes a two-layer ML-based architecture with a central ML engine and ML-based threat detectors placed at the edge nodes.

The use of this advanced distributed architecture, supported by Google Remote Procedure Call Protocol (gRPC) telemetry data and network flows, will help to detect and mitigate the above-mentioned threats. In some specific data plane attacks, it is necessary to deploy distributed detection engines at the edge, with specific inference ML models that should be developed, to solve the attacks close to the origin.

The use of a simulated environment based on real NFV/SDN telecom infrastructures allows to generate traffic, train models, and deliver accurate inference ML engines to the edge and to the Cybersecurity net application for early mitigation.

To ensure the resilience of TeraFlowSDN ML models against adversarial attacks, multiple libraries have been included for designing defences in ML-based components and testing them against sophisticated adversarial attacks.

Finally, TeraFlowSDN includes un-, semi-, and supervised learning approaches for multi-layer network security monitoring, with additional embedded intelligence, using standard interfaces, containerization, and load balancing, while paving the way towards carrier-grade deployment of ML-based security monitoring [13].

5.4 Design Guidelines for NIFs

Artificial intelligent systems are irreversibly set on the evolutionary path of every future vertical as well as of every object and service we humans will interact with in the near future. This trend is motivated by the need to support elastic and demanding real-world use cases such as automated and cooperative mobility, e-health, gaming, entertainment, etc.

In this scenario, it is acknowledged that the telecom operators will have the opportunity to fill a central role in providing innovative solutions for application and service developers who need to combine the advanced capabilities of B5G and 6G networks with the fluid cloud-based application development processes emerged in the last decade (such as platform as a service (PaaS), continuous integration/continuous delivery (CI/CD) pipelines, and micro-service/serverless-based designs). A significant example of such an ecosystem are AI-enabled applications, which have become a major innovative force in almost any vertical and are being foreseen as one of the pillars that will boost the fourth industrial revolution. While great progress has been made during the last few years with respect to the accuracy and performance of AI-enabled applications, their integration into potentially autonomous decision-making systems or even critical applications requires E2E quality assurance, ubiquitous availability, and low latency. In the following, the reference model for NIFs is specified, as well as the implementation of some NI functions.

5.4.1 Reference Model for NIFs

The orchestration of such a new generation of AI-enabled application requires the introduction of new abstractions into the network service orchestration domain. Here, a novel concept of NIFs is introduced to refer to the AI-enabled E2E application sub-components that can be deployed across edge-enabled B5G and 6G networks. A new generation of AI-enabled applications will be obtained through the chaining of various NIFs across the various levels of the edge architecture. This will trigger the need for novel network automation platforms supporting all aspects of network and service management, including the deployment and scaling of NIFs over a distributed facility as well as the various ancillary tasks that are to be performed to deploy such application, e.g., the creation of a new network slice.

To achieve this breakthrough, innovation is needed in two different areas: (i) the reference models for NIFs and (ii) the E2E orchestration of NIFs.

Reference models for NIFs: There is a need to define a reference model for NIFs, capable of capturing and representing the heterogeneity of NIFs at different levels of the technology stack. This can be developed as a network of interconnected modular ontologies, implemented in standard knowledge representation languages (e.g., OWL), following well-known state-of-the-art ontology engineering methodologies. Special attention needs to be put into describing NIFs not only from a functionality point of view, as available in other catalogues, but also considering other capabilities (e.g., computation, communication, storage, and hardware acceleration) that complement those of the NIFs, as well as any other aspects related to constraints necessary to support their dynamic orchestration. Similarly, the reference model shall also offer capabilities to describe methods and approaches to support the provisioning of the AI-enabled applications.

Building applications by chaining NIFs calls for the definition of a reference model for NIFs capable of capturing and representing the wide diversity of NIFs that will be deployed in B5G and 6G networks. This model is necessary to provide a common packaging for AI functions and a virtual marketplace for standardized AI components. There are several approaches in the literature to model artefacts (including datasets, pieces of software, algorithms, models, computational resources, etc.) to facilitate their understanding and reuse, as well as their potential chaining in complex applications. A relevant example is the workflow infrastructure conservation using semantics ontology network [15], focused on documenting computational scientific infrastructures, describing their relevant capabilities, and how they can be deployed, with four different ontologies: software stack, hardware specs, scientific virtual appliances, and workflow execution requirements. Other similar works in this direction are the software ontology (SWO) [16], a model for describing the software involved in the storage and management of data, with a strong focus on the biomedical domain or OntoSoft-VFF (ontology for software version, function and functionality) [17], focused on the description of the functionality and evolution through time of any software used to create workflow components. In addition to this, the state-of-the-art provides general-purpose metadata profiles and vocabularies that have been used for cataloguing different types of assets/entities: Dublin Core for all types of digital objects, DCAT and DCAT-AP for open government datasets, as well as its extensions, the DataCite metadata schema, and the research object model for scientific workflows.

E2E orchestration of NIFs: Deploying large-scale AI-enabled applications calls for the provisioning of resources across multiple administrative and technological domains. Therefore, it is of paramount importance to partition applications composed of multiple NIFs across multiple domains according to the requirements of each NIF. The proposed approach aims to cover more extreme provisioning scenarios in terms of hardware and software resources by starting from "de facto" standards for the orchestration of cloud and edge services, such as Docker and Kubernetes, and their emergent variants (e.g., FaaS). Acknowledging the heterogeneity and complexity of the currently available edge computing platforms, especially when it is as broad as the one foreseen in B5G and 6G networks, research innovative solutions for the E2E orchestration of the AI-enabled applications is needed. This goal can be achieved by leveraging novel network automation platform, in which monitoring solutions tackling the specific requirements of optimized hardware, edge devices, communication infrastructures, and cloud services will be taken into account. The monitoring subsystem will collect raw information to obtain the QoS indicators that provide insights regarding the correct behaviour of multiple NIFs orchestrated and linked to create complex AI-enabled applications. The quality indicators will cover traditional metrics for IT systems (e.g., performance) but also specific items to assess that NIFs operate according to their initial design. At the same time, the collected metrics and the quality indicators will ensure the capacity to oversee the operation of orchestrated NIFs and even to integrate AI-based functionalities to detect abnormal situations or to implement predictive maintenance.

While the NIFs paradigm opens the door for a number of opportunities for the customers, it also poses numerous challenges for the infrastructure providers including the need for supporting properties such as QoS, scaling, and fault-tolerance for the AI-enabled applications. This problem is exacerbated by the need for supporting stringent QoS requirements such as ultra-low latency and high data rate of novel applications and services envisioned in B5G and 6G networks. The AI-enabled applications force service orchestrating mechanisms to take into account additional factors, apart from the code and data locality, in order to satisfy diverse latency and data rate requirements by making appropriate placement decisions at MEC servers, which are expected to play a pivotal role in curtailing the round-trip delay of the applications. Extensive work on orchestration has been done in the past years by the service community and by the network community. An extensive survey on an E2E MANO in multi-technology and multi-operator 5G networks can be found in [18]. A study on multi-domain network and service orchestration is conducted in [19] highlighting its challenges in data-centre networks as well as in cellular networks.

5.4.2 Customized AI Techniques that Empower Practical NI

6G network will need to rely on customized instances on AI, which natively embody the NI (in the NIP). In the following, two exemplary solutions are discussed.

5.4.2.1 Avoiding the loss-metric mismatch in NI

Loss functions drive the training process of supervised machine learning models. In the vast majority of cases, loss functions are designed to be generic enough to work well in a wide range of scenarios. In regression problems, including forecasting tasks, mean absolute error (MAE), mean square error (MSE), and mean squared logarithmic error (MSLE) are common choices for expressing the loss.

However, in many practical cases in network management, such traditional losses do not characterize well the target performance metric of forecasting tasks. For instance, in anticipatory resource allocation problems encountered across mobile network infrastructure domains, the goal is to anticipate a capacity that is *sufficient* to accommodate the future traffic demand. Indeed, under-provisioning of capacity leads to the disruption of the offered service and violations of the service-level agreements (SLAs) with the service providers, while overprovisioning causes a more affordable squandering of resources. There, it is critical that the predictor learn to forecast a minimum quantity that is always above the demand.

Using a traditional loss function to perform forecasts in cases such as those outlined above results in a so-called loss-metric mismatch, where the regression objective, represented by the loss to be minimized, does not correspond to the optimization of the actual performance metric (Figure 5.14). As a result, the AI model that implements the regression model does not learn predictions that are aligned with the expected network management objective.

Thus, the NI native architecture defined above shall *support the use of customized loss functions that are carefully developed based on expert system knowledge*, i.e., a deep understanding of the network engineering or management task at hand, as well as of the variables that affect it and how they do so. This illustrates how the tailored design of loss functions for NI shall occur.

In the left plot (a), a pure traffic predictor is trained using a legacy loss, e.g., MAE or MSE for regression. The resulting forecast cannot be used as is, but serves as an

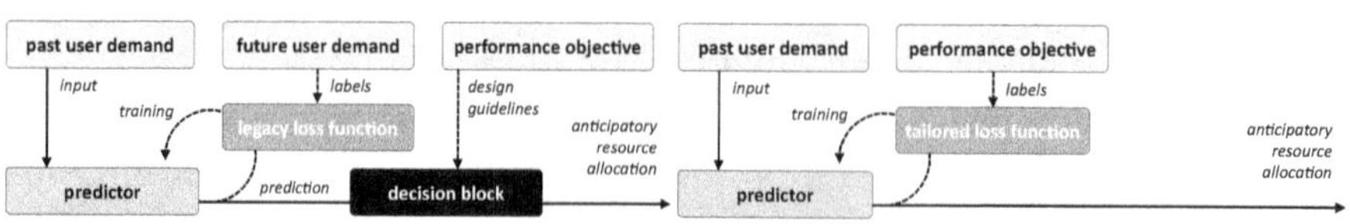

Figure 5.14. Different approaches for solving the loss-metric mismatch.

input to the actual decision block, which is manually designed by human experts to output the anticipatory MANO actions so that the target network performance objective is met. Yet, the decision block is agnostic of the inherent accuracy of the predictor and just trusts the forecast it receives. In the right plot (b), the novel approach proposed here is outlined: expert knowledge is used to directly design a dedicated loss that encodes the relationship between the prediction and the performance objective. As a result, the predictor is trained to produce forecasts that optimize the performance and can be directly used to drive the MANO actions. Importantly, the action decision is now aware of the unavoidable prediction error (e.g., lower accuracy in predicting small traffic volumes) and automatically compensates for it (e.g., by taking more conservative actions to accommodate small-traffic future demands).

5.4.2.2 Loss meta-learning for NI

The performance metric to be optimized by anticipatory MANO actions is not always known a priori by the network operator. This is the case, for instance, when the performance must be measured at the application layer (i.e., in the service provider domain) or when it concerns end user satisfaction (e.g., if it relates to mean opinion scores or quality of experience). In these situations, designing tailored loss functions is not possible, since the human expert (e.g., network manager or system engineer) does not know the exact relationship between the forecast and target performance.

6G networks shall leverage on innovative guidelines to deal with NI design in the complex situations described above. Specifically, instead of imposing a predefined expression of the loss function used to train the predictor, *a design of forecasting models is proposed that is free to meta-learn the loss function that best suits the network management objective at hand.* In practise, this is realized by combining a loss-learning block with the actual predictor, as shown in Figure 5.15. This block is responsible for learning the loss function, or, equivalently, capturing the relationship between the forecast produced by the predictor and the target management objective. Once

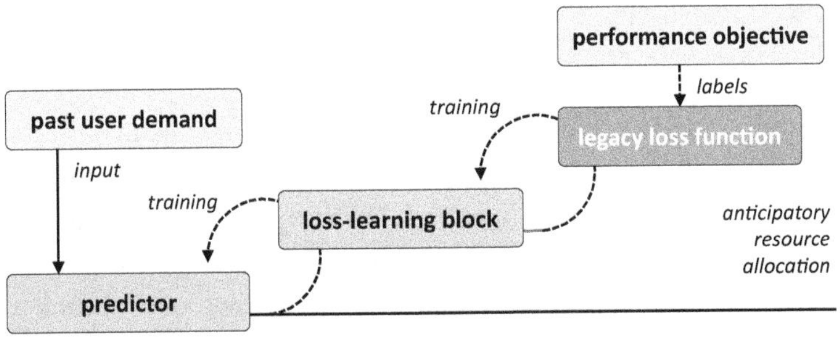

Figure 5.15. Loss meta-learning for NI.

ready, the loss-learning block can operate as a tailored loss function: it receives the output of the predictor and determines its quality for the precise management task. Therefore, it can be employed to train the predictor so as to steer the optimization of its parameters towards minimizing the actual MANO objective. The full design of the proposed framework is specified in [20].

5.4.3 Sustainable Decentralized AI Solutions

Sustainable decentralized AI-native architectures involve using energy-aware AI techniques for designing and optimizing 6G network slicing MANO in a sustainable and scalable way. The invoked architecture enables the next innovations:

- A FL-based AE is considered to reduce the amount of raw data exchange between local AEs and E2E AEs, making data analysis and prediction more energy-efficient. This technique avoids transferring the local dataset to a remote server, which leads to a high communication overhead and higher energy consumption.
- Decentralized deep reinforcement learning (D-DRL) techniques are considered in the DEs, such as multi-agent DRL (MA-DRL) and federated DRL (F-DRL). Such type of decentralized AI techniques can be utilized in the DEs in several ways: (1) To perform cross-domain joint VNF placement and energy control. The energy cost could be added in the denominator of the DE multi-objective reward function along with latency, while the throughput can be maximized by plugging it in the numerator with the costs. The multi-objective weights are fine-tuned according to operator/tenant priorities that depend on the slice type and business strategy. (2) Traffic-aware local decision agents dynamically placed in the network. These federated decision entities tailor their resource allocation policy according to the long-term dynamics of the underlying traffic, defining specialized clusters that enable faster training and communication overhead reduction.

5.4.4 Implementation of Intelligent Distribution from the Computation Perspective

From a computational point of view, the network can be seen as an entity consisting of a large number of computing nodes interconnected via communication links. The nodes are not homogeneous but range anything from high-performance data centres and edge servers down to user equipment and miniature IoT computing devices. Some of them have a more complicated internal architectural structure, featuring multiple processor cores, CPUs, processor cards, and clusters of them with internal communication channels, as well as memory and input/output devices. Also, communication links are typically heterogeneous. In addition, neither the computational workload of the network nor its hardware components are constant but alter more or less frequently all the time, depending on the prevailing user data traffic distributions and acts of service providers. Let us call computation here parallel if it happens simultaneously in multiple units within a node and distributed if multiple nodes are involved.

In order to get the best performance out of the network, the computation needs to occupy intra-node units and be distributed among the network nodes so that the overall computing capacity would be maximized and the resource usage (e.g., time, energy, design effort, etc.) and cost would be minimized while ultimately supporting the designated use cases. A typical network node alone has more than a thousand design parameters, not to mention the plurality of nodes and fitting the subsets of the computation into them. It is therefore evident that determining the optimal distribution is a demanding multi-target Nondeterministic Polynomial (NP) complete problem, which cannot be solved with the current technology. The current 5G systems approach distribution by relying on simple offloading functionality from UE to the network, virtual machines allowing flexible allocation of resources in the cloud, containers reducing the state of computation, multicore CPUs allowing (constrained) parallel processing, and a programming paradigm relying mostly on independent sequential components and asynchronous execution of threads.

To boost the more macroscopic 6G network-level architectural design decisions, the hidden capability of low-level architectural techniques would be worth considering. For more flexible and efficient placement of computation, an architecture utilizing the following techniques to overcome the limitations of 5G network and processing solutions could be potentially applied (see Figure 5.16).

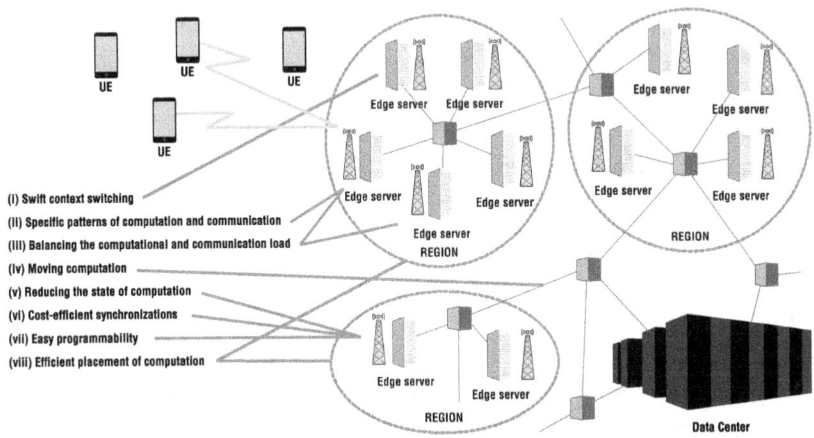

Figure 5.16. Intelligent computation distribution with low-level architectural techniques.

To support parallel and distributed execution, a programmer and the system need to divide the computation and related communication into parts (threads, processes, components, and subservices) that execute in parallel or concurrently in multiple execution units and communicate with each other. The number of computational (software) parts often exceeds the number of (hardware) execution units, introducing the need for multitasking, i.e., sharing units between the parts. In order to avoid efficiency issues, multitasking requires (i) swift context switching between the parts targeted to the execution unit. Context switches can be made faster, e.g., via multithreading [21] that uses on-chip resources to retain multiple contexts and allows for instant switching between them. If computational parts of a single service or task are executed in multiple execution units, there is a need to perform intercommunication between these units that can form (ii) specific patterns of computation and communication. Execution of these patterns can be accelerated by techniques such as multicasting and multioperations [22]. Multioperations are primitives of parallel computation where threads perform reductions, e.g., additions, on values provided by multiple threads into a single value in a constant number of steps. Often, in the case of non-uniformly distributed computing tasks, there is a need for (iii) balancing the computation and communication load between the execution units in order to manage hot spots and overcome the capacity limits of a single unit. Techniques including work sharing and work stealing between the nodes, e.g., edge servers, close to each other can be considered requiring (iv) moving computation between the execution units. Since current techniques using virtual machines and containers require moving a substantial amount of data from the source unit to the target unit as well as interrupting the execution, saving the state, and restoring it after the movement, techniques such as grouping a number of neighbouring nodes into a distributed computer would be useful to reduce the need for data block movements. Since the delay caused by moving computation is proportional to the amount of data moved, (v) reducing the state of computation can increase the performance. This is possible, e.g., by grouping homogeneous computations into flexible vector-like entities with one control flow but multiple data flows, such as thick control flows [23]. If there exist dependencies between the parallel parts, one needs to perform synchronizations to obtain the correct results. This is expensive in current multicore CPUs. A possible way to support (vi) cost-efficient synchronizations is to use the wave synchronization mechanism overlapping synchronizations with computations.

Distribution of computation can be investigated from the perspective of usability or programmability, which is now receiving more attention as a KPI for emerging 6G systems. Methods of (vii) easy programmability for parallel functionalities within the nodes include, e.g., shared memory emulation [24, 25], supporting unified and synchronous memory access with strict memory consistency and eliminating many complications of current multicore CPU programming. Finally, the algorithms for (viii) efficient placement of computation contribute to efficient computation distribution, which makes the right data to be in the right place at the right time with minimal cost. Up to the node level, the placement of functionality and data could be done explicitly by the programming or with the help of simple hardware mechanisms, such as symmetric multiprocessing. If the node is a high-capacity data centre, more complex techniques are needed for optimum allocation of computational resources due to limitations on the scalability of the legacy techniques and the communication bandwidth between nodes.

To avoid excessive synchronization and communication delays, only nearly independent components should be distributed to separate nodes. A good strategy could be based on simple greedy algorithms placing the computation into a node with the lowest workload within the region of neighbouring nodes or more complex ML-based solutions. In this case, although computation and data placement throughout the entire network would not be feasible, the hierarchical placement for a certain range of networks (e.g., regions of edge servers, subnetworks of a certain size) would be possible in a given environment (i.e., energy limit, the number of users, existence of physical defects, and traffic congestion).

5.5 A Multi Agent Reinforcement Learning Framework

This section provides an overview of the integration of RL, a subset of machine learning, into radio networks. The advancements in device and communication protocols in 6G have enabled greater automation in managing the Core-RAN-Edge network component. The objective of RL is to maximize long-term performance by automating a complex and dynamic system. This involves multiple network segments with different scales and characteristics, including the nature of the controller, decision-making periods, and centralized or decentralized nature. As a result, the MA-DRL [26] scheme may be utilized to support the learning and deployment of RL-based agents in managing 6G systems.

5.5.1 Design Concepts of the MA-DRL Scheme

Reinforcement learning (RL) is one of the main machine learning paradigms (Figure 5.17). An RL agent is an actor in a dynamic system/environment (for instance, the RAN in 6G); it can interact with and alter the system and eventually learn through interaction data with the system. Once trained, the RL agent continuously acts with the objective of maximizing a long-term performance. RL

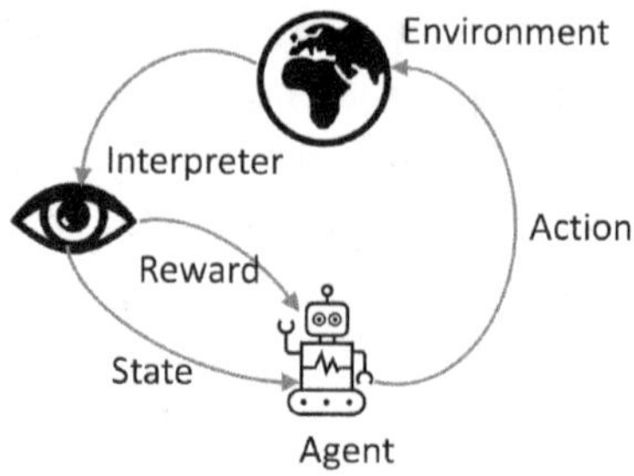

Figure 5.17. The reinforcement learning interaction loop.

is researched in many domains such as games (board games and video games), robotics, telecommunications, etc.

RL models the system as a Markov decision process (MDP). The configuration of the system is represented as a state in a state space S. The set of decision/actions the agent can take in the environment is A. A transition model, $P_a(s)$, describes the effect of the agent's action $a \in A$ on the environmental state $s \in S$. A reward signal, $R_a(s)$, represents the positive or negative feedback the agent receives after it takes an action a in a state s. In order to be applied to 6G, the system of application will need to be described as a MDP by defining the set of S, A, P, and R.

RL is composed of various families of approaches that take into account various characteristics of the problem.

- Multi-agent RL (MA-RL) deals with systems where multiple actors simultaneously compete or cooperate to solve a task. The multi-agent setting is motivated by problems where the real world imposes some communication or sensory constraints that force the agents to act in a decentralized way once deployed in the world.
- Specific RL methods deal with systems where the state is only partially observable. In a 6G system, some components might need to take decision without having a full and complete information about the system.
- Off-policy methods can learn an RL agent while executing the decisions of another fixed, predefined agent (that may not be RL related). This is of interest if one wants to learn about RL agents without having them interact in a hazardous way with a sensitive system while learning.

Traditional RL algorithms apply to small and discrete problems but do not scale when the number of observations or actions increases. The combination of RL and deep learning has somewhat reduced this issue. DNNs are a powerful and flexible tool to learn different levels of abstraction from data. The ability of DNNs to deal with high-dimensional data is game-changing to scale RL to more complex and eventually real-world problems.

5.5.2 The MA-DRL Scheme

This section outlines the functional implementation of the MA-DRL scheme in a 6G system. The proposed logical architecture, as already discussed above, divides the system functions into multiple major logical elements, including the cognitive plane, the monitoring plane, and the control plane [26].

The *monitoring plane* collects and monitors functional and performance parameters, including KPIs concerned. In this monitoring plane, there are two RL components: the *RL reward monitor* and the *resource analytics*. The RL reward monitor translates the feedback from the 6G system into a reward understandable by the RL scheme. The resource analytics is a component that analyses the feedback from the 6G system and is responsible for detecting special events such as non-stationarities.

Based on the inputs from the monitoring plane, the *cognitive plane* analyses the system and makes decisions to offer intelligent operational strategies such as optimizing resource allocation and deployment inputs to the relevant RL-driven modules such as RAN slicing, cell-free (CF) scheduling and beam steering. Most of the RL elements are therefore located in that cognitive plane. The RL simulator implements the training environment (mimicking the 6G system). The *RL agent training host* performs the RL training. The *RL agent designer* is a database of trained agents and training configurations. Finally, the *RL engine* manages the three previous blocks.

The cognitive plane outputs its intelligent decisions to other planes for actions to be applied. The control plane is called on demand to enforce RT actions decided by the cognitive plane via the *RL policy adapter*. The RL adapter is in charge of translating the MDP actions into proper control actions specific to the RL application.

For reference, the components of the MA-DRL scheme within the proposed logical architecture are reported in Figure 5.18.

On-line versus off-line learning

This section presents a comparison between the offline and online training workflows. The offline workflow involves agent training on a simulated environment, whereas the online workflow involves the agent directly training in the actual infrastructure.

The online and offline deployments emphasize some challenges of the application of RL to 6G systems, such as:

- RT control: the agent must compute its action at the required control frequency.
- Safety: the agent's actions must not be harmful to the overall system. In the online deployment, the agent should be able to perform safe exploration.

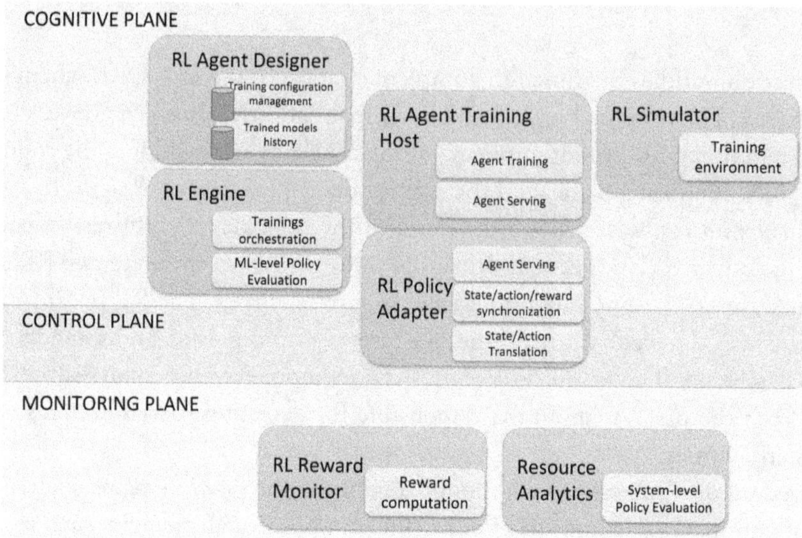

Figure 5.18. Overview of the MA-DRL scheme components.

- Robustness: the MA-DRL agents should be robust to inputs with small deviations (e.g., noisy sensors) to their training data.
- Dynamic environment: the agent should adapt to inputs whose dynamics may vary over time (see non-stationarity).

Non-stationarity, trust, and domain of validity

In addition to the above discussion, the building of a tool to instill confidence in the deployed RL agent is explored. This is accomplished by equipping the MA-DRL scheme with monitoring functions.

The validity of the trained policy deployed in the 6G system needs to be assessed by defining system-level policy evaluation rules, which will trigger alerts to the RL engine. The resource analytics component should implement these rules. If the rules exploit KPIs collected from the monitoring plane, the interface between the resource analytics component and the relevant monitoring components needs to be implemented. In some cases, the reward used during training can provide valuable information during exploitation that could be used to raise alerts or for monitoring purposes. Therefore, the reward computation from the monitoring plane needs to be implemented in the RL reward monitor component.

Finally, the handling of these alerts by the RL engine and RL agent designer will be defined. A simple approach could involve the agent designer randomly selecting trained agents until one is found that no longer triggers alerts. A more advanced approach would entail identifying a matching agent trained in a similar environment configuration based on the KPIs reported by the alerts. In this case, the relevant KPIs and potentially threshold/tolerance values for configuration matching need to be specified.

5.5.3 MA-DRL for Joint Slicing Scheduling

This section provides an illustrative example of how the MA-DRL scheme can be utilized in an RL application by demonstrating how a joint-slicing scheduler problem can be formulated as an RL problem and incorporated into the MA-DRL scheme (Table 5.3).

Table 5.3. System components of the MA-DRL scheme.

Components	**Specifications**
MDP	The state space consists of three stats at each UE, namely: *WBCQI*, *Queue Length*, and *Delay of the First Packet.* The action space is the number of physical RBGs assigned to each UE, constrained by the total number of physical RBGs. The reward is a weighted sum of *Throughput*, *Latency*, and *Packet Loss* with the weights subjected to further sensitivity analysis.
Policy Adapter	• Translating between JSONs on the system side from-and-to the tensors on the agent side of the gym interface. • Time synchronization so that the corresponding <*States, Actions, Rewards*> are not messed up.
Reward Monitor	Need for tools to analyse the distribution of rewards over time under different weights for sensitivity analysis.
Training Host	Use vanilla policy gradients with pure MLPs for policy representation. All training is done offline.
Simulator	The simulator consists of: • OpenAI's gym interface sitting between the DRL agent and the RabbitMQ message queue. • The OAI's system with FlexRIC wrapping OAI's 5GNR sitting behind RabbitMQ • A telnet that programs OAI's 5G-L2SIM to replay the provided CQIs and traffics trace as UEs to the OAI's 5GNR base station. The four components (DRL agent, gym interface, RabbitMQ, and FlexRIC) are interacting as described in the "training scheme" section. **Interfaces** The gym interface will provide callable Python functions (e.g., **.step()**) on the agent side to enable the use of standard DRL implementations. The gym interface and the OAI's system will communicate through the RabbitMQ message queue.

This use case addresses connectivity in factories in the context, where more machines with diverse operational modes will be integrated together, making QoS-guaranteed wireless connectivity desirable. To this end, they address the challenge of supporting multiple connectivity "slices" with diverse requirements and ask how to do it optimally.

This RL application is tailored towards a deployment in a Bosch factory where the technical requirements of the two slices are diverse: the PLC offloading and video processing offloading:

- Slice PLC: Delay requirement is tight while data rate is normal.
- Slice video processing: Delay is not as tight but uplink data rate is high.

In the formalization, an RL problem will be solved for different slicing configurations, and the best configuration will be then selected. The specific components of the system in the following table.

5.6 AI-Driven Air Interface Design

AI-driven air interface design is a novel methodology for designing RANs, and it is envisioned to be one of the enabling technologies of the 6G intelligent networks. It relies on artificial intelligence techniques to learn single functionality or multiple functionalities in a transmitter or receiver of a radio communication link from training data. This chapter provides sample AI-driven air interface designs targeting hardware impairment compensation or mitigation, spectral efficiency enhancement, channel estimation, and radio resource allocation. The section is organized as follows: sub-section 5.6.1 presents an AI-driven receiver method for radio frequency (RF) hardware impairment compensation; sub-section 5.6.2 presents a fully learned receiver; sub-section 5.6.3 is about AI-native air interface design with constellation shaping to mitigate hardware impairments; sub-section 5.6.4 addresses AI-native channel estimation; sub-section 5.6.5 presents AI-driven channel estimation for reconfigurable intelligent surface (RIS)-enabled networks; and sub-section 5.6.6 presents AI-driven resource allocation methods in cell-free MIMO networks.

5.6.1 AI-driven Receiver Methods for RF Hardware Impairment Compensation

Data traffic over cellular communication networks has an increasing trend, which imposes requirements on the 6G radio cellular networks to support it even higher than their 5G counterparts. However, developing techniques to fulfil these requirements in the presence of limitations in RF hardware is challenging. Signal transmission is subject to distortions due to RF hardware impairments which have an even higher impact in high-throughput scenarios. For example, the wideband transmission is severely impacted by power amplifier (PA) non-linearity; high-band transmission (e.g., sub-THz communication) is degraded by oscillator phase noise; and high-throughput transmissions are limited by the quantization noise of digital-to-analogue converters (DACs). Therefore, it is required to develop techniques to compensate the impact of RF hardware impairments and enable high-throughput communication under extreme conditions.

Several solutions have been investigated to reduce the impact of hardware impairments, e.g., improved oscillator circuit design to reduce the phase noise which increases overall hardware cost and adjusting the operating point of the PA by applying power backoff which reduces the energy efficiency of the transmitter and reduces coverage. In addition, there have been techniques to linearize the transmitter, e.g., by performing digital pre-distortion (DPD). These techniques usually require an estimate of PA characteristics based on a feedback loop using a sampling receiver circuit at the transmitter side to measure the PA's output and down convert it to a baseband signal for signal processing to estimate PA's characteristics. The DPD performs pre-distortion of the signal based on the estimated PA's characteristics so that the overall RF-chain resembles approximately linear characteristics. The DPD processing is overall complex and energy consuming task, especially for wide-band signal transmissions.

Artificial intelligence techniques enable a new design paradigm, where instead of mitigating RF hardware impairments or trying to linearize the transmitter, a receiver can be trained to learn signal detection in the presence of RF hardware impairments. An AI-empowered receiver based on artificial neural networks (ANN) is proposed in [27]. The radio transceiver architecture with AI-empowered receiver is outlined in Figure 5.19. The AI-empowered receiver learns the optimized demapper to compute soft bits as input to the soft decoder (low-density parity-check (LDPC)) in the presence of RF hardware impairments (e.g., phase noise or PA non-linearity). The demapper can be implemented using a fully connected NN with real

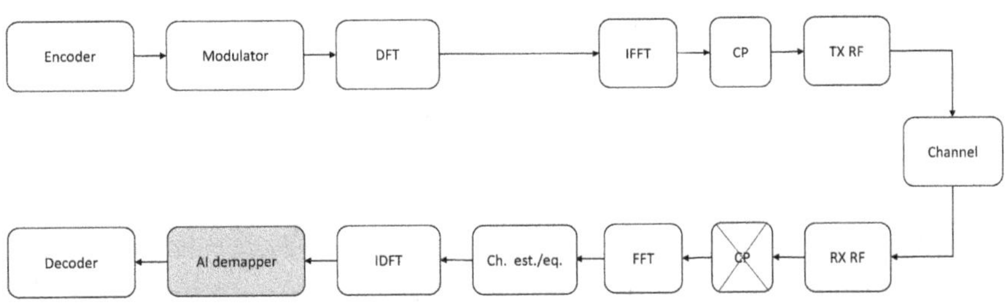

Figure 5.19. The radio transceiver architecture with AI-empowered receiver.

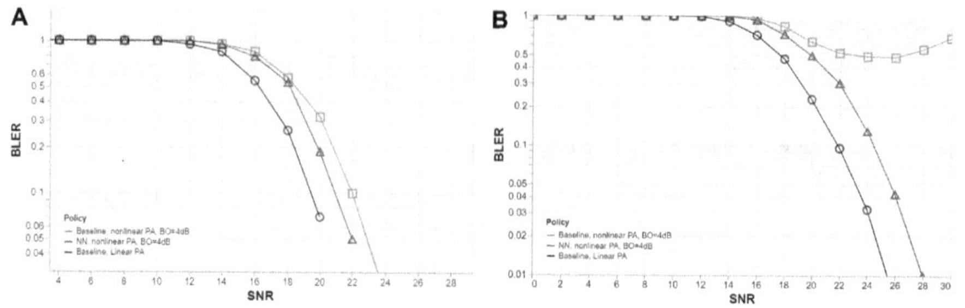

Figure 5.20. Performance of the AI-empowered receiver.

and imaginary parts of the equalized received signal and signal-to-noise ratio (SNR) estimate as inputs and the log likelihood ratio (LLR) values corresponding to the transmitted information bits as outputs.

The simulation results for the block error rate (BLER) performance of the AI-empowered receiver with a demapper composed of NN with five hidden layers each with 64 neurons are shown in Figure 5.20. For 64QAM-modulated signals with modulation and coding scheme (MCS) index 19, the AI-empowered receiver achieves roughly 1 dB performance gain compared to the legacy receiver method at 10% BLER operating point. For 256QAM-modulated signals, even for the lowest defined code rate in the standard (MCS index 20), the legacy receiver cannot successfully detect the signals and fails to reach 10% BLER. However, the AI-empowered receiver provides satisfactory results without any visible error-floor. This confirms that AI-empowered receiver can enable using higher-order modulations while providing signal reception with desired reliability, hence they can be used in high-throughput scenarios where higher-order modulations are desired.

The achievable throughput for the AI-empowered receiver and the legacy receiver with link adaptation at the transmitter is shown in Figure 5.21. The legacy receiver in the presence of linear PA provides an upper bound on throughput. In the presence of PA non-linearity, the legacy receiver achieves lower throughput compared to the upper bound, and the performance gap becomes larger at high SNR. The AI-empowered receiver achieves higher throughput compared with legacy receiver. The performance gain is larger at higher SNR values, and the achieved throughput saturates at the same level as the one for the upper bound on throughput. Hence, the AI-empowered receiver can partially compensate the impact of distortions due to PA non-linearity of a transmitter and can increase throughput and/or extend the coverage of a communication link.

The AI-empowered receiver can be used to relax the requirements on inband distortions (e.g., the requirements on maximum error vector magnitude, EVM) and the linearization methods to be performed at the transmitter side, e.g., DPD. The receiver method can be used in uplink communication of a cellular network,

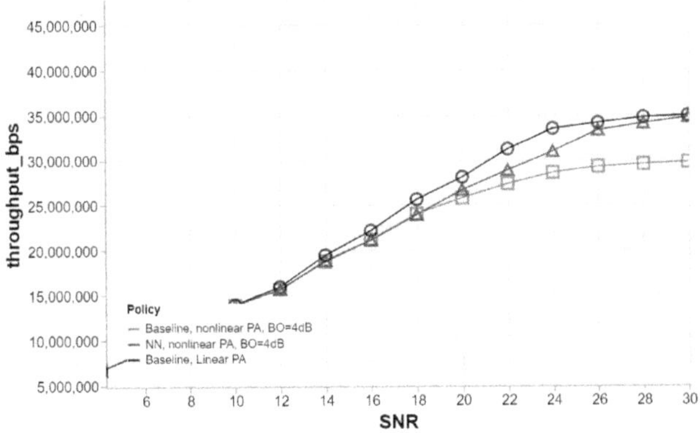

Figure 5.21. Achievable throughput with adaptive modulation and coding and for AI-empowered receiver.

which is usually coverage limited, and performing linearization techniques (e.g., DPD) at the UE is challenging, and it is desired to increase the energy efficiency of the UE to increase battery lifetime. Alternatively, the proposed method can be used in a downlink scenario to increase throughput and/or extend the coverage area of the base station (BS), enhancing the energy efficiency of BSs.

5.6.2 DeepRx: A Fully Learned Air Interface Receiver

A key component of the air interface is the receiver. Traditionally, the receiver has many processing blocks, including raw channel estimation from the reference signals, smoothing and interpolation, equalization, demapping, and decoding. The traditional receiver therefore splits the processing of the channel estimate (via known and received reference signals) from the processing of the rest of the signal and only combines them in the equalization step. This means that the receiver's performance is limited by the accuracy of the channel estimation, which is inherently limited, e.g., for high mobility UEs. This limitation can be circumvented by training the receiver algorithm from the antenna signal to the uncoded bits, as demonstrated in the DeepRx receiver shown in Figure 5.22. By also using the unknown received data, such an ML receiver learns an algorithm that can track the changes in the channel response in time and frequency, even in the case of sparse reference signals, leading to improved performance, as shown in Figure 5.23.

A similar DeepRx-type receiver can also be extended to support MIMO detection. Conventional MIMO detection algorithms contain an equalization block that relies on input-input multiplications. In order to train a NN to replace such algorithms efficiently, it is necessary to equip the NNs with the ability to represent input-input multiplications. Otherwise, the NN must learn to approximate them, which can be inefficient. The DeepRx MIMO [29] has different ways to represent

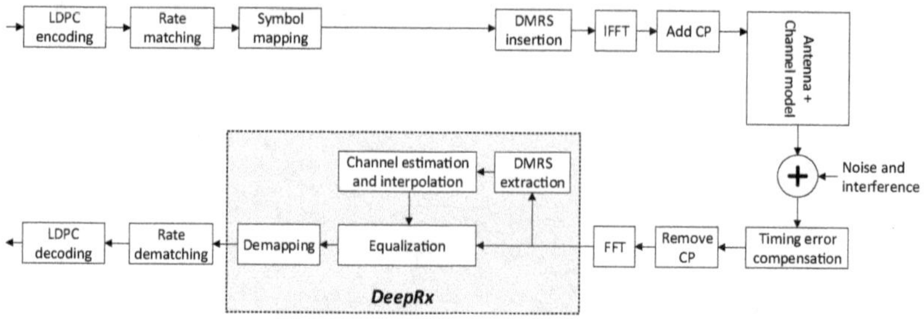

Figure 5.22. DeepRx replaces most of the processing blocks in the frequency domain receiver processing in the physical layer. The blocks inside the DeepRx box describe the traditional receiver processing blocks that DeepRx replaces with a trained neural network.

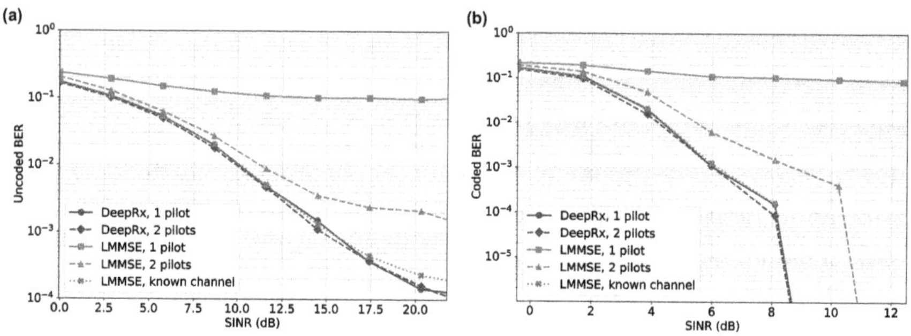

Figure 5.23. DeepRx SIMO reaches linear minimum mean square error (LMMSE) with known channel performance, even for highly mobile UEs of 0–130 km/h as in these figures.

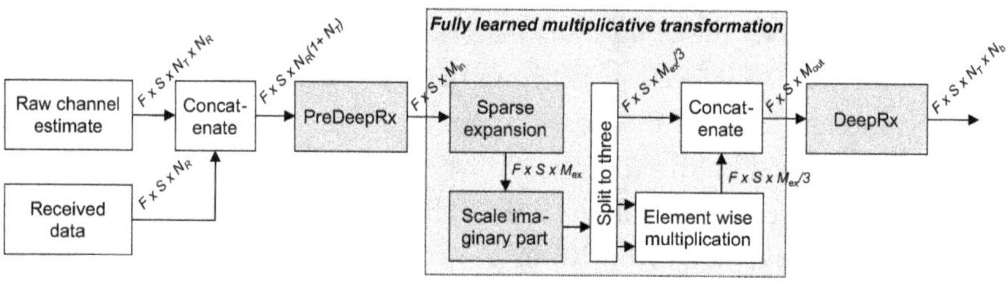

Figure 5.24. DeepRx can be extended to MIMO by adding the capability of input-input multiplication to the neural network, which allows more efficient learned equalization.

the multiplications. One example is shown in Figure 5.24, where the network learns which inputs should be conjugate multiplied and which should be multiplied. The performance results of such a DeepRx MIMO network are shown in Figure 5.25.

DeepRx shows that it is possible to learn state-of-the-art receiver algorithms based on data. Since the learned method relies on very few assumptions on the actual waveforms, it is possible to utilize this kind of receiver for other types of waveforms, e.g., ones that utilize novel constellation shapes.

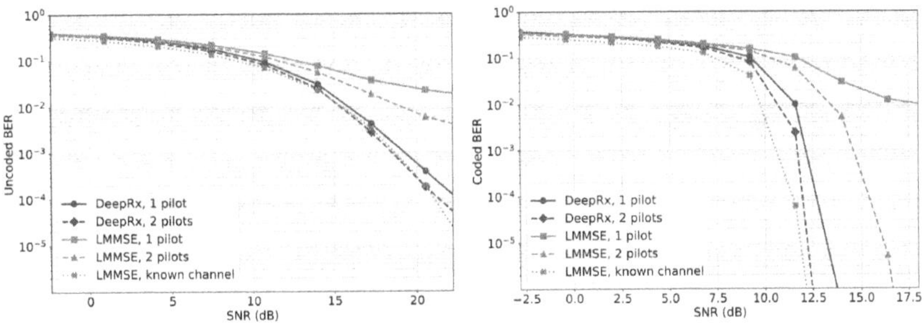

Figure 5.25. DeepRx MIMO can surpass the LMMSE receiver in TDL-E channels and reach near LMMSE with known channel performance.

5.6.3 AI-native Air Interface Design with Constellation Shaping and Hardware Impairment Mitigation

One step towards a truly AI-native air interface is to allow ML to design certain aspects of the physical layer. In practise, this means that the chosen properties of the air interface will be natively tailored for ML-based processing, e.g., on the receiver side. Ideally, this will result in enhanced spectral efficiency as the air interface is optimized with the help of ML.

In this approach to air interface design, ML is employed to learn the constellation shape and a transformation function for the time-domain waveform. These elements are learned in conjunction with an ML-based receiver, which has a similar architecture to DeepRx [30]. The benefits of such an approach are twofold. First, as opposed to 5G and earlier network generations, where the receiver must be provided so-called demodulation reference signals (DMRSs) for channel estimation and consecutive equalization, the proposed AI-based approach with learned constellation shape can operate without any pilots. This naturally removes the overhead they incur, as all the resource elements (REs) can be used to carry useful data. Second, the learned transformation of the time-domain waveform makes the whole system more resilient against PA-induced nonlinear distortion. Namely, using such a transformation, the power of out-of-band emissions can be reduced, while at the same time facilitating more accurate detection at the receiver side.

Figure 5.26 illustrates the overall approach on a conceptual level. In this example, an otherwise conventional orthogonal frequency-division multiplexing (OFDM) transmitter is assumed, with the exception that the bits are mapped to symbols using the learned constellation. In addition, as mentioned above, a convolutional neural network (CNN) is added to the transmitter's inverse fast Fourier transform (IFFT) output in order to ensure resilience against PA-induced nonlinearities.

The complete transmission link is trained E2E, which can be done with supervised learning. This requires a differentiable model between the bit source and the receiver output, including the transmitter, multipath channel, and the receiver.

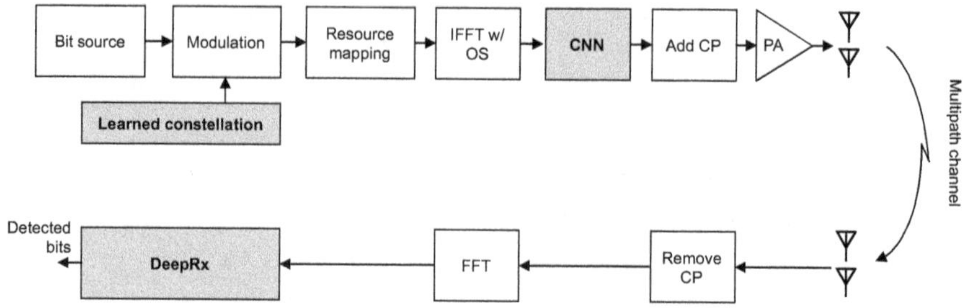

Figure 5.26. E2E learning for constellation shaping and out-of-band emission reduction.

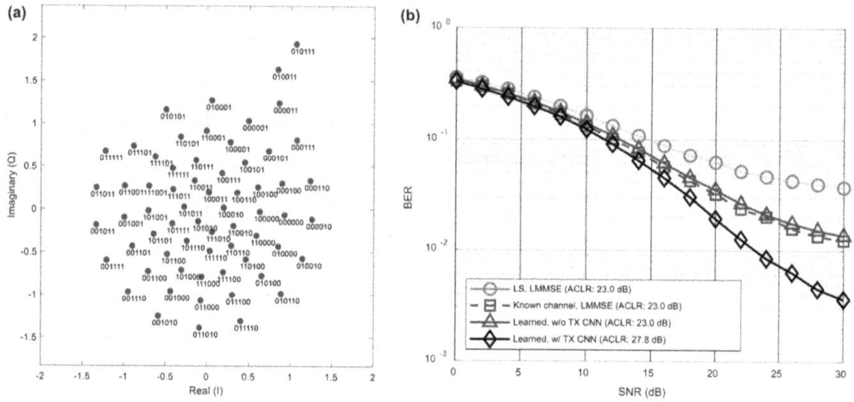

Figure 5.27. (a) An example of a learned constellation shape, and (b) the corresponding BER under a nonlinear PA.

Moreover, to ensure that a generic scheme is learned, the training is done using several randomized PA responses. Binary cross entropy between the transmitted bits and the detected bit estimates is used as the primary training loss function, which is equivalent to maximizing the throughput between the transmitter and receiver. In addition, the out-of-band emissions at the PA output are also added to the loss term, which incentivizes the transmitter to produce a waveform with as little emissions as possible. With this, the transmitter and receiver are incorporated into a single E2E model and can thereby learn jointly the constellation shape, transmission scheme for minimizing out-of-band emissions, and the required receiver-side processing for detecting the modulated bits.

Figure 5.27 shows an example of a learned constellation for a multipath scenario, as well as the achieved uncoded bit error rate (BER). The training is carried out with the TensorFlow library, where the full link is implemented as an E2E model. This also includes the channel model, which is based on precomputed channel coefficients generated under an urban micro (UMi) simulation scenario. In these results, the transmitter PA is operating near saturation. The proposed ML-based solution is compared to two baselines:

- A practical receiver, which estimates the channel from pilots using least squares and interpolates it linearly over the whole slot (extrapolation of the channel estimate beyond the last pilot symbols is done with the nearest neighbour rule). The symbol estimates are obtained with LMMSE equalization, after which the soft bits are calculated using the log maximum A-posteriori (log-MAP) rule.
- A genie-aided receiver, which has perfect channel knowledge and performs LMMSE equalization followed by log-MAP demapping.

Both benchmarks utilize a conventional quadrature amplitude modulation (QAM)-OFDM waveform with two OFDM symbols per slot (14 OFDM symbols) dedicated to pilot transmissions. It is evident from Figure 5.27(b) that the learned pilotless scheme is superior to the benchmark schemes in terms of the achievable spectral efficiency, as it achieves a lower BER with less overhead. The pilotless operation is made possible by the learned asymmetric constellation shape, shown in Figure 5.27(a), which the receiver can use for blind detection of transmitted symbols. The lower BER of the learned scheme is mostly due to the nonlinear distortion within the RX signal, which the baseline schemes are not capable of suppressing.

To investigate the impact of PA-induced nonlinearities, Figure 5.28(a) shows first the BER of the different schemes under a fixed SNR for varying PA input backoff values. It is evident that the learned system with the TXCNN achieves the lowest BER when the backoff is small (this is the most nonlinear operating point

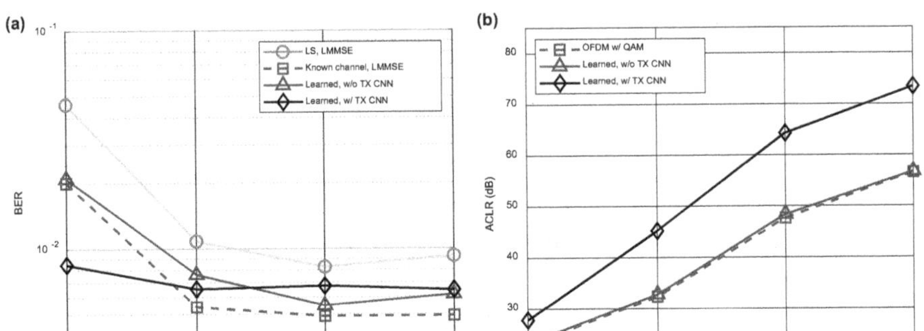

Figure 5.28. (a) BER of the different schemes with respect to the PA input backoff when the SNR is fixed at 24 dB, and (b) ACLR of the different schemes with respect to the PA input backoff. Note that here the PA backoff is defined with respect to unit variance, i.e., it represents the power of the PA input signal (not to be confused with PA backoff with respect to 1 dB compression point).

of the PA). However, when the nonlinearities are less severe, the benefit of the CNN diminishes, as expected. Figure 5.28(b) shows the adjacent channel leakage ratio (ACLR) of the learned waveform, compared to a conventional QAM OFDM waveform, measured at the PA output. Here, ACLR is defined as the ratio of the power of the in-band desired signal to the power of the out-of-band emissions. The effect of the TX CNN is again clear, as it achieves a clearly superior ACLR compared to the other schemes.

Altogether, these findings demonstrate the potential of an AI-native air interface in reducing BER, improving the spectral efficiency and reducing the signalling overhead over the current systems. Moreover, the benefits of ML-based processing when dealing with hardware impairments are also evident. The proposed approach facilitates more accurate detection of distorted signals while reducing the out-of-band emissions.

5.6.4 AI-driven Channel Estimation

Channel estimation (ChE) is among the first functions of a radio receiver that has been enhanced by AI/data-driven methods [57]. The ChE problem has been targeted by many tools stemming from AI.

For ChE problem, the minimum mean squared (MMSE) estimator is optimal in the Neyman-Pearson-Lemma sense. However, in practise, MMSE is not utilized widely, due to the following reasons: the first one is computational complexity. The MMSE requires inversion of the covariance matrix of the first input, which can grow with o(n^3). The second reason is the run-time sample complexity, i.e., the number of samples required to construct an accurate sample covariance matrix. For ChE, the processing delay that's required to accumulate pilots to build an accurate sample covariance matrix is not tolerated. Thus, an NN inspired by the mathematical representation of the MMSE estimator is proposed [57–59]. This NN consists of two layers that represent the MMSE estimator and a multiplication layer that passes the input to the estimator.

The proposed NN architecture is shallow and therefore more manageable for mathematical analysis, although it may not scale well with input size. If this NN architecture implicitly implements the MMSE, the well-known covariance matrix decomposition can be utilized, in which the large covariance matrix is represented in terms of Kronecker product smaller matrices. To handle data with spatial (vertical and horizontal), temporal, and frequency dimensions, the large input covariance matrix is decomposed into four smaller matrices. This enables the use of the same NN architecture on these four slices of data. At each stage, a slice of the 4-dimensional input vector is fed to an NN tailored for that slice of data. This has been portrayed well in Figure 5.29.

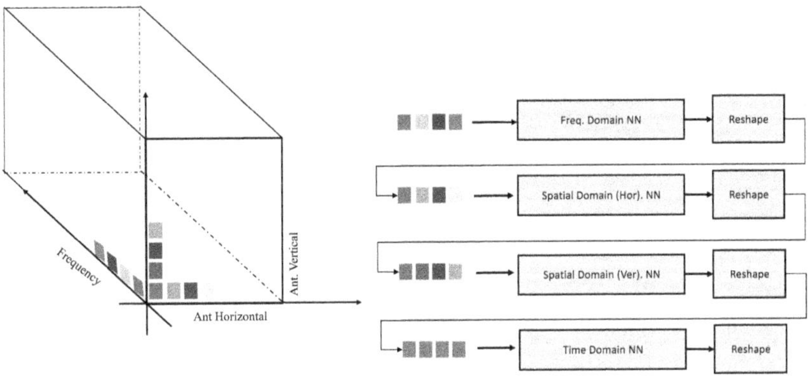

Figure 5.29. Data slices in the frequency and spatial directions of vertical and horizontal.

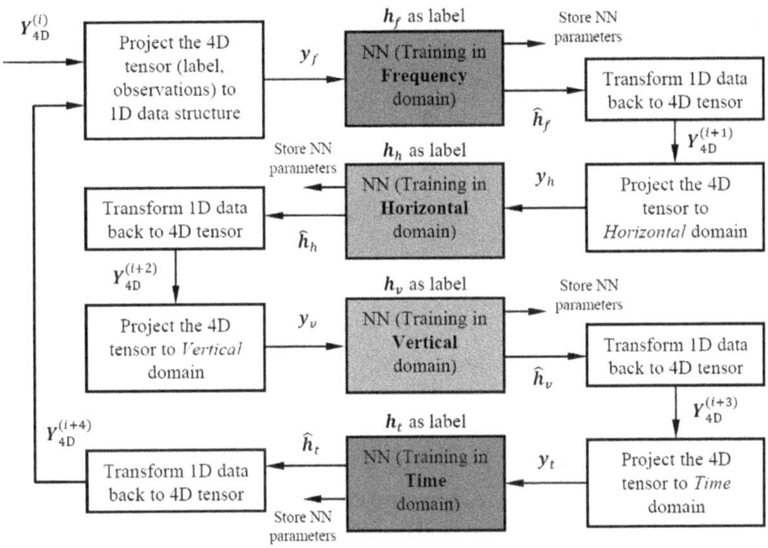

Figure 5.30. The neural network architecture consists of multiple smaller neural networks with reshaping operations in between. This design allows for processing large input vector size, without the need to large NNs.

The overall NN structure is depicted in Figure 5.30. For each of the NNs, the input data have to be reshaped to fit the NN. This reshaping acts as an additional re-shuffling, which further helps the generalization of the learned weights.

The performance of this NN is compared to that of the optimal MMSE solution. In Figure 5.31, the coloured graphs are the result of turbo-AI. The 3D turbo-AI illustrates three stages of turbo-AI, namely frequency, horizontal, and vertical, while the 4D turbo-AI benefits from all the dimensional processing. This further shows the trade-off for computational complexity versus performance in terms of normalized mean squared error (NMSE).

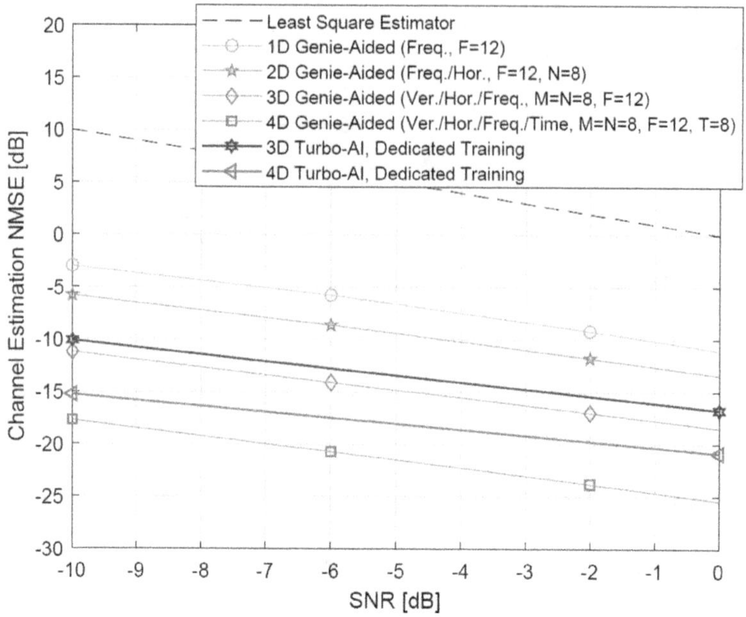

Figure 5.31. Comparison of the proposed ChE schemes, turbo-AI, with the statistically optimal solution MMSE-based estimator. The MMSE estimator for the above results included a 6144 × 6144 matrix inversion operation.

5.6.5 AI-based Sparse Channel Estimation for RIS-aided Communications Networks

A reconfigurable intelligent surface is composed of a large number of passive elements, each of which can reflect back incoming signals while incurring a phase shift. The wireless propagation environment can be reconfigured based on this property of RISs, and the phases of individual elements can be configured to optimize a certain objective function that corresponds to some aspect of system performance. However, most optimization solutions need channel state information to be known at the entity controlling the RIS (the assumption is that the RIS is controlled by the BS, where BS estimates the channel, calculates optimal phase shifts, and sends control signals to the RIS to configure the phase shifts). However, channel estimation is difficult as RIS consists of a large number of passive elements, which requires measurements at the BS with long pilot sequences due to large dimensionality. Also, it is necessary to estimate both the direct channel and the reflected channel through RIS.

The channel estimation for the reflected channel through the RIS requires the design of phase configuration of the RIS elements. It has been studied in [31], and an optimal codebook is proposed based on a minimum variance unbiased estimator. The dimensionality of the parameter space can be reduced by considering the

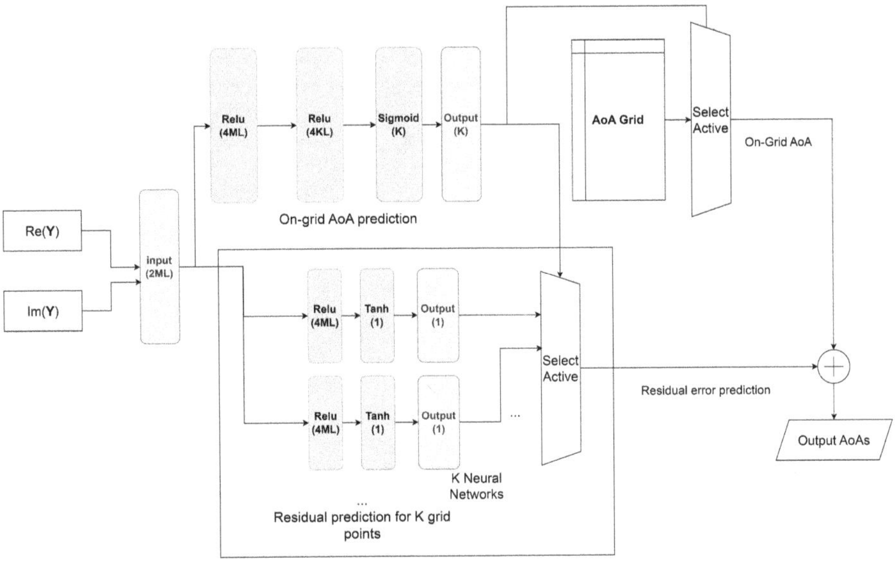

Figure 5.32. NN architecture for AoA prediction.

angular domain sparsity of the mmWave channels, resulting in improved accuracy with reduced pilot overhead. In [32], a sparse representation of the concatenated BS-RIS-user channel is derived, where channel estimation is converted into a sparse signal recovery problem. A double-structured orthogonal matching pursuit (DS-OMP)-based cascaded channel estimation scheme is proposed in [33], based on the double-structured sparsity of the angular cascaded channels. In this work, a channel estimation method based on the sparse representation of the channel is proposed, and angular parameters are estimated using a NN. The NN-based solution is a one pass method compared to iterative traditional sparse estimation techniques, while results show better accuracy compared to [31].

The uplink channel estimation of an RIS-assisted mmWave system is considered, where the direct link is assumed to be non-line-of-sight (NLoS) and the RIS has line-of-sight paths with both the user and BS. A compact representation for the RIS channel is derived, where the angle of arrivals (AoAs) is discretized. First, the case where AoAs lie exactly on the discrete grid (on-grid) is considered, and a sparse estimation method is proposed based on OMP. Also, the results are compared with NN-based predictions. Then, the case where AoAs can take any discrete value deviating from the discrete grid (off-grid) is considered, where the residual error from on-grid points is expressed separately. The NN architecture shown in Figure 5.32 is used for the prediction of AoAs, which consists of several NNs. The top NN is used to predict the on-grid AoA points, and the residual error is calculated by the NNs shown in the bottom. This work has been published in [34].

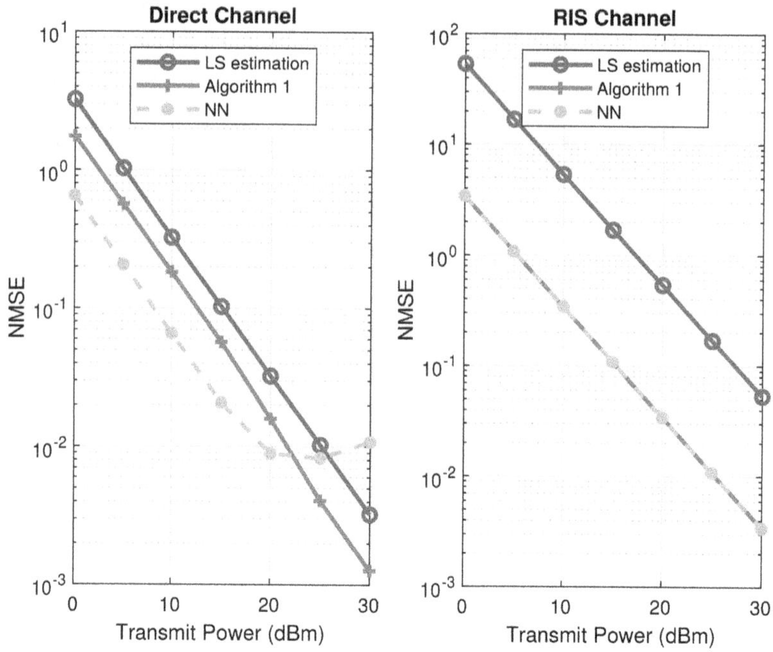

Figure 5.33. Comparison of performance of proposed methods with LS estimation for both direct and reflected channels, in the off-grid case.

Numerical simulations are performed to evaluate the results, and channel estimation error is considered as the KPI. Comparison of the performance of the proposed methods is shown in Figure 5.33, where the NMSE of direct channel and cascaded channel through RIS are considered separately. The performance of the NN is compared with least squares estimation and sparse estimation algorithms with and without perfect AoAs. Results show that NN outperforms all the other methods; however, a saturation effect is seen with increased transmit power due to leakage from grid imperfections.

5.6.6 AI-based Radio Resource Allocation for Cell-free Massive MIMO Networks

Proper radio resource management (RRM) is of much importance in improving the system performance in communication networks. However, the classical optimization or heuristic-based resource allocation algorithms have several challenges, such as high computational complexity, suboptimal solutions in complex and non-convex problems, lack of flexibility and parameter sensitivity, and inaccuracy of the model-based resource allocation methods [37]. Novel communication architectures such as cell-free massive MIMO networks and high-frequency communication systems have an increased system complexity due to the large number of antenna elements in the transceivers and the increased AP deployment density for a high-frequency radio access technology. RRM becomes more challenging in such systems due to the increased system complexity and high dimensionality of the resource allocation problems. In recent literature, the learning capability of ML algorithms is exploited to overcome the above challenges associated with conventional approaches in complex communication systems [36–40]. Most of the existing studies focus on a supervised learning approach where a model is trained to learn the mapping between the inputs (user locations or channel statistics) and the optimal outputs (power allocations) obtained by an optimization algorithm.

On the other hand, an unsupervised learning-based resource allocation approach does not require the optimal resource allocations to be known during model training as in supervised learning; hence, it alleviates the need of generating a large dataset with thousands of samples by solving the computationally complex optimization problem [38]. Thus, it makes the data preparation and model training simpler, more practical, and more flexible since the deep learning model can be easily retrained in a changing environment over time. An unsupervised learning-based resource allocation scheme for a cell-free massive MIMO network is proposed here to learn the resource allocations in a data-driven manner with lower computational complexity than an optimization-based algorithm. Specifically, joint optimization of user power allocations and fronthaul capacity allocations (between CSI and data) to maximize the network sum throughput in an uplink of a limited-fronthaul cell-free massive MIMO network is considered. A DNN is directly trained using a custom loss function to optimize the sum throughput objective instead of training with labelled data. The large-scale channel coefficients between the users and the access points are used as the DNN input.

The loss function for model training is defined as the negative value of the sum rate, which is a function of large-scale channel coefficients and the trainable parameters θ of the DNN. It is differentiable with respect to the trainable parameter set θ which allows training the model via mini-batch gradient descent method. In each iteration of the training, a set of channel realizations is generated from its distribution, and the average loss is calculated over the mini-batch. During training, the model learns parameters θ to minimize the loss that maximizes the sum rate and outputs the power allocations and fronthaul capacity allocations. Furthermore, the DNN could be used in two modes: (1) offline training mode, where the model is trained offline for a large dataset with different channel instances; and (2) online training mode, where the offline trained model is retrained in each channel instance, allowing further customization and fine-tuning of model parameters based on large-scale channel inputs in each channel realization, to further optimize the sum rate performance.

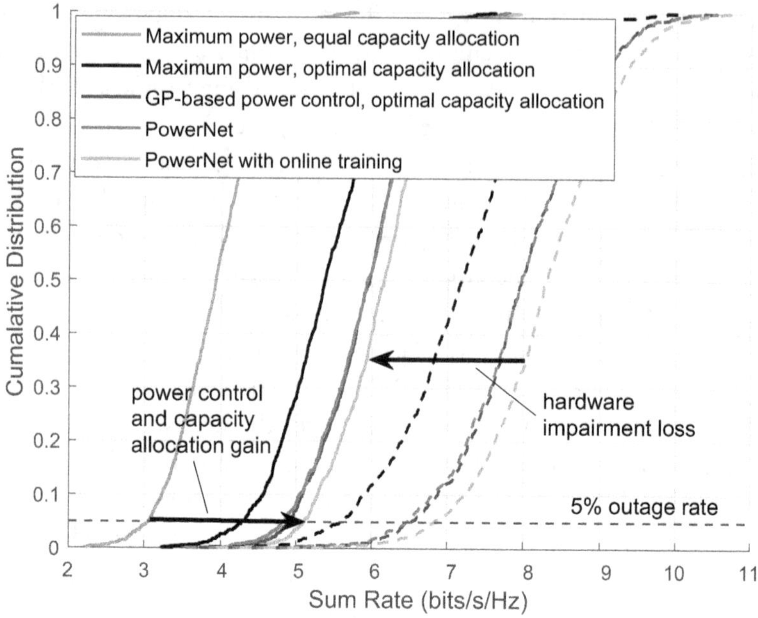

Figure 5.34. Sum rate performance comparison between proposed method (PowerNet) and baseline (optimization-based) with and without hardware impairments in the transceivers. Dashed lines: Ideal transmitters and receivers without hardware impairments. Solid lines: With hardware impairments in the transmitters and receivers.

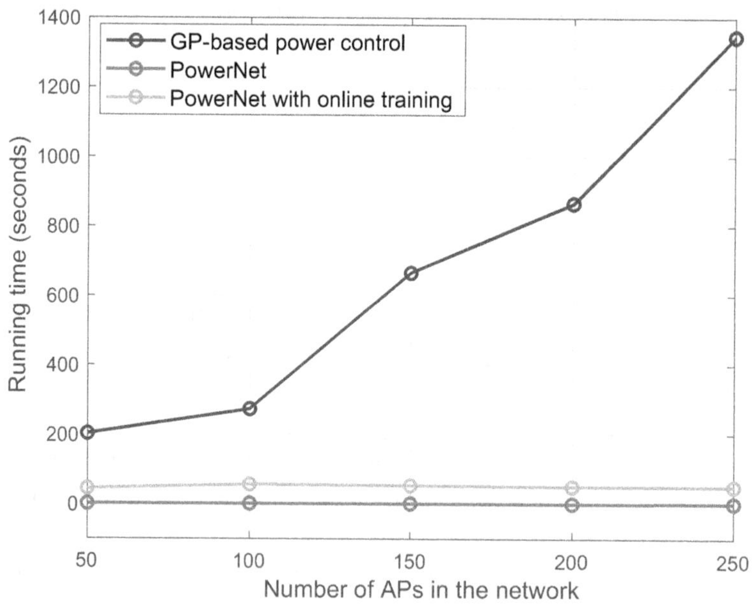

Figure 5.35. Recorded CPU timing for the CVX solver and the PowerNet to produce outputs for 100 channel realizations for different network configurations.

As seen from Figures 5.34 and 5.35, numerical simulations have shown the proposed ML approach to learn resource allocation vectors, resulting in similar sum throughput performance compared to optimization-based baseline while having lower computational complexity. The time complexity of the DNN does not drastically scale with the number of users and APs, as the optimization-based algorithm does. While the time complexity of the optimization-based algorithm exponentially scales when the number of APs or users is increasing, the time complexity of the DNN only increases slightly when the system parameters are scaled. In contrast to the iterative algorithms used in the optimization-based approach, the ML approach only requires one-shot output calculation, which only involves matrix multiplication and addition operations to produce the DNN outputs; hence, it has a significantly reduced time complexity of the DNN.

5.7 Statistical Federated Learning for Resource Provisioning

A paramount feature of 6G systems is its massive and highly heterogeneous network slicing, which enables tenants to not only target specific industries, but also bring digitalization to end users through the expected widespread adoption of novel, advanced, and diverse digital services (e.g., holographic communications, next-generation extended/virtual reality applications, tactile internet, industry 5.0). This fact will require 6G networks to manage a large number of slices, which could potentially span multiple technology domains, such as the RAN, edge, cloud, and core. This fact poses a major challenge to traditional centralized MANO methods, which are vulnerable to single points of failure and require extensive monitoring. This results in a large communications overhead and delayed, heuristic-based decisions, leading to increased energy costs and reduced scalability and sustainability for the network. To address this issue, sustainable and scalable 6G network slicing must rely on distributed AI architectures, where local computation tasks are performed closer to the monitoring points, reducing energy costs and exchanging only some local updates (e.g., the model weights or compressed/preprocessed data) or compressed statistics through federated and multi-agent learning strategies.

5.7.1 AI for SLA Management in RAN

AI techniques will be a crucial factor in the automation of resource allocation for dynamic network slicing and its associated SLA. In order to reach both scalability and sustainability in network slicing, AI-enhanced analytic algorithms must be brought closer to the distributed monitoring points throughout the network. This approach significantly reduces the raw data exchange, allowing only certain AI model parameters to be transmitted for coordination or collaboration purposes. In such a scenario, FL-based analysis is an attractive solution that enables: (i) local learning at each network element, whether virtual or physical, by exchanging only the weights of its model with the aggregation server; and (ii) addressing the absence or inadequate distribution of local datasets by capitalizing on the knowledge of other elements participating in the FL task. The challenge in this case is twofold: first, to ensure that the outputs of the models learned offline conform to a predetermined statistical distribution, such as the slice-level SLAs, when deployed in a testing scenario; and second, to ensure that these models converge when faced with non-independent and identically distributed datasets in live network environments.

This section presents a new decentralized management and analytics framework designed to handle a massive number of dynamic slices in Beyond5G and 6G scenarios, enhancing scalability, sustainability, and responsiveness of self-management and self-configuration of network slices for ZSM. The framework aligns with both ZSM [42] and enhanced network management interface (ENI) objectives [43]. Additionally, the framework is implemented in three technological domains: cloud, edge, and RAN [44].

5.7.2 Statistical FL-based Policy for RAN

To demonstrate the scalability and ZTM capabilities of the aforementioned architecture, consider the deployment solution illustrated in Figure 5.36. It shows a disaggregated RAN architecture with a central unit (CU)-distributed unit (DU) functional split, in which each transmission/reception point (TRP) is co-located with its DU, which is connected to the corresponding CU using a fronthaul link. Each CU runs a VNF at the edge on top of commodity hardware and includes a co-located MS and an AI-enhanced AE, which are instantiated per slice. For each central unit i and slice j, the MS (i, j) performs the data collection to build a local (small) dataset D_(i,j) of size Di,j, consisting of a set of input features vector metrics x_(i,j)^((n)) (e.g., over-the-top traffic patterns, average channel quality indicator (CQI), average number of active users, etc.) and the corresponding outputs y_(i,j)^((n)) (e.g., physical resource blocks (PRBs) occupancy, CPU load, etc.). These local datasets are generally insufficient to train precise analytical models. As a result, local AEs participate in a FL task, where an E2E slice-level AE acts as an aggregation server. Note that TRPs are often deployed in areas with varying traffic patterns, both spatially and temporally, which are strongly affected by the user distribution and depend on the context in which the TRPs are deployed (e.g., residential areas, commercial districts, and entertaining events such as concerts or football matches). Furthermore,

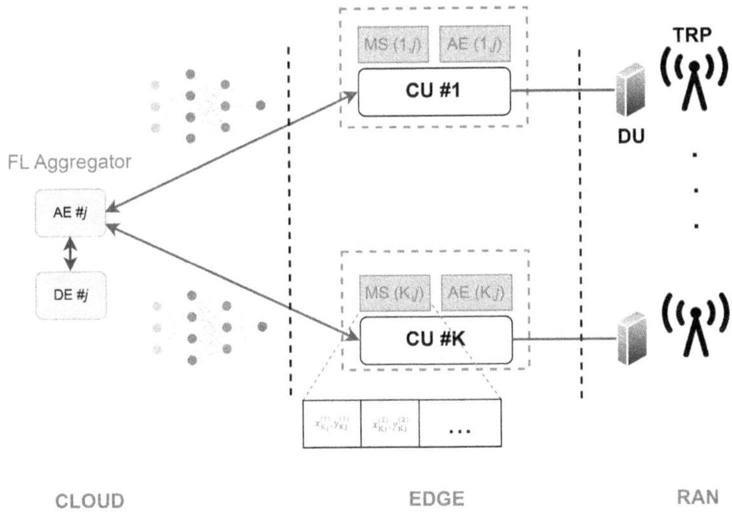

Figure 5.36. Cross-domain statistical FL-based with decentralized MS/AE.

the radio performance indicators are associated with time-dependent channel conditions. As a result of these aforementioned aspects, the local mini-data datasets are inherently non-independent and identically distributed.

The benefits of the proposed management framework can be demonstrated through a use case of slice-level resource prediction with SLA constraints [45]. The main objective is to respect the SLA violation rate for each slice while minimizing the total communication management overhead. A typical SLA is established between the tenant of slice j and the infrastructure provider, to ensure that the utilization of slice resources (e.g., CPU load or number of occupied PRBs) does not exceed the range $[\alpha_j, \beta_j]$ with a probability given by a specified upper-bound threshold δ_j. These SLAs can be defined in terms of percentiles and the empirical cumulative density function (ECDF) constraints and solved via a proxy-Lagrangian and two-player game [45]. In practise, the infrastructure provider and the tenant of slice j may agree that the Q-th percentile of utilization a specific resource must be lower than some given agreed bound to ensure isolation, where the Q-th percentile refers to the value below which *Q%* of the samples of this resource are distributed.

In order to prevent the exchange of raw monitoring data with the cloud domain, each AE performs SLA-constrained resource prediction locally. This involves learning an AI-based resource provisioning model, under long-term SLA constraints, considering space-time-varying input features such as slice traffic and radio conditions, which depend on the CU locations at the RAN and slice type. Given that the local MS datasets are not exhaustive, the local AEs participate in a FL task to enhance their predictions, acting as local clients. Only the j-th slice model weights and SLA violation rates achieved by the local models are communicated to the E2E

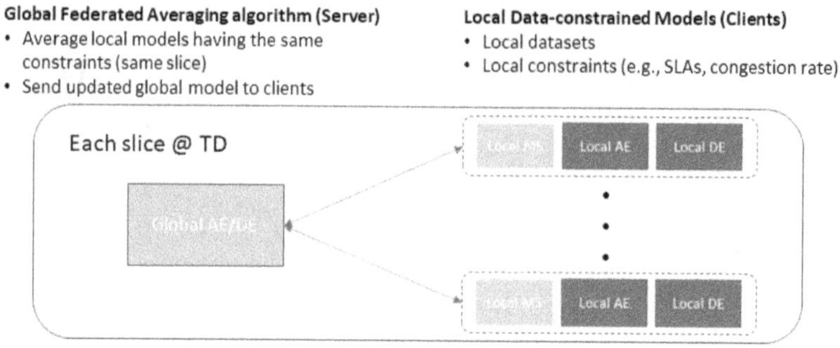

Figure 5.37. FL-based decentralized AEs for slice n.

slice-level AE, located at the cloud, which plays the role a FL aggregator. At each FL round, the E2E AE averages the received model weights and broadcasts the result to the local AEs that update their corresponding local resource provisioning models. The whole process is illustrated in Figure 5.37.

The algorithmic statistical FL (StFL) architecture and algorithmic framework for decentralized resource allocation have been presented in [46] and [45], respectively. As shown therein, the StFL solution enables SLA enforcement, while significantly reducing the overall network overhead, when compared to the centralized counterpart or with FedAvg algorithm [46]. Furthermore, as analysed in [44], the StFL algorithm achieves more than $\times 10$ times energy efficiency, paving the way for 6G sustainable massive slicing networks. To further reduce the communications overhead, minimize the computation and management of data, and ensure scalability under massive slicing, a subset of the active AEs can be selected in each FL round. Towards this goal, a stochastic policy has been proposed in [47] and [48] to select a subset of m clients out of the total set of K (where $m < K$) local AEs to take part in the FL optimization, based on their received SLA metrics. This approach uses these metrics to compute the violation rate probability distribution over the local AEs based on a softmin activation layer, and a stochastic policy is considered to randomly select the clients in the next round of the FL training based on the violation rate distribution. As a result, AE that reach lower SLA violation rates can participate in the FL round with a higher probability. In practise, the whole process can be summarized as follows: during each FL round t, the E2E DE selects stochastically m local AEs to participate in the FL task, based on the violation rate probability distribution, which is obtained using the softmin activation layer. Then, the E2E DE provides feedback in the form of an activation bit to inform the participating AEs. In this scenario, only the AEs selected would participate in the training process and transmit their weights to the FL aggregator that will average them. Finally, the resulting global model is broadcast to all the local AEs involved in the training. This procedure is repeated until convergence.

5.8 Network Slicing-Driven by Deep Reinforcement Learning

Network slicing enables multiple virtual networks to be instantiated and customized to meet heterogeneous use-case requirements over 5G and beyond network deployments. However, most of the solutions available today face scalability issues when considering many slices, due to centralized controllers requiring a holistic view of the resource availability and consumption over different networking domains. To address this challenge, a hierarchical architecture is designed to manage network slice resources in a federated manner [49]. Motivated by the advancement in DRL schemes and the open RAN paradigm, a set of traffic-aware local decision agents (DAs) is proposed to be dynamically placed in the RAN. These federated decision entities tailor their resource allocation policy according to the long-term dynamics of the underlying traffic, defining specialized clusters that enable faster training and communication overhead reduction.

Emerging use cases such as vehicle-to-everything (V2X) communication, the Internet of Things (IoT), and augmented/virtual reality (AR/VR) are some of the examples of 5G/6G use cases that need to co-exist on a shared physical infrastructure. However, the diverse demands for bandwidth, latency, and reliability increase the need for effective orchestration solutions to manage these services effectively and efficiently. Network slicing holds great potential as a solution to this challenging scenario. Network slicing creates an all-encompassing environment to underpin a plethora of network services by running fully or partly isolated logical networks, namely slices, on the same physical infrastructure. The zero-touch network is conceived as a next generation of network management that leverages the principles of NFV and SDN to be the cornerstone for supporting fully automated operations and on-demand configuration without the need for fixed contractual agreements and manual intervention. To handle these radical changes, the ZSM framework reference architecture has been designed by the ETSI [50]. It is designed to support fully automated network and service management. The architecture supports a set of architectural design principles, including:

- Modularity aspects for creating self-contained and loosely coupled services to prevent monoliths and tight coupling.
- Extensibility enables the network to extend new services and service capabilities.
- Scalability fulfils increasing or decreasing demands to deploy managed entities, and modules can be independently scaled.
- Resilience aspects cope with the degradation of the infrastructure and other management services as well as simplicity makes minimal complexity while still meeting the functional and non-functional requirements.

The modular characteristic is paired with intent-based interfaces, closed-loop operations, and AI/ML techniques to empower the full automation of the management operations. One of the essential building blocks in the ZSM is intelligent decision-making elements.

The fluctuating nature of traffic demand dramatically complicates the process of resource planning and allocation, particularly in the RAN domain. Resource allocation decisions, such as bandwidth allocation, must consider the added variability of the wireless channel and mobile users. Traditional RAN slicing solutions envision a centralized controller with a comprehensive, RT view of the network and its resources. However, this approach needs to work on scalability issues in actual implementations due to the large volume of monitoring information that must be exchanged and the high number of BSs. As a result, it becomes challenging to implement optimal resource allocation strategies in a timely and efficient manner.

Despite the innovative approach, it is still necessary to determine an effective way of managing slicing scenarios with many vertical services. In this regard, a hierarchical architecture is proposed for network slice resource orchestration. The framework considers the variable distribution of mobile traffic demands and sets up a network of local decision agents as virtual software instances within the NRT RAN intelligent controller (RIC). These DAs can access local RAN monitoring information and extract local knowledge without relying on a centralized entity for decision-making.

The framework utilizes a dynamic agent selection mechanism based on local traffic conditions similarity, allowing for more efficient information exchange and collaboration among local DAs. The benefits of this approach include: (i) resource allocation at the edge of the network, leading to more timely and accurate information; (ii) reduced control information that needs to cross the network to reach the central controller, reducing overhead towards the core network; and (iii) provisioning of FL schemes that enhance the capabilities of DAs. DAs will not only learn from a local observation space but also from information from other RAN nodes, thus improving the generalization of the learning process. The main innovation of this approach is based on utilizing the distributed RAN information to create a new type of specialized agents that work together in homogeneous clusters through a federation layer. This results in a scalable and stable decision-making process under rapidly changing traffic conditions. The proposed framework is also compatible with O-RAN.

5.8.1 Framework Overview

This solution is based on the slicing concept in mobile networks, where multiple network tenants share a portion, namely slice, of the common mobile network infrastructure. Each slice has dedicated networking resources to meet its service level agreement (SLA). The focus is on the RAN domain, where the SLAs are expressed in terms of maximum slice throughput and transmission latency. The term transmission latency refers to the average time that traffic belonging to a slice must wait within the BS transmission buffers before being processed due to inter-slice scheduling procedures.

The mobile network infrastructure consists of a set of BSs with slices deployed on them. Each BS has a capacity expressed regarding PRBs of fixed bandwidth. The available PRBs must be divided among the slices according to their RT traffic demand and SLA requirements. It is assumed that each network slice has predefined latency and throughput requirements as part of the SLA between the network operator and the slice owner. The focus is on the RAN domain, and the latency is considered as the queuing delay experienced by traffic while passing through the scheduling processes of each BS. The system operates in time slots, which represent a decision interval. The PRB allocation decisions can only be made at the beginning of each decision interval, with an interval determined by the infrastructure provider ranging from a few seconds to several minutes.

The allocation of radio resources to end users is considered a two-step process, assuming the existence of a preliminary admission and control mechanism. First, once network slices are admitted into the system, the infrastructure provider schedules the allocation of radio resource slots for each tenant. Then, based on the slice's resource availability, each tenant may enforce its own scheduling solutions for its end users, depending on their use case or business requirements.

An F-DRL-based architecture is leveraged to handle the RAN slicing scenario. This architecture involves the local DAs operating as software instances within each BS, as depicted in Figure 5.38. The agents are responsible for enforcing slice PRB allocation decisions based on the local monitoring information received from the network MS. However, the distributed nature of RAN deployments and the varying spatiotemporal behaviour of mobile traffic traces make it challenging for an agent trained on complex and multi-variate monitoring metrics. To address the aforementioned challenges, an FL layer has been introduced to enable inter-agent communication and expedite the learning process through knowledge sharing.

The quest for intelligent and optimal control in massive telecommunication environments has aroused intensive research on the applications of DRL methods. The DRL can provide a promising technique to be incorporated into network slicing and solve the control and optimization issues. DRL combines RL and DNN to extract knowledge based on experience gained by interacting with the environment (network slicing). To solve slice issues such as admission control, congestion control, energy efficiency, resource management, service creation, dynamic network configuration, anomaly and fault detection, security, and reliability, the agent of network slicing should update the Q-function for optimal actions. Deep Q-learning

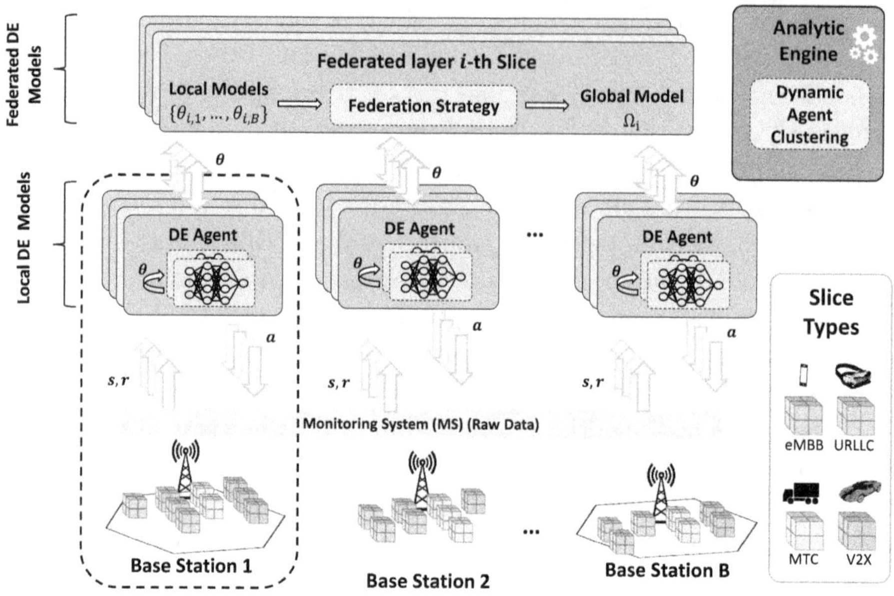

Figure 5.38. Generic federated DRL architecture for RAN slicing [49].

benefits from a NN to approximate the Q-value function. The state can be considered as input, and the Q-value of all possible actions is generated as the output. An agent (NN) in DRL generates the dataset on-the-fly, continuously interacts with a network, and is well-suited to problems with numerous possible states with high dimensions. This model-free method stores the experiences of each interaction with the environment in a replay buffer, which enables the network parameters to be updated without any prior knowledge of the environment's statistics. Figure 5.39 illustrates the local double DQN (DDQN) agent workflow.

5.8.2 Federated DRL for RAN Slicing

FL enables the training of machine learning models across decentralized entities that only have limited data access. Unlike multi-agent RL, which requires a centralized data source for all agents to observe, select actions, and receive rewards, FL allows for a collaborative learning process by aggregating multiple model updates. It results in a shared prediction model that is more refined, improves the learning rate, protects privacy, and offers better generalization. In the F-DRL-based framework, each agent trains a local DDQN model and shares its experience in the form of hyperparameters with other entities in the same federation layer. This iterative process leads to a global updated model that is stored in a cloud platform or an edge platform for faster feedback. To enhance efficiency and minimize communication overhead, the federation layer collects and shares the local models only every

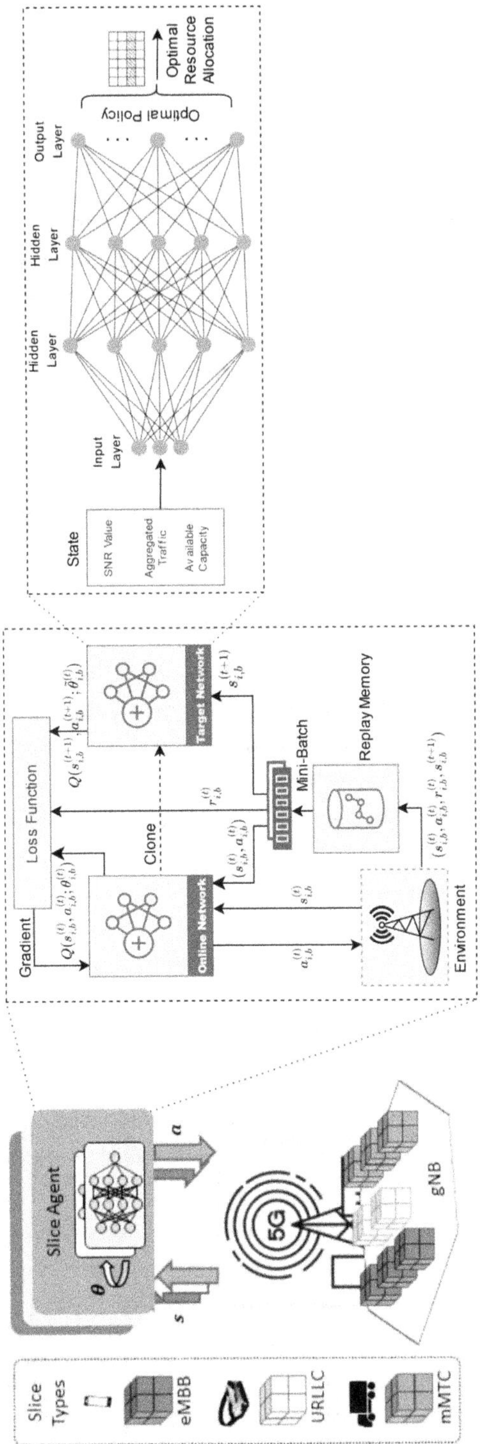

Figure 5.39 An illustration of local DDQN workflow.

T decision interval, referred to as a "federation episode." Different strategies can be applied to derive the global federated model.

Mobile traffic demands follow repetitive patterns linked to human activities and have a spatiotemporal nature. Characterizing these patterns can lead to improved network utilization forecasting and more effective resource allocation planning. However, more than simply relying on the geographical locations and proximity of BSs is needed to accurately understand traffic demands, as land usage may vary even within BS in the same area. It presents a challenge in this framework, as not all federated agents should exchange information, and it should not be limited to just nearby entities. To address this issue, a clustering algorithm is proposed that utilizes network monitoring traces and their similarity to guide the formation of DA subsets.

5.9 Analytics Engine and Interpretable Anomaly Detection

Network slicing will remain a fundamental concept in 6G networks, with an even more increased scale. While slicing increases the flexibility of the system in order to support different types of deployments and services, it also increases the complexity of the management solutions manyfold. Current standardization directions have focused on centralized management of such virtualized systems, such a solution is becoming extremely complex and hard to scale to the number of slices that are envisioned. Rather than a centralized solution, as it is considered by both ETSI NFV [51] and 3GPP management architectures [52], a solution based on the three decentralized components, namely MS, AE, and DE, offers many advantages, as described in the first section. These components are deployed on the different entities taking part in the management process, i.e., MANO, OSS/BSS, NSMF, and the in-slice management plane, distributing the management functions among these entities, in order to ensure a scalable management system.

The AE is designed to analyse the status of network slices using telemetry data collected by the MS, predict slice KPIs, and detect anomalies in KPI trends. The AE operates per slice, supporting the distributed architecture defined above. Its outputs, i.e., predictions of slice KPIs or anomaly results, are then reported to DE as key indicators to learn from and infer actionable decisions to maintain and optimize the slice performance as defined in SLAs. AE exploits the proposed distributed and scalable MANO architecture in Figure 5.40 to push the analysis close to the data collection MS in each domain (i.e., RAN, edge, and cloud), minimizing the need to transfer raw slice performance and configuration data across the different network domains and slices. The E2E slice KPIs to be predicted by the AE

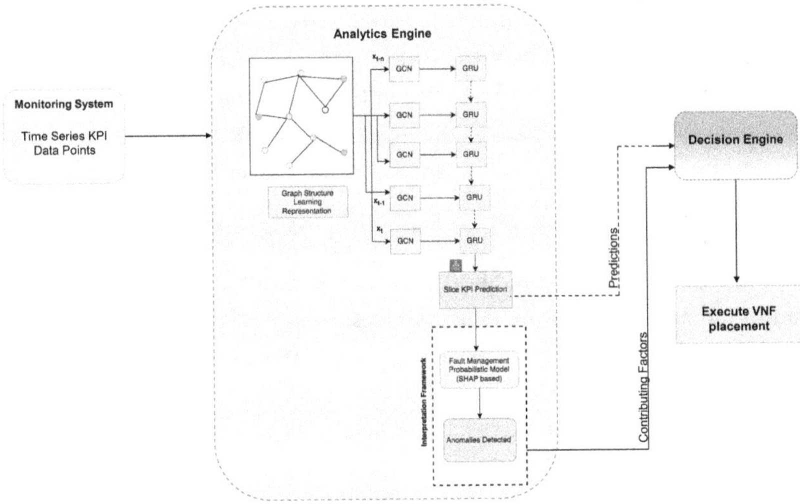

Figure 5.40. Proposed graph-based interpretable anomaly detection (G5IAD) framework.

have been defined in line with standards specifications [53] and reflect the slice performance, including, for example, uplink/downlink throughput for network slice instance (NSI), average E2E uplink/downlink delay, virtualized resource utilization per NSI, etc. Compared to KPI prediction, focus is placed on considering the unique characteristics of 5G and 6G systems when implementing KPI prediction for a slice level AI engine, as opposed to previous systems.

5.9.1 Fault Management Probabilistic Model

Multivariate time-series analysis is an important model that simultaneously analyses multiple measurements to study the behaviour of time-dependent data, and forecasts future values depending on the history of variations in the data. Figure 5.40 illustrates the AE workflow, which incorporates the input data points from MS and is further trained using graph theory to produce the slice KPI predictions and perform fault detection on predictions. In the following, the AE and design of an interpretation framework are introduced, as shown in Figure 5.40.

Graph neural networks model the relationship between a set of objects (nodes or vertices V) and their connections/interrelationships (given by a set of edges E linking the respective nodes/vertices) in non-Euclidean space (data points, which may or may not have the same underlying domain, i.e., same graph structure defined by an adjacency matrix). A combination of graph CNN and recurrent connections is proposed, which uses multiple time series as input to the network. This approach helps in capturing both the spatial and temporal relationships among the network attributes in a slice. The work in this proposed framework adopts the hybrid graph theory and recurrent neural network-based architecture to learn meaningful network information in an unsupervised manner. To discover hidden associations among network nodes (representing the input features), a graph learning layer computes the graph adjacency matrix, which is later used as an input to all graph convolution modules. The graph learning layer learns a graph adjacency matrix that captures the hidden relationships among the multitude of network slice KPIs. This is then fed into the gated recurrent units to capture the temporal dependencies. The advantages of both of these methods are drawn upon to model both temporal and intra-feature dependencies.

The model receives time series input slice KPIs as input, along with time sequences of slice KPI values for the KPI to be predicted (or analysed for anomalies). The sequences for the input model, as shown in Figure 5.40, are passed through the recurrent neural network GRU layers, while the correlation matrices are processed by graph convolutional networks (GCN). Here, the initial node features (time series data points) are provided as an input to the GCN, and then, the node embeddings are computed by applying the series of convolutional modules. The proposed method utilizes historical time series data as input and employs a graph CNN to capture the topological structure of the network, thereby obtaining relational dependencies. Then, the time series data with relational dependencies are input into gated recurrent units to capture the temporal features. The final results are obtained through a fully connected layer, which predicts the slice KPIs.

Recurrent neural networks and hidden Markov model are used to estimate the probability of a sequence (e.g., sliceLatency KPI) occurring using these probability distributions. Note that $p(x_i|x_{1:\,i-1})$ is the probability of the integer x_i occurring after the sequence $x_{1:\,i-1}$. A language model for sequences specifies a probability distribution for the next in a sequence, given the set of previous sequences. Using a training set of well-known normal sequences, the probabilistic NN is then trained to generate this probability distribution, as shown in Figure 5.41. Given a set of

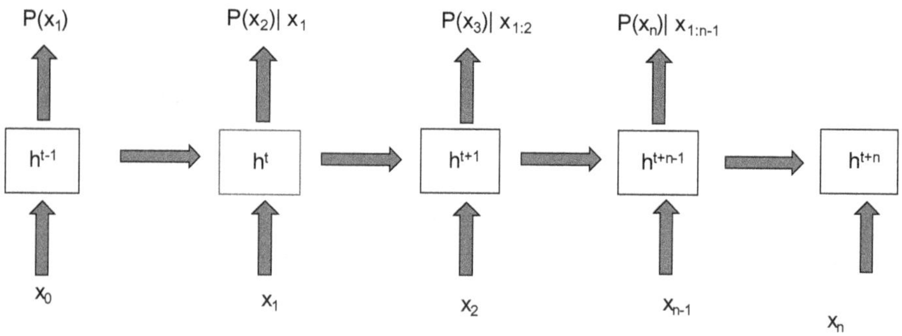

Figure 5.41. Probabilistic classifier for predicting anomalous sequences.

normal sequences, the anomaly detection algorithm evaluates the test instances as anomalous or normal. As the algorithm implemented is cost efficient (reduced training and testing times), it consumes much less computational resources in comparison to the traditional DNN algorithms.

5.9.2 Interpretation Framework

Most DNNs are "black box" that cannot provide easily interpretable insights into the relationship between input and output. Particularly when there is high-dimensional data and multiple layers in the NN, there is a need for a method (post-hoc, after training) that can be used to provide the interpretability of models on the datasets where the ground truth of interpretation results is not available. To overcome this, a relatively new technique in machine learning, known as a supports the interpretation of the NN, or any complex machine learning model, by determining how input features contribute to the value of output features. The SHAP framework unifies methods such as LIME and DeepLIFT [54] under the class of additive feature attribution methods. Reference [55] demonstrates the game theory-based SHAP framework, and the authors in [56] explained how Shapley values could be used for explaining the anomalies detected by a trained NN-based model.

An interpretation framework is provided to enable better insight and automation for the AE, as it is based on deep learning. The DE can use this information to make appropriate decisions on the next steps to take, such as reconfiguring the slice based on the features that have the most impact on the KPI that needs to be brought back into an acceptable range.

The deep explainer SHAP function takes as input the model, an instance of anomalous input x, and a set of background instances. For a particular feature x'i, it calculates a set of SHAP values that measure the importance of each of the features $x_1, x_2, \ldots, x_n$ in predicting x'i. This local interpretability can be represented graphically by using force plots, decision plots which enable us to pinpoint the SHAP value of features with respect to each other. In the anomalies interpretation procedure, the anomalous sequences are first filtered based on the probability score. Then, for each anomalous row, the top features with the lowest probability score are selected. For each feature in the list of top features, the model weights are set, and the Deep SHAP explainer is used to calculate the SHapley values. The deep explainer SHapley function outputs allFeatures, shapValue, and inputValue for anomalous KPI sequences. The shapValue corresponds to the impact of each feature on all other features in the input. The features are sorted based on the shapValue in descending order. Then, the positive shapValue features and their corresponding inputValue are stored as effect (featureEffect) and value (featureValue), respectively.

5.10 Summary and Outlook

The integration of NI into the network architecture will enable the realization of the 6G vision, which aims to provide unprecedented connectivity, capacity, and low latency. To achieve this goal, standardization work has been performed to support analytics coming from network data, and a plethora of artificial intelligence-based algorithms have been proposed. However, there is still a need for a novel architectural framework to integrate NI into major network architectures and enable specific AI-based enablers for the network operation towards 6G. In this chapter, a common architectural framework for the native integration of NI has been proposed towards the fully automated and scalable operation of 6G networks.

This novel NI framework includes an NI stratum that embraces different parts of the network architecture, from access to core, from infrastructure to management, and ensures that AI algorithms are handled with specific criteria related to data acquisition, policy enforcement, and decision making.

Also, several algorithms have been discussed in this chapter, highlighting the capabilities of such an approach based on AI. Once accomplished, AI and 6G networks will be integrated together as a single technology to meet the 6G vision.

References

[1] Hexa-X "D4.2 AI-driven communication & computation co-design: initial solutions". Accessed: April 6, 2023. [Online]. Available: https://ec.europa.eu/research/participants/documents/downloadPublic?documentIds=080166e5e1787205&appId=PPGMS.

[2] Ivan Paez, Luca Cominardi, Miguel Camelo, Paola Soto-Arenas, Nina Slamnik-Krijestorac, Marco Fiore, Marco Gramaglia, Albert Banchs, Maria Molina, Ana Hernandez, Danny de Vleeschauwer, Chia-Yu Chang, Andres Garcia-Saavedra, Andra Lutu, Lidia Fuentes, Joaquín Ballesteros, Aspa Skalidi, Alexandros Kostopoulos, & Georgios Iosifidis. (2021). DAEMON Deliverable 2.1: Initial report on requirements analysis and state-of-the-art frameworks and toolsets. Zenodo. https://doi.org/10.5281/zenodo.5060979.

[3] DAEMON "D3.1: Initial design of real-time control and VNF intelligence mechanisms (Version 1)," 2021. Accessed: April 6, 2023. [Online]. Available: https://doi.org/10.5281/zenodo.5745433.

[4] O. Gheibi, D. Weyns, and F. Quin, "Applying machine learning in self-adaptive systems: A systematic literature review," In *ACM Transactions on Autonomous and Adaptive Systems (TAAS)*, vol. 15, no. 3, pp. 1–37, 2021. Accessed: April 4, 2023. [Online]. Available: https://doi.org/10.1145/3469440.

[5] M. Camelo, L. Cominardi, M. Gramaglia, M. Fiore, A. Garcia-Saavedra, L. Fuentes, D. De Vleeschauwer, P. Soto-Arenas, N. Slamnik-Krijestorac, J. Ballesteros, C. Y. Chang, G. Baldoni, J. M. Marquez-Barja, P. Hellinckx, and Latré, "Requirements and Specifications for the Orchestration of Network Intelligence in 6G," In *2022 IEEE 19th Annual Consumer Communications & Networking Conference (CCNC)*, Las Vegas, NV, USA, pp. 1–9, 2022. Accessed: April 4, 2023. [Online]. Available: https://doi.org/10.1109/CCNC49033.2022.9700729.

[6] S. Kukliński, L. Tomaszewski, R. Kolakowski, A. M, Bosneag, A. Chawla, A. Ksentini, S. Ben Saad, X. Zhao, L. A. Garrido, A. Dalgkitsis, B. Bakhshi, and E. Zeydan, "AI-driven predictive and scalable management and orchestration of network slices," In *ITU Journal on Future and Evolving Technologies – Integrated and autonomous network management and control for 6G time-critical application*, vol. 3, no. 3, pp. 570–588, 16 November 2022. Accessed: April 6, 2023. [Online]. Available: https://doi.org/10.52953/IPUI5221.

[7] *Management of network slicing in mobile networks; Concepts, use cases and requirements*, Technical Specification (TS) 28.530, v17.1.0, 3GPP, Apr. 2021.

[8] C.Y. Chan, N. Nikaein, O. Arouk, K. Katsalis, A. Ksentini, T. Turletti, and K. Samdanis, "Slice Orchestration for Multi-Service Disaggregated Ultra-Dense RANs.", In *IEEE Communications Magazine*, vol. 56, no. 8, pp. 70–77, 2018. https://doi.org/10.1109/MCOM.2018.1701044.

[9] MonB5G, "D2.4, Final release of the MonB5G architecture (including security)," 2019. Accessed: April 6, 2023. [Online]. To appear in: https://cordis.europa.eu/project/id/871780/results.

[10] G. Choudhury, D. Lynch, G. Thakur, and S. Tse, "Two use cases of machine learning for sdn-enabled ip/optical networks: Traffic matrix prediction and optical path performance prediction," In *Journal of Optical Communications and Networking*, vol. 10, no. 10, pp. D52–D62, 2018. Accessed: April 6, 2023. [Online]. Available: https://doi.org/10.1364/JOCN.10.000D52.

[11] R. Vilalta, R. Munoz, R. Casellas, R. Martínez, V. López, O. G. de Dios, A. Pastor, G. P. Katsikas, F. Klaedtke, P. Monti, A. Mozo, T. Zinner, H. Øverby, S. Gonzalez-Diaz, H. Lønsethagen, J. M. Pulidot, and D. King, "Teraflow: Secured autonomic traffic management for a tera of sdn flows," In *2021 Joint European Conference on Networks and Communications & 6G Summit (EuCNC/6G Summit)* pp. 377–382, 2021. Accessed: April 6, 2023. [Online]. Available: https://doi.org/10.1109/EuCNC/6GSummit51104.2021.9482469.

[12] S. Orlowski, M. Pioro, A. Tomaszewski, and R. Wessaly, "SNDlib 1.0–Survivable Network Design Library," In *Proceedings of the 3rd International Network Optimization Conference (INOC 2007)*, Spa, Belgium, April

2007. Accessed: April 6, 2023. [Online]. Available: https://doi.org/10.1002/net.20371.

[13] M. Furdek, C. Natalino, F. Lipp, D. Hock, A. Di Giglio, and M. Schiano, "Machine learning for optical network security monitoring: A practical perspective," In *Journal of Lightwave Technology*, vol. 38, no. 11, pp. 2860–2871, 2020. Accessed: April 6, 2023. [Online]. Availabe: https://doi.org/10.1109/JLT.2020.2987032.

[14] *Management of network slicing in mobile networks; Concepts, use cases and requirements*", Technical Specification (TS) 28.530, v17.1.0, 3GPP, Apr. 2021.

[15] I. Santana-Perez, R. F. da Silva, M. Rynge, E. Deelman, M. S. Pérez-Hernández, and O. Corcho, "Reproducibility of execution environments in computational science using semantics and clouds," In *Future Generation Computer Systems*, vol. 67, pp. 354–367, 2017. Accessed: April 6, 2023. [Online]. Available: https://doi.org/10.1016/j.future.2015.12.017.

[16] J. Malone, A. Brown, A. Lister, J. Ison, D. Hull, H. Parkinson, and R. Stevens, "The Software Ontology (SWO): a resource for reproducibility in biomedical data analysis, curation and digital preservation", In *Journal of Biomedical Semantics*, vol. 5, no. 25, 2014. Accessed: April 4, 2023. [Online]. Available: doi:10.1186/2041-1480-5-25.

[17] L. Carvalho, D. Garijo, C. B. Medeiros, and Y. Gil, "Semantic Software Metadata for Workflow Exploration and Evolution," In *Proceedings of the Fourteenth IEEE International Conference on eScience*, Amsterdam, The Netherlands, 2018. Accessed: April 6, 2023. [Online]. Available: https://doi.org/10.1109/eScience.2018.00132.

[18] R. Guerzoni, I. Vaishnavi, D. Pérez-Caparrós, A. Galis, F. Tusa, P. Monti, A. Sganbelluri, G. Biczók, B. Sonkoly, L. Toka, A. Ramos, J. Melian, O. Dugeon, F. Cugini, B. Martini, P. Iovanna, G. Giuliani, R. Figueiredo, and L. Murillo, "Analysis of end-to-end multi-domain management and orchestration frameworks for software defined infrastructures: an architectural survey," In *Transactions on Emerging Telecommunications Technologies*, vol. 28, no. 4, 2017. Accessed: April 6, 2023. [Online]. Available: https://doi.org/10.1002/ett.3103.

[19] K. Kostas, N. Nikaein, and A. Edmonds, "Multi-domain orchestration for NFV: Challenges and research directions," In *Proc. of IEEE International Symposium on Cyberspace and Security*, Granada, Spain, 2016. Accessed: April 6, 2023. [Online]. Available: https://doi.org/10.1109/IUCC-CSS.2016.034.

[20] A. Collet, A. Banchs, and M. Fiore, "LossLeaP: Learning to Predict for Intent-Based Networking," In *IEEE INFOCOM 2022 – IEEE Conference on Computer Communications*, pp. 2138–2147, 2022. Accessed: April 6,

2023. [Online]. Available: https://doi.org/10.1109/INFOCOM48880.2022.9796918.

[21] J. Kowalik, S. Janusz, *Parallel MIMD computation: the HEP supercomputer and its applications,* MIT Press, Cambridge, 1985.

[22] M. Forsell, J. Roivainen, V. Leppänen, and J. L. Träff, "Implementation of multioperations in thick control flow processors," In *IEEE International Parallel and Distributed Processing Symposium (IPDPS) Workshops*, 2018. Accessed: April 6, 2023. [Online]. Available: https://doi.org/10.1109/IPDPSW.2018.00121.

[23] V. Leppänen, M. Forsell, and J-M. Mäkelä, "Thick Control Flows: Introduction and Prospects", In *Proceedings of the 2011 International Conference on Parallel and Distributed Processing Techniques and Applications (PDPTA'11),* 2011.

[24] J. Keller, C. W. Keßler, and J. L. Träff, *Practical PRAM Programming,* John Wiley & Sons, 2001.

[25] J. Jaja, *Introduction to Parallel Algorithms.* Addison-Wesley, Reading, 1992.

[26] 6G BRAINS "D2.3, Multi-agent Deep Reinforcement Learning Scheme Specification," 30 September 2021. Accessed: April 6, 2023. [Online]. Available: https://doi.org/10.5281/zenodo.5786348.

[27] H. Farhadi and M. Sundberg, "Machine learning empowered context-aware receiver for high-band transmission," In *IEEE Globecom Workshops,* Taipei, Taiwan, Dec. 2020. Accessed: April 6, 2023. [Online]. Available: https://doi.org/10.1109/GCWkshps50303.2020.9367518.

[28] H. Farhadi, J. Haraldson, and M. Sundberg, "A deep learning receiver for non-linear transmitter," In *IEEE ACCESS,* vol. 11, pp. 2796–2803, Oct. 2022. Accessed: April 6, 2023. [Online]. Available: https://doi.org/10.1109/ACCESS.2023.3234501.

[29] D. Korpi, M. Honkala, J. M. J. Huttunen, and V. Starck, "DeepRx MIMO: Convolutional MIMO detection with learned multiplicative transformations," In *Proc. IEEE International Conference on Communications (ICC),* Jun. 2021. Available: https://arxiv.org/abs/2010.16283.

[30] M. Honkala, D. Korpi, and J. M. J. Huttunen, "DeepRx: Fully convolutional deep learning receiver," IEEE Transactions on Wireless Communications, vol. 20, no. 6, pp. 3925–3940, Jun. 2021. Accessed: April 6, 2023. [Online]. Available: https://doi.org/10.1109/TWC.2021.3054520.

[31] T. L. Jensen and E. De Carvalho, "An Optimal Channel Estimation Scheme for Intelligent Reflecting Surfaces Based on a Minimum Variance Unbiased Estimator," In *ICASSP 2020 – 2020 IEEE International Conference on Acoustics, Speech and Signal Processing (ICASSP),* pp. 5000–5004, 2020. Accessed: April 6, 2023. [Online]. Available: https://doi.org/10.1109/ICASSP40776.2020.9053695.

[32] P. Wang, J. Fang, H. Duan, and H. Li, "Compressed Channel Estimation for Intelligent Reflecting Surface-Assisted Millimeter Wave Systems," In *IEEE Signal Processing Letters*, vol. 27, pp. 905–909, 2020. Accessed: April 6, 2023. [Online]. Available: https://doi.org/10.1109/LSP.2020.2998357.

[33] X. Wei, D. Shen, and L. Dai, "Channel Estimation for RIS Assisted Wireless Communications – Part II: An Improved Solution Based on Double-Structured Sparsity," In *IEEE Communications Letters*, vol. 25, no. 5, pp. 1403–1407, May 2021. Accessed: April 6, 2023. [Online]. Available: https://doi.org/10.1109/LCOMM.2021.3052787.

[34] D. Dampahalage, K. B. Shashika Manosha, N. Rajatheva, and M. Latva-Aho, "Supervised Learning Based Sparse Channel Estimation For RIS Aided Communications," In *ICASSP 2022 – 2022 IEEE International Conference on Acoustics, Speech and Signal Processing (ICASSP)*, pp. 8827–8831, 2022. Accessed: April 6, 2023. [Online]. Available: https://doi.org/10.1109/ICASSP43922.2022.9746793.

[35] T. V. Chien, T. N. Canh, E. Björnson, and E. G. Larsson, "Power control in cellular massive MIMO with varying user activity: A deep learning solution," In *IEEE Transactions on Wireless Communications*, vol. 19, no. 9, pp. 5732–5748, 2020. Accessed: April 6, 2023. [Online]. Available: https://doi.org/10.1109/TWC.2020.2996368.

[36] C. D'Andrea, A. Zappone, S. Buzzi, and M. Debbah, "Uplink power control in cell-free massive MIMO via deep learning," In *2019 IEEE 8th International Workshop on Computational Advances in Multi-Sensor Adaptive Processing (CAMSAP)*, pp. 554–558, 2019. Accessed: April 6, 2023. [Online]. Available: https://doi.org/10.1109/CAMSAP45676.2019.9022520.

[37] F. Hussain, S. A. Hassan, R. Hussain, and E. Hossain, "Machine learning for resource management in cellular and IoT networks: Potentials, current solutions, and open challenges," In *IEEE Communications Surveys & Tutorials*, vol. 22, no. 2, pp. 1251–1275, Second quarter 2020. Accessed: April 6, 2023. [Online]. Available: https://doi.org/10.1109/COMST.2020.2964534.

[38] N. Rajapaksha, K. B. S. Manosha, N. Rajatheva, and M. Latva-aho, "Deep learning-based power control for cell-free massive MIMO networks," In *ICC 2021 – 2021 IEEE International Conference on Communications (ICC)*, 2021. Accessed: April 6, 2023. [Online]. Available: https://doi.org/10.1109/ICC42927.2021.9500734.

[39] L. Sanguinetti, A. Zappone, and M. Debbah, "Deep learning power allocation in massive MIMO," In *2018 52nd Asilomar Conference on Signals, Systems, and Computers*, pp. 1257–1261, 2018. Accessed: April 6, 2023. [Online]. Available: https://doi.org/10.1109/ACSSC.2018.8645343.

[40] Y. Zhao, I. G. Niemegeers, and S. H. De Groot, "Power allocation in cell-free massive MIMO: A deep learning method," In *IEEE Access*, vol. 8, pp. 87 185–87 200, 2020. Accessed: April 6, 2023. [Online]. Available: https://doi.org/10.1109/ACCESS.2020.2992629.

[41] G.A. Reina, A. Gruzdev, P. Foley, O. Perepelkina, M. Sharma, I. Davidyuk, I. Trushkin, M. Radionov, A. Mokrov, D. Agapov, J. Martin, B. Edwards, M.J. Sheller, S. Pati, P. Moorthy, W. Narayana S. Shih-han, P. Shah, and S. Bakas, "OpenFL: An open-source framework for Federated Learning", In *arXiv*, 2021. Accessed: April 6, 2023. [Online]. Available: https://doi.org/10.48550/arXiv.2105.06413.

[42] ETSI, "Zero Touch Network and Service Management (ZSM)," 2023. Accessed: April 6, 2023. [Online]. Available: https://www.etsi.org/committee/ZSM.

[43] ETSI, "Experiential Networked Intelligence (ENI)," 2023. Accessed: April 6, 2023. [Online]. Available: https://www.etsi.org/technologies/experiential-networked-intelligence.

[44] H. Chergui, L. Blanco, L. A. Garrido, K. Ramantas, S. Kukliński, A. Ksentini, and C. Verikoukis, "Zero-Touch AI-Driven Distributed Management for Energy-Efficient 6G Massive Network Slicing," In *IEEE Network*, vol. 35, no. 6, pp. 43–49, November/December 2021. Accessed: April 6, 2023. [Online]. Available: https://doi.org/10.1109/MNET.111.2100322.

[45] H. Chergui, L. Blanco, and C. Verikoukis, "Statistical Federated Learning for Beyond 5G SLA-Constrained RAN Slicing," In *IEEE Transactions on Wireless Communications*, vol. 21, no. 3, pp. 2066–2076, March 2022. Accessed: April 6, 2023. [Online]. Available: https://doi.org/10.1109/TWC.2021.3109377.

[46] H.-B. McMahan, E. Moore, D. Ramage, S. Hampson, and B. A. Y. Arcas, "Communication-Efficient Learning of Deep Networks from Decentralized Data." In *20th International Conference on Artificial Intelligence and Statistics (AISTATS'2016)*, 2016.

[47] H. Chergui, A. Ksentini, L. Blanco, and C. Verikoukis, "Toward Zero-Touch Management and Orchestration of Massive Deployment of Network Slices in 6G," In *IEEE Wireless Communications*, vol. 29, no. 1, pp. 86-93, February 2022. Accessed: April 6, 2023. [Online]. Available: https://doi.org/10.1109/MWC.009.00366.

[48] S. Roy, H. Chergui, L. Sanabria-Russo, and C. Verikoukis, "A Cloud Native SLA-Driven Stochastic Federated Learning Policy for 6G Zero-Touch Network Slicing," In *ICC 2022 – IEEE International Conference on Communications*, May 2022. Accessed: April 6, 2023. [Online]. Available: https://doi.org/10.1109/ICC45855.2022.9838376.

[49] F. Rezazadeh, L. Zanzi, F. Devoti, H. Chergui, X. Costa-Pérez, and C. Verikoukis, "On the Specialization of FDRL Agents for Scalable and Distributed 6G RAN Slicing Orchestration," In *IEEE Transactions on Vehicular Technology*, 2022. Accessed: April 6, 2023. [Online]. Available: https://doi.org/10.1109/TVT.2022.3218158.

[50] *Zero-touch Network and Service Management (ZSM)*, Reference Architecture ETSI GS ZSM 002, 2019.

[51] ETSI, "Network Functions Virtualisation (NFV)," 2023. Accessed: April 6, 2023. [Online]. Available: https://www.etsi.org/technologies/nfv.

[52] *Management and orchestration; 5G end to end Key Performance Indicators (KPI)*, Technical Specification (TS) 28.554, v18.0.1, 3GPP, March 2023.

[53] 3GPP, "System architecture for the 5G System (5GC)," 3GPP TS 29.520 v17.1.0, Dec 2020.

[54] U. Schlegel, H. Arnout, M. El-Assady, D. Oelke, and D. A. Keim, "Towards a rigorous evaluation of XAI methods on time series". In *2019 IEEE/CVF International Conference on Computer Vision Workshop (ICCVW)*, pp. 4197–4201, Oct. 2019. Accessed: April 6, 2023, [Online]. Available: https://doi.org/10.1109/ICCVW.2019.00516.

[55] L. Merrick and A. Taly, "The explanation game: Explaining machine learning models using shapley values," In *Machine Learning and Knowledge Extraction: 4th IFIP TC 5, TC 12, WG 8.4, WG 8.9, WG 12.9 International Cross-Domain Conference, CD-MAKE 2020*, Dublin, Ireland, August 25–28, 2020, Proceedings 4 (pp. 17–38). Springer International Publishing.

[56] L. Antwarg, R. M. Miller, B. Shapira, L. and Rokach, "Explaining anomalies detected by autoencoders using Shapley Additive Explanations," In *Expert systems with applications*, vol. 186, pp. 115736, 2021.

[57] D. Neumann, T. Wiese, and W. Utschick, "Learning the MMSE Channel Estimator," In *IEEE Transactions on Signal Processing*, vol. 66, no. 11, pp. 2905–2917, June 1, 2018, doi: 10.1109/TSP.2018.2799164.

[58] Y. Chen, J. Mohammadi, S. Wesemann, and T. Wild, "Turbo-AI, Part I: Iterative Machine Learning Based Channel Estimation for 2D Massive Arrays," In *2021 IEEE 93rd Vehicular Technology Conference (VTC2021-Spring)*, Helsinki, Finland, 2021, pp. 1–6, doi: 10.1109/VTC2021-Spring51267.2021.9449026.

[59] Y. Chen, J. Mohammadi, S. Wesemann, and T. Wild, "Turbo-AI, Part II: Multi-Dimensional Iterative ML-Based Channel Estimation for B5G," In *2021 IEEE 93rd Vehicular Technology Conference (VTC2021-Spring)*, Helsinki, Finland, 2021, pp. 1–5, doi: 10.1109/VTC2021-Spring51267.2021.9448950.

DOI: 10.1561/9781638282396.ch6

Chapter 6

Towards Sustainable Networks

By Agapi Mesodiakaki, Arifur Rahman, et al.[1]

6.1 Introduction

Today's society faces major challenges such as the pandemic and global warming that need to be addressed while creating innovation-led opportunities for economic prosperity and job creation in a circular, green, and digital economy. Sustainability is a *holistic concept* covering environmental, social, and economic aspects while being a key element for circular economies and thoroughly developed into the globally implemented United Nations (UN) Sustainable Development Goals (SDGs) [1].

For the 6G vision, the fundamental network design paradigm must be extended from established performance-oriented to performance- and value-oriented parameters. This leads to a new class of evaluation criteria, defined as key-value indicators (KVIs) [2], e.g., *sustainability, inclusiveness, and trustworthiness*, as detailed in Section 2.1. Sustainability, which is the main focus of this chapter, is explicitly

1. The full list of chapter authors is provided in the Contributing Authors section of the book.

considered from two perspectives: (i) 6G itself needs to be sustainable and mapped to network energy efficiency as well as material efficiency (circularity) and environmental footprint (*Sustainable 6G*) and (ii) 6G as an enabler for sustainable growth in other markets and value chains, e.g., by promoting smart transportation, manufacturing, and agriculture. For instance, connected vehicles such as driverless electric ones enabled by 6G can not only promote clean energy but also reduce emissions by optimizing traffic flow (*6G for Sustainability*).

It is also worth noticing that the proposed categorization is in accordance with the International Telecommunication Union (ITU), which organizes the Information and Communications Technology (ICT) sustainability into three orders of effects: (i) *first-order effects* that denote the life cycle impacts of goods, networks, and services (i.e., its footprint), (ii) *second-order effects* that denote the impacts in other sectors due to the use of ICT, and (iii) *other (higher-order) effects* that denote higher order effects such as those associated with behavioural changes [3]. In the proposed categorization, the *first-order effects* of 6G are mapped to *Sustainable 6G*, while *6G for sustainability* corresponds to the *second-* and, to some extent, *higher-order effects*.

Unlike technical capabilities (targeted in previous chapters), quantifying societal values such as sustainability is challenging, as it may depend on factors like background, culture, and even ideology of individuals as well as operating locations, regulatory environment, and stakeholder structures. Although the UN SDGs are considered the most comprehensive approach to determining sustainability as a whole, it is commonly agreed that these goals and indicators have to be customized to the sector in focus. For the ICT sector, ITU, Global Enabling Sustainability Initiative (GeSI), and Science-Based Targets initiative (SBTi) have jointly developed directions to reduce the ICT environmental footprint by 45% between 2015 and 2030 to decarbonize in line with the 1.5°C limit of the Paris Agreement [4].

To achieve this goal, it is of high importance to note that globally, the energy consumption from *using devices represents more than 40% of the ICT power consumption while networks and data centres shared the remaining consumption roughly equally between them* [3]. *From a carbon emission and life cycle perspective, devices again dominate and represent more than half of the overall emissions, while networks represent around 25%.* Overall, the majority of networks and data centre emissions are associated with the use stage, while devices use stage and embodied emissions require as much attention. However, this balance looks different for other impact categories, so it is important to consider the environmental impacts of the full life cycle also for networks and data centres (including aspects such as lifetime, recyclability, materials efficiency, etc.).

Taking into consideration the paramount importance of climate change as well as the KVIs and the constraints mentioned above, the following sustainability key performance indicators (KPIs) for 6G as part of the future ICT sector are identified [5, 6]:

- **Enablement effect on other sectors:** enabling reductions of emissions of more than 30% CO_2eq in 6G powered sectors of society (*6G for Sustainability*); the enablement effect refers to the CO_2 emission reduction between a baseline scenario and a scenario with an applied solution that reduces greenhouse gases (GHG) emissions. The main challenge in this case is to define a clear methodology for evaluating the "enablement" impact of ICT on other sectors, with ITU working currently towards this direction [7]. Additional challenges include that the overall effect of 6G is beyond reach as the total use of 6G with the magnitude of unknown use cases cannot be foreseen. Therefore, for the time being, the evaluation of 6G can only be scenario-based and refer to specific use cases.
- **Economic target:** total cost of ownership (TCO) reduction by more than 30% compared to current networks (*Sustainable 6G*); the TCO refers to the sum of capital expenditure (CAPEX), i.e., one-time costs, and operational expenditure (OPEX), i.e., recurring costs. To achieve this KPI, it is of high importance that in a typical mobile network today, CAPEX is ~30% and OPEX is ~70% of the TCO over a 10-year period, with the radio access network (RAN) being the biggest cost component in both CAPEX (~50%) and OPEX (~65%) [8], then followed by transport, core network, energy, and other network costs (e.g., people, network management and maintenance, etc.). A breakdown of RAN CAPEX shows that the largest cost components are site construction, spectrum, and equipment, while a breakdown of RAN OPEX shows that the largest contributors are power consumption, site rentals, and operations.
- **Enhance energy efficiency:** reducing energy consumption per bit in networks by more than 90% (*Sustainable 6G*); *energy efficiency can be expressed as the ratio between the energy consumed per hour measured (in watt/hour) and the data volumes (in bytes) sent during the same time period.* The latter definition is in accordance with the European Telecommunications Standards Institute (ETSI)'s assessment [9], which suggests the use of MWh/Tb. Alternative definitions refer to *the amount of bits successfully sent over the network divided by the total energy consumed* or equivalently to *the total network throughput divided by the total power consumption, measured in bps/W or equivalently bits/joule.*

To complement the 6G sustainability targets, it is fundamental to jointly take into account all the sustainability aspects of networking, including hardware, planning, deployment, operations, and the entire equipment life cycle.

The heterogeneity of resources and services, comprising communication, computing, control, sensing, and so on, naturally calls for an *ever-deeper end-to-end (E2E) cross-layer design and optimization*, taking into consideration the entire 6G network (core, transport/aggregation, access, and user equipment/devices) during its entire life cycle with the parallel need of *energy efficiency being an integrated network design criterion.*

6.2 Technology Enablers for Network Sustainability

To achieve the aforementioned sustainability targets, 6G networks will employ a number of key sustainability enablers, presented in this chapter, which can be divided into four categories based on the way they achieve sustainability: (i) the enablers *at the deployment level*, presented in Section 6.2.1, that include architectural innovations (disaggregated and virtualized RAN) or hardware innovations (energy-neutral devices), (ii) the enablers *at the management/orchestration level* (of algorithmic nature) that target at network operation efficiency maximization (sustainable resource allocation), presented in Section 6.2.2, (iii) the ones *at service/application layer* (application-aware networks), presented in Section 6.2.3, and (iv) the *cross-layer sustainability enablers* (sustainable radio-aware digital twin), presented in Section 6.2.4, that include innovations in two or more layers, respectively.

6.2.1 Sustainability Enablers at the Deployment Level

Sustainability enablers at the deployment level include novel technologies that have been designed in a sustainable way, such as, for instance, the use of highly efficient electronic components. The latter may refer to all the electronic layers that impact global consumption including the baseband unit (computation part) as well as the radio unit (radio frequency (RF) amplifying part). The computation part is mainly dependent on Moore's law and microchips integration, while the RF amplifying part is driven by power amplifiers technology improvements and materials, e.g., gallium instead of silicon for its good performance at high frequencies. Additional sustainability enablers *at the deployment level* that are highlighted in this section include, e.g., a novel disaggregated and virtualized RAN (vRAN) architecture enabling elastic edge computing. Other promising sustainability enablers *at the deployment level* include wireless power transfer (WPT) and energy harvesting, targeting at energy-neutral devices with advanced characteristics, i.e., devices that are powered through WPT or energy harvesting, being detailed in Section 6.2.1.2.

6.2.1.1 Sustainable virtual elastic edge computing architecture

The 5G service-based architecture (SBA) has adopted edge computing as a key paradigm and has defined the necessary interfaces (e.g., Mp2 reference point) and enabling technologies (e.g., traffic steering at edge sites through the User Plane Function, UPF) that allow edge deployments to be fully integrated with the 5G Core. Moreover, via the EDGEAPP initiative [10], a new application architecture for verticals was defined, allowing over-the-top (OTT) applications to be deployed at local area data networks (LADNs) at the network edge.

Building on the work of the [11] for the 5G Non-StandAlone (NSA) mode and the aforementioned edge computing enablers, future 6G networks intend to leverage an existing Multi-access Edge Computing (MEC) platform (e.g., StarlingX) that participates in ETSI's plug tests and being fully integrated with the 5G network functions virtualization infrastructure (NFVI) [12] by supporting coordinated resource allocation for MEC applications and 5G network functions.

This is accomplished by coordinating two different management and orchestration sub-systems (i.e., the NFV orchestrator (NFVO) and MEC orchestrator, or MEO) which interact through the Mm1 reference point, defined as part of the ETSI MEC specifications [13].

While 5G early-stage approaches adopted a common virtual machine (VM)-based technology stack for MEC and NFV, 5G/B5G approaches focus on Cloud-Native technologies (i.e., Docker Containers, Kubernetes Virtual Infrastructure Managers (VIMs), and the Helm Container Management Framework) which are widely regarded as the future of vertical application development [14] with enhanced flexibility and increased application performance.

While support for Kubernetes VIMs is gradually emerging in MEC platforms (e.g., in StarlingX) and is discussed in [15], *there is currently a gap in supporting disaggregated Cloud-Native apps*. To fill this gap, extensions to the MEO are proposed to support the disaggregation of application functions, which will be defined as collections of helm charts, both horizontally (i.e., across edge sites) and vertically (i.e., from the cell site towards the core cloud).

The proposed sustainable virtual elastic edge computing architecture shown in Figure 6.1 will consider the deployment of the vRAN components either at the "bare metal" of the MEC hosts' NFVIs or within virtual network functions (VNFs), as proposed in [15], to offer compatibility with the network slice as a service (NSaaS) sub-system. Moreover, future 6G networks focus on extending the Mobile Edge (ME) platform at the host level, to allow MEC apps to be accessed by any user equipment (UE), irrespective of physical location. This will be accomplished through the interaction of the ME platform with the UPF deployed at the radio edge node, via the Mp2 reference point. The radio edge UPF (right side of Figure 6.1) will act as a branching point, steering user plane flows towards the

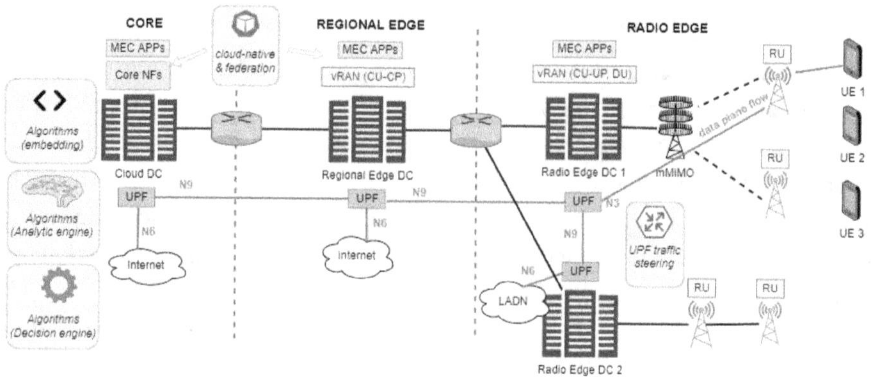

Figure 6.1. Overview of the sustainable virtual elastic edge computing architecture [17].

targeted MEC host, where they are served by the corresponding MEC application. For example, the user plane flows that correspond to ultra-reliable low latency communications (URLLC) traffic will be generally terminated at the same radio edge node (and served by its local MEC host), with near-zero latency [16]. The placement of these application functions to any edge data centre (i.e., the radio, regional, and cloud edges) where they can be accessed by any UE will allow the optimized distribution of latency budgets. Moreover, dynamic virtual network embedding algorithms (left side of Figure 6.1) will be explored for 6G networks, to determine the optimal disaggregation of application functions at any edge data centre, considering compute, networking, storage, and latency constraints as defined by the MEC application manifests. Hence, increased resource utilization via inter-data centre load balancing can be achieved. Therefore, a MEC platform is instantiated based on ETSI MEC functionalities implementing the required extensions in line with the vision of [16] of a virtual elastic infrastructure.

The virtual elastic edge computing architecture, depicted in Figure 6.1, leverages the computational resources of the different domains (cloud, regional, and edge) and *enables sustainability at the deployment level* by means of involving:

(i) *distributed analytic engines* (at the left side of Figure 6.1) *at all tiers of the edge computing architecture,*
(ii) *decision engines at the two core-tier orchestration subsystems (i.e., the NFVO, which manages E2E slices via the NSaaS module, and the Mobile Edge Application Orchestrator (MEAO)),* and
(iii) *UPF traffic steering mechanisms to monitor the status of the MEC hosts' resources.*

The decision engine decides the proactive or reactive allocation of the elastic edge computing architecture and networking resources to serve the respective demands. The decision engine of the elastic edge computing architecture can integrate the resource allocation algorithms discussed in Section 6.2.2 to enable sustainable deployment by implementing the decide part of the envisioned closed-loop control based on the principles of observe, decide, and act.

In the context of involving analytic engines and decision engines, the Radio Edge UPF may act as a branching point, steering user-plane flows towards the targeted MEC host, where they are served by the corresponding MEC application. Thus, user plane flows from URLLC traffic will be generally terminated at the Radio Edge node (and served by its local MEC host), with near-zero latency [16]. The UPF traffic steering mechanisms as shown in Figure 6.1 explore the status of neighbouring MEC hosts resources, which may also be determined not only by end-user workloads and traffic demands but also by RAN workloads (e.g., in charge of running the distributed unit and centralized unit).

Consequently, the virtual elastic edge computing architecture enables *sustainability at the deployment level* while considering the aspects of *optimal disaggregation of application functions, considering compute, networking, storage, and latency constraints* as defined by MEC application manifest deployed at the cloud, regional, and edge of the networks.

The architecture shown in Figure 6.1 leverages the subset of ETSI MEC/NFV blocks that are most relevant to enable the sustainable virtual elastic edge computing infrastructure [18]. The selected blocks from [18] are illustrated in Figure 6.2, where the *blue and red blocks* (i.e., the VIM, MEC platform, MEC platform manager, MEC Orchestrator, NFV orchestrator, and MEC Apps) refer to the MEC and NFV architectures, respectively. The VIM comprises the functionalities that are used to control and manage the interaction of a VNF/Cloud-native

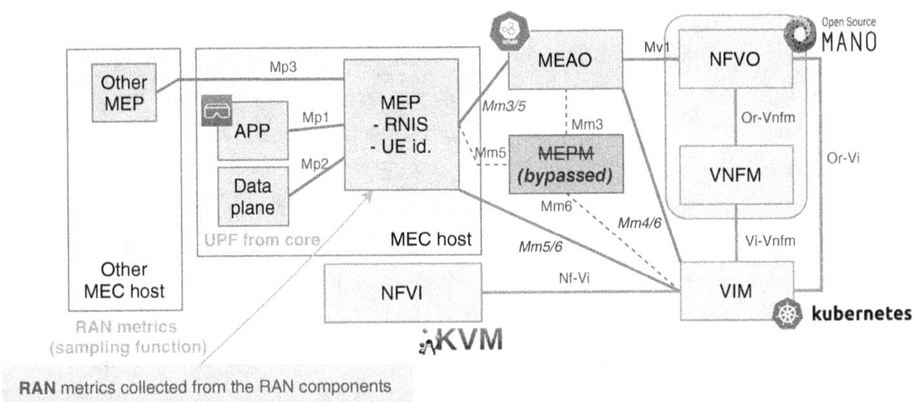

Figure 6.2. Elastic MEC framework to enable sustainable virtual elastic edge computing architecture [17].

network function (CNF) with computing, storage, and network resources under its authority, as well as their virtualization [19]. The MEC platform will offer an environment where the MEC applications can discover, advertise, consume, and offer MEC services, receive traffic rules from the MEC platform manager and applications, provide instructions to the data plane, and receive DNS records from the MEC platform manager, respectively. The MEC Platform Manager (MEPM) is responsible for managing the life cycle of MEC applications, including the reporting of events to the MEAO, and for managing service authorizations, traffic rules, and DNS configuration [18]. The MEO is responsible for maintaining a global view of the resources, the applications, and the users and for triggering the instantiation/termination of applications. However, when deploying MEC in the NFV environment, MEC applications appear as VNFs/CNFs towards the ETSI NFV MANO components. The MEC applications run as virtualized applications, such as a virtual machine (VM) or a containerized application, on top of the virtualization infrastructure provided by the MEC host and can interact with the MEC platform to consume and provide MEC services [19]. MEC applications are packaged by application developers (or in some cases also by MEC operators). The impact of sustainability of the virtual elastic edge computing architecture is further boosted by leveraging sustainable resource allocation algorithms (as the ones presented in Section 6.2.2) as part of the decision engine while collecting the analytics from the architecture.

6.2.1.2 Energy-neutral devices

Apart from the sustainable architectural design, described in Section 6.2.1.1, technology enablers at the deployment level include the design of sustainable devices. In this context, energy-neutral devices are studied in this section, detailing also how they can have a positive impact on the sustainability of wireless networks. An energy-neutral device is defined as "*a passive or active device with a guaranteed continuity of use through a Wireless Power Transfer (WPT) or energy harvesting link that offers sufficient energy. In other words, the device experiences no net negative effect on its potential energy resources from its harvesting* (E_{in}) *and consumption operations* (E_{cons})" [20].

Energy-neutral devices can play an important role in the sustainable realization of the 6G trend and vision of a massive Internet of Things (IoT) deployment including many low-power connected devices. Given that massive IoT-enabled applications have often been considered in view of the "5G/6G for sustainability" context, it is anticipated that environmental and mobility challenges could benefit from wireless sensors of such types in the future.

However, the deployment of massive IoT could come with an undesired impact, namely, the ecological footprint of batteries required to operate the IoT nodes.

Hence, specific attention is needed to reduce battery usage, which typically contains toxic materials.

Consequently, in order to perform a fair analysis regarding energy, a life cycle assessment (LCA) regarding also the batteries themselves is required. In general, *LCA covers the entire life cycle of the studied system (from raw materials extraction to end-of-life treatment) and can deliver results for multiple environmental impact indicators (e.g., climate change, abiotic resource depletion, and water consumption).* Although LCA has been standardized for ICT by the ITU [3], application of such standards already during technology development is challenging as the basis for any LCA is the use of resources and release of emissions.

The awareness that LCA is essential has grown over the past years, and studies on different batteries' modes and usages have been published [21–23]. The main findings in terms of battery efficiency are summarized in Table 6.1, providing both the energy storage capacity of the batteries, typically expressed in Wh, and their *cumulative energy demand (CED, in joule), a term widely used in LCA to consider the direct and indirect energy use (e.g., indirect energy may account for the energy required to generate materials used in the product) over a product's entire life cycle.* The ratio between these two parameters gives an indication of the *efficiency of the energy 'invested' in a battery versus what it delivers.* For further studies that have performed LCA for IoT devices as a whole, the reader may be redirected to [24].

Interactive applications have been identified in different domains that could benefit substantially from the interaction with energy-neutral devices [26], which are expected to be more efficient compared to battery usage (presented in Table 6.1), while having a much longer lifespan, with toxicity and waste generation being greatly reduced. It is also anticipated that distributed large antenna infrastructures, such as the RadioWeaves technology under development [25], have the potential to power sufficiently devices that need to perform only simple operations and within the coverage range so as to enable energy-neutral operation [26].

Table 6.1. Reported LCA for the impact of batteries [25]. The cumulative energy Demand (CED) and energy storage capacity E_c are used to compare the energy efficiency E_c/CED, of battery usage of non-energy-neutral devices.

Year	Ref.	Type	E_c	CED	E_c/CED
2015	[21][a]	Lithium-ion Manganese Oxide battery (LMO)	1 Wh	0.85 MJ eq	4.2×10^{-3}
2016	[22][b]	Alkaline	1.68 Wh	0.7 MJ eq	8.6×10^{-3}
2017	[23][c]	Li-ion	1 Wh	328 Wh	3.0×10^{-3}

[a]Parameters used in the table were taken from plotted data of the environmental impact comparison for solid state and laminated cells; [b]parameters used in the table come from the analysis for a AAA battery collected with a car; [c]contrary to the others, these batteries are rechargeable. Only one charge cycle is considered here. Multiple charge cycles reduce the CED.

In this context, the deployment of distributed large antenna array systems opens interesting opportunities for interaction with energy-neutral devices:

(a) The massive number of antenna elements makes it possible to increase the efficiency of the wireless power transfer significantly through a (very) large antenna array gain. As further commented below, when operating in a near field, energy can be focused in a spot.

(b) The distributed deployment of the antenna array systems increases the probability for energy-neutral devices to be in the proximity of one or more contact service points (CSPs) [27], with charging capabilities. It is evident from the basic Friis radio equation that proximity is a major benefit for wireless power transfer. Indeed, while RF-based wireless power transfer constitutes an attractive solution for remote applications, the efficiency of the transfer is typically very low and limits the use cases for which it can bring an adequate charging solution [28].

When considering sub-10 GHz frequencies, massive-element arrays are physically large, and the energy-neutral devices to be powered are in many use cases located in the array near field. This opens the possibility to focus power on a focal point rather than a beam. This enables efficient WPT and moreover avoids high radiation levels at unintended locations.

A typical simulation result is shown in Figure 6.3 for a room environment served by a RadioWeaves infrastructure operating at 2.4 GHz [25]. The multiple-input single-output (MISO) path loss (PL) was evaluated on a cutting plane (at z = 1 m, perpendicular to the centre of the RadioWeaves panel) through the simulated room,

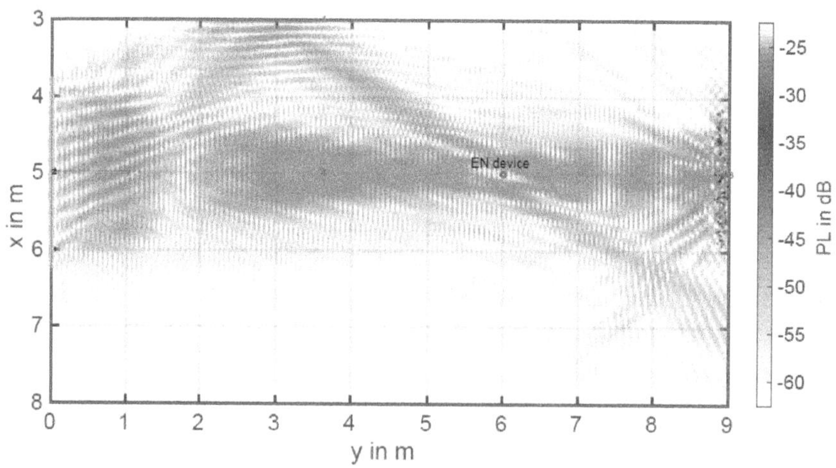

Figure 6.3. Illustration of the spot focusing of distributed beamforming towards the energy-neutral device.

for a carrier frequency $f_c = 2.4$ GHz with perfectly known channel state information (CSI). The results demonstrate the spot-focusing effect towards the energy-neutral device.

To analyse the transfer mechanisms in the near field, it is not possible to rely on relatively simple planar wavefront assumptions, yet spherical wavefronts should be computed. A simulation environment has been prepared to investigate specific geometric settings and to analyse the link budgets and architecture requirements for energy-neutral devices in realistic scenarios. This framework is also used to develop and evaluate signalling schemes and signal processing algorithms for interacting with energy-neutral devices. Moreover, adequate channel models are being developed for the distributed large array topologies, which will be essential in the assessment of performance and efficiency of further technological developments.

Another challenge in interacting with energy-neutral devices relies on the initial access, as explained in [25]: how to power up a battery-less device while no information is available yet on its channel state information, and hence no array gain can be leveraged yet. Indeed, the energy-focusing solutions rely on CSI. Solutions for initial access under development can, e.g., make use of a site-specific model of the propagation environment to realize robust beam sweeping towards potential device locations [29]. In addition, special attention is given to ensure that the proposed solutions can operate within the regulatory constraints.

Finally, although the system design study for energy-neutral devices is in progress together with the development of signalling and signal processing, additional results can be found in [30] and [31].

6.2.2 Sustainability Enablers at Network/Management Level

The combination of different sustainability enablers can result in even higher gains. For instance, the disaggregated and virtualized RAN of Section 6.2.1.1 (enabler *at the deployment level)* can result in significant network energy efficiency gains if accompanied by energy-efficient E2E network, compute, and storage resource allocation (enabler *at the management/orchestration level).*

This is due to the fact that connect-compute-control services entail in general periodic exchange (transmission) and processing of a large amount of capillary data that are continuously collected by heterogeneous devices, to perform complex (collaborative) tasks and/or enable edge intelligence operations. As computing tasks become more complex, it becomes infeasible to run them locally on the device. At the same time, distant central clouds, although extremely powerful, typically entail long delays to be reached. To this end, MEC represents a promising solution, by bringing a secure information technology environment [32], with storage and computing resources, close to the end service consumers.

While this is clearly beneficial from several perspectives, it also comes with non-negligible challenges, among which one can identify: (i) a massive deployment of computing and storage units, which are resource- and energy-hungry, and (ii) the explosion of data traffic (including the uplink) due to continuous exchange of sensors/devices' data [33]. While the former could contribute to increasing the energy consumption of wireless networks, the latter contributes to increased communication energy consumption and resource utilization.

Furthermore, electromagnetic field (EMF) exposure, although not considered an issue [34], should be monitored due to the ever-stricter recommendations/impositions by regulation bodies and state laws [35].

Deployment strategies, resource management, and orchestration can help mitigate such effects. Hence, in Sections 6.2.2.1–6.2.2.4, the focus is on radio and compute resource orchestration for computation offloading services, with the following KPIs and KVIs taken into account: (i) data offloading rate (data offloaded per unit time), (ii) E2E delay (including communication and computation), (iii) communication and computation power consumption, and (iv) EMF exposure [36]. To this end, *a holistic view of the system* is proposed, in order to dynamically and jointly orchestrate and manage heterogeneous resources to optimize offloading performance.

In particular, this section focuses on the *sustainability enablers at the network/management level*, highlighting different *multi-objective resource allocation techniques*, while explaining how they can promote sustainability. In all selected algorithms, a novel architecture, as depicted in Figure 6.4 (in accordance with the virtual elastic architecture of Figure 6.1), is assumed, which consists of a set of computing resources operating in the three layers of the proposed network architecture (e.g., radio-edge or MEC, regional-edge or fog, and cloud, see also Figure 6.5).

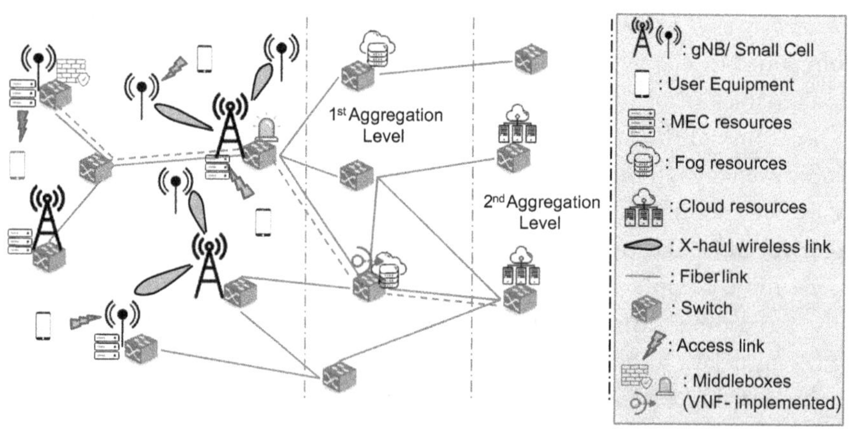

Figure 6.4. Setup overview (in accordance with Figure 6.1) [37].

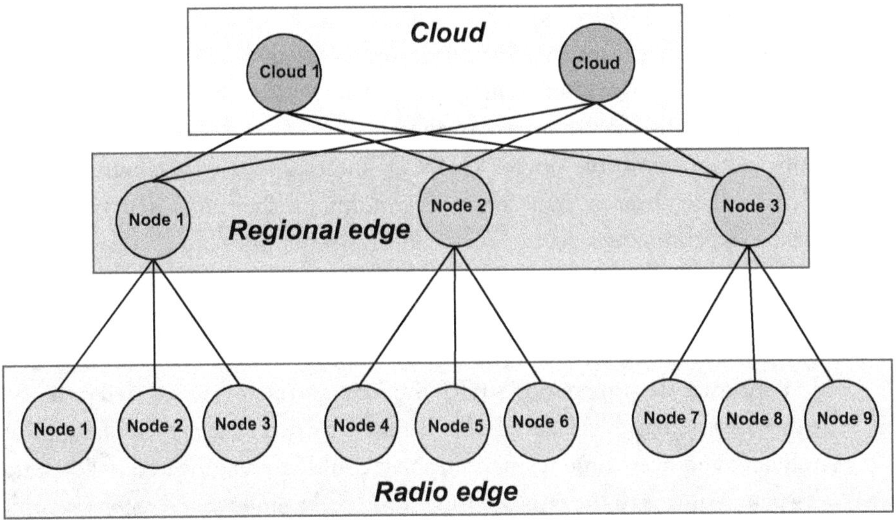

Figure 6.5. The multi-layered computing infrastructure under study.

The edge resources can provide rapid storage and processing of data, extending cloud resources to the network periphery. The edge ranges from the on-device and on-premises edge, to the "far edge" (i.e., within a range of a hundred kilometre from users and premises) and the "near edge" (typically some hundreds of kilometres from the users) and to regional data centres, which together with the traditional cloud resources form an edge-cloud continuum. These computing resources are interconnected with a variety of networking mediums (wired and wireless) and technologies. Subsequently, the following four problems are studied:

(i) cost-efficient computational resource allocation (Section 6.2.2.1),
(ii) cost-efficient joint communication, compute, and storage resource allocation (Section 6.2.2.2),
(iii) energy-efficient joint communication, compute and storage resource allocation in virtual elastic infrastructures (Section 6.2.2.3), and
(iv) EMF-aware joint communication and compute resource allocation (Section 6.2.2.4).

Additional *sustainability enablers at network/management level* may include:

- the *efficient selection of bandwidth and signal characteristics*; the signal properties such as frequency carriers, aggregation capabilities, bandwidth specification, as well as the physical layer features of the signal (e.g., modulation and coding schemes, the waveform type, and precoding) should be efficiently managed to improve the spectral efficiency, thus leading to sustainable solutions,

- the *introduction and optimization of sleeping periods* for sustainable network management; apart from the main improvements achieved by the orthogonal frequency division multiplexing (OFDM) structure which allows for rapid sleep modes (discontinuous transmission, DTX and reception, DRX), the multiple-input multiple-output (MIMO) configuration now allows to switch off part of the antenna transceivers depending on the traffic demand. However, new technics as lean carrier and deep sleep modes are also needed that will not sacrifice the quality of service (QoS) or user experience, and
- leveraging artificial intelligence (AI) (being discussed in detail in Chapter 5) for sustainable network management/orchestration; for instance, employing AI to optimize sleeping periods of RF modules and adapt the needed resources to the user demand. AI could also be used to detect energy consumption anomalies and over-dimensioned sites that could be reengineered to adapt the network resources to the targeted QoS. Finally, AI-empowered receivers could perform signal detection in the presence of power amplifier non-linearities, hence, enabling the operation of the power amplifier with lower back-off, leading to higher energy efficiency, and consequently, sustainability.

6.2.2.1 Cost-efficient computational resource allocation in virtual elastic infrastructures

In such multi-layered computing infrastructures, as the ones under study that target at a diverse set of objectives (e.g., minimization of the cost and average latency per slice), the amount of the required computations grows exponentially with the number of agents (i.e., the number of computing nodes). For this reason, in this section, novel rollout and multi-agent reinforcement learning (MARL) algorithms are proposed for the infrastructure shown in Figure 6.5 in which the decision at each stage is made by executing a local rollout algorithm for each agent that uses a base policy, together with coordinating information from other agents. In this way, the local computation required by each agent is independent of the number of agents, while the amount of the total computation grows linearly with the number of agents. In general, the provided computing capacity increases when moving from the edge to the cloud; however, at the same time, performance limitations in terms of processing latency and available bandwidth also appear. The workloads of the multi-layered infrastructure can be executed in the form of either single tasks or a set of tasks, where each task's computational workload can be infinitely decomposed for execution to the available resources. These workloads can be baseband processing workloads that correspond to distributed unit and centralized unit workloads or generic application workloads. The main parameter of interest is the total execution time of each task or set of tasks, along with the efficient utilization of resources, serving the tasks' time (e.g., deadline) constraints. Tasks are assigned, respectively: ephemeral and low-latency required computations on the edge, while complex computations are assigned at the cloud. Based on the above, the proposed methodology, namely, virtual-elastic resource allocation, allocates tasks to the available resource, deciding the decomposition into smaller subtasks in a way so as to serve the task time requirements. Two "extreme scenarios" can be considered:

- A *highly time-critical application*: in this case, the application's workload is executed at the edge. This, of course, may not be feasible due to the limited capacity of the edge resources and as a result, horizontal and vertical disaggregation will be necessary, where part of the application's workload is assigned to edge resources in the same layer or in higher-layer resources.
- A *non-time-critical application*: in this case, the workload can be executed at the central cloud or in any other resource that is available.

A multi-agent rollout algorithm has been developed for the virtual-elastic resource allocation problem, where computing slices are decided to serve application workload demands, as depicted in Figure 6.6. These slices correspond to reservations in multiple resources and in multiple layers. This methodology is one of the most reliable reinforcement learning (RL) methods, and it is based on the idea of policy iteration, i.e., starting from some policy and generating an improved policy from the set of feasible solutions. The simulation parameters to evaluate the performance of the proposed algorithm are listed in Table 6.2.

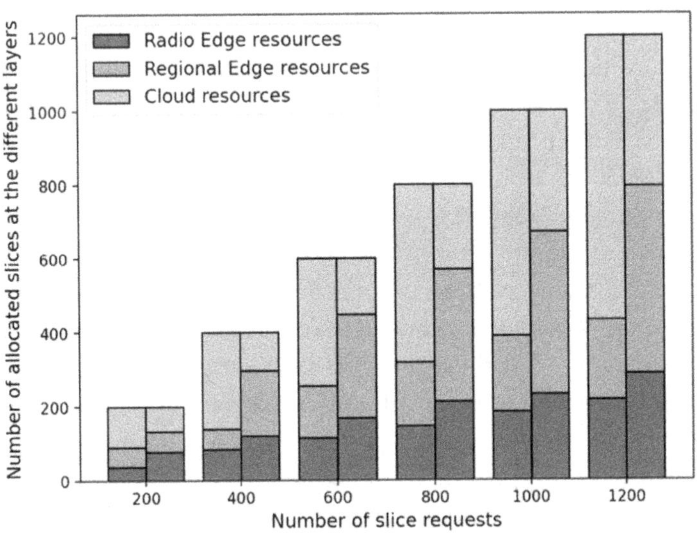

Figure 6.6. Number of allocated slices at the different computing layers for the proposed methodology when the objective is the minimization of the cost (left bar) and the minimization of the average latency per slice (right bar) for various number of slice requests.

Table 6.2. Simulation parameters.

Layer	Number of Nodes	Number of Cores Per Node	Latency from Data Generation to the Layer Latency Unit (l.u.)	Cost Per Layer Cost Unit (c.u.)
Radio edge	9	40	[0,2]	1.96
Regional edge	3	300	[3,5]	1.44
Cloud	2	1000	[6,10]	1

Several simulation experiments of the mentioned multi-agent rollout mechanism have been performed, in which a varying number of slice requests (bag of tasks) [200–1200] need to be served by the proposed multi-layered computing infrastructure shown in Figure 6.5. As the capacity of edge resources is limited and increases for higher layers, it is assumed that the cost of utilizing a resource decreases by 40% from the edge to the cloud (1 c.u. for cloud resources). As shown in Figure 6.6, when the objective is the minimization of the cost (left bar), more slices are served in the cloud. Also, the utilization of cloud resources increases with the increase in the served slice requests, taking advantage of the wide availability of cloud resources compared to the resources in other layers. The lower cost of using cloud resources increases further their utilization. For this reason, the utilization of the regional edge resources is lower and increases significantly when the optimization criterion is the minimization of the average latency per slice. In this case, radio edge and regional edge resources are highly utilized, showcasing their ability to efficiently serve time-critical workloads.

Hence, when the main optimization criterion is the cost, cloud resources are the most adequate ones to serve demanding applications with relaxed latency requirements. On the other hand, radio and regional edge resources can be effectively combined to serve the time-critical baseband processing requirements for which the latency constraints are strict.

In Figure 6.7, the effect of the different optimization criteria is examined in the form of experienced latency. When the only optimization criterion is the processing cost, the demands whose latency constraint is strict are allocated at the lower layers of the infrastructure and at the higher layers when the latency constraint is more relaxed. Hence, there are not any significant variations in the experienced latency as the number of slice requests that are served increases. On the other hand, when the main optimization objective is the minimization of latency, the experienced latency per slice increases with the increase in the number of slices that are served because the resources of the lower layers, which are limited, are fully utilized and demands are served by higher layer resources (regional edge and cloud resources).

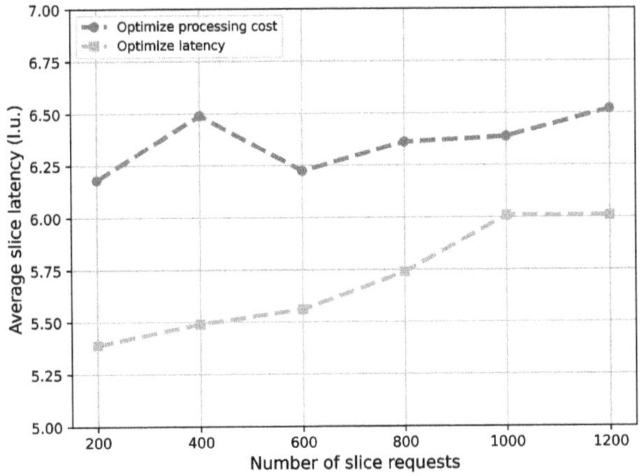

Figure 6.7. Average slice latency when the objective is the minimization of the cost and the minimization of the average latency per slice, for various number of slice requests.

6.2.2.2 Cost-efficient joint network, compute, and storage resource allocation in virtual elastic infrastructures

This section focuses on the development and evaluation of near-real-time dynamic network slicing reconfiguration mechanisms by considering both the fixed and wireless parts of the network. The approach considered is based on the development of novel network slicing algorithms by decomposing a global team objective into many sub-objectives.

Due to the three-tier nature of the network architecture considered (with core, regional edge, and radio edge tiers [17]), different agents can be considered at each tier, where the high-level multi-agent reinforcement learning will be considered in the NFVO. Therefore, the challenge is to decompose the global team objective into many sub-objectives, targeting to optimize network parameters (e.g., maximize the spectral efficiency).

To this end, selecting a latent variable (a series of primitive actions with a cumulative reward) is targeted by each agent, with each agent selecting assignments independently. Each assignment is a policy conditioned by the latent variable for an extended duration, generating a behaviour of which the latent variable is decidable. Thus, a specific set of latent variables can be defined, each one corresponding to a slice type, e.g., considering five types for the further enhanced mobile broadband (FeMBB), ultra-massive machine-type communications (umMTC), extremely ultra-reliable and low-latency communications (eURLLC), long-distance and high-mobility communications (LDHMC), and extremely low-power communications (ELPC), respectively, in accordance with Section 2.1.

Based on the above-mentioned analysis, the following parameters are defined: **N**: the number of agents (n ∈ [N]), **a**: a joint action, and $\boldsymbol{\mu}$: a joint high-level policy that learns to select slices to optimize an extrinsic team reward function that maps global state and joint action to a scalar reward. All agents have the same observation space and action space, while all agents take individual actions based on individual observations. In parallel, a low-level multi-agent RL can be considered at the regional edge (e.g., at the near-real-time RAN Intelligent Controller, RIC). These agents receive the chosen slice type from a high-level agent. Further definitions include the following: π: a joint low-level policy: learns to choose primitive actions (choice of remote units (RUs) to cluster and choice of physical resource block (PRB) resources to assign) to produce useful and decodable behaviour (trajectory) by optimizing a low-level reward function. $D = \{(z, \tau)\}$: data set of slice-trajectory pairs: each pair consists of the slice type chosen by a high-level policy and the corresponding trajectory (peak data rate, mobility, transmission power, latency, and whatever defines the slice types) generated by the low-level policy. Also, a slice decoder is considered, which calculates the probability of realization of the slice type given the trajectory produced by an agent in the low-level policy.

Thereby, a reward system of this two-level hierarchical approach is defined, where each agent considers:

- at a high-level policy, *an extrinsic system reward*, to conduct centralized training of high-level policies for cooperation and
- at a low-level policy, *a combination of an intrinsic reward and an extrinsic system reward*, to conduct decentralized training of low-level policies with independent RL.

To this end, five agents are considered that correspond to the five types of slices, i.e., FeMBB, umMTC, eURLLC, LDHMC, and ELPC. Each agent tries to find the proper slice type according to the request. The main objective of the developed solution is to maximize the use of available resources, which results in the maximization of the network capacity and sustainability aspects, e.g., energy efficiency. Figure 6.8 presents the reward of the multi-agent set. The reward system presents both agent rewards and the total reward. The general goal for the mixing network is to maximize the number of assigned resources. Therefore, the ascending plot shows the assigned resources, and therefore the network capacity is increased. The variation in the plot is due to the fact that, in this simulation, if the requests repeat asking for the same slice type which can be unavailable after a while, agents will receive their rewards for not assigning the wrong slice type, but the total reward may decrease as there are free resources.

It should be noted that the reward is considered to guarantee that trajectories are useful for the overall performance, while the intrinsic reward is considered to promote the association of latent variables with predictable behaviour. In other words, the intrinsic reward encourages the generation of distinguishable behaviour

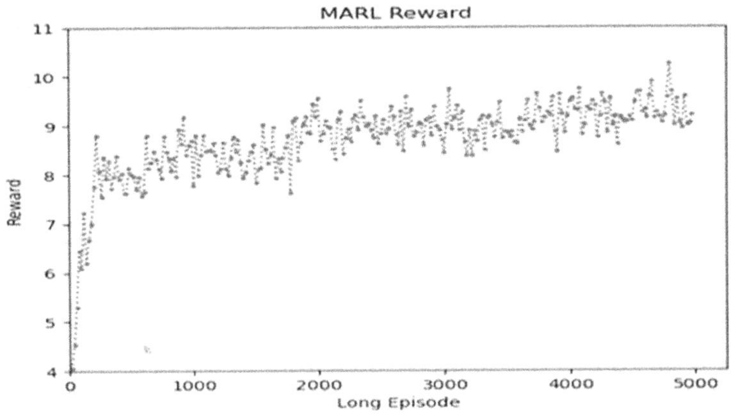

Figure 6.8. Reward of the multi-agent set.

for different slices (only by doing so can the low-level policy produce sufficiently distinct "classes" in the dataset for the decoder to achieve high prediction performance).

The aforementioned approach offers a hierarchical decomposition in two key dimensions over agents, and across time, which is able to simultaneously address the difficulty of learning cooperation at the noisy low-level actions in stochastic environments, as well as the difficulty of long-term credit assignment due to highly delayed rewards. Hierarchical approaches may also reduce the computational complexity to address the exponential increase in sample complexity with a number of agents.

6.2.2.3 Energy-efficient joint network, compute, and storage resource allocation

Extending the multi-objective problem description presented in Section 6.2.2.2, to further optimize service deployment and operation in terms of sustainability, the E2E path of the traffic should be jointly optimized with the network, compute, and storage resource allocation. To that end, each service request corresponds to a service function chain (SFC) consisting of an ordered set of VNFs that should be efficiently deployed in the computational nodes of the network, ensuring that the VNFs are deployed in the exact order specified by the SFC and the capacity constraints of the nodes and involved links are not violated [38]. In addition, these VNFs can be also deployed in different forms, i.e., in VMs, containers, or unikernels, and different network locations, i.e., in MEC units for reduced latency, in data centres for reduced cost or in intermediate locations, as shown in Figure 6.4.

Hence, as already mentioned in previous sections, there is a trade-off between delay and cost that requires further study. On the one hand, MEC nodes (followed by regional fog nodes) are closer to the UE and thus their selection should be prioritized over cloud data centres for delay-intolerant services, while at the same time, their limited resources in addition to the need for a compact size (due to their location very close or onto the antenna) make them expensive (high CAPEX) and also increase their OPEX. On the contrary, data centres are mainly located in distant locations, with very low rental prices and abundant available space, which enable deployments consisting of hundreds (or more) of server racks, thus leading to inexpensive abundant computational resources. However, their long distance from city centres constraints their use mainly to delay-tolerant services only.

The 6G transport network interconnecting the computation and communication nodes will consist of front/mid/backhaul (X-haul) links of different technologies and capabilities (i.e., both fibre and wireless links), calling for joint consideration of the access network and transport in the modelling process, i.e., optimization of both user association and traffic routing problems. For wireless X-haul links, an attractive solution lies in the use of millimetre-wave (mmWave) frequencies, due to their wide spectrum bands and high antenna gains compensating for the increased path loss in these bands [39]. In the access network, a combination of gNodeBs (gNBs) and a dense overlay of small cells (SCs) employing 5G-New Radio (5G-NR) frequencies, including mmWave, is expected.

In this context, resource allocation becomes challenging due to the high network and resource heterogeneity, resulting in a large number of strongly coupled decision variables. Hence, efficient resource allocation strategies should:

(i) jointly consider all different types of resources, i.e., communication, computational and storage, and technologies, e.g., 5G-NR, mmWave, fibre, as well as their constraints,
(ii) take into account the E2E network path from the traffic's source to the destined UE to guarantee E2E optimality,
(iii) induce low computational complexity to enable near real-time decisions, while meeting the E2E delay target, and
(iv) achieve high energy efficiency.

Developing energy-efficient solutions serves a twofold goal: reducing the OPEX of the involved stakeholders (e.g., mobile network operators, infrastructure providers, etc.) and leading to environmentally friendly solutions by limiting the associated carbon footprint. In this context, energy-efficient online resource allocation solutions are needed to jointly solve the user association, traffic routing, and VNF placement problems while ensuring SFC chaining and guaranteeing the QoS of the service requests in 6G networks.

Therefore, in the following, a Heuristic for Energy-efficient VNF placement, traffic Routing, and user assOciation (HERO) is proposed aiming at maximizing

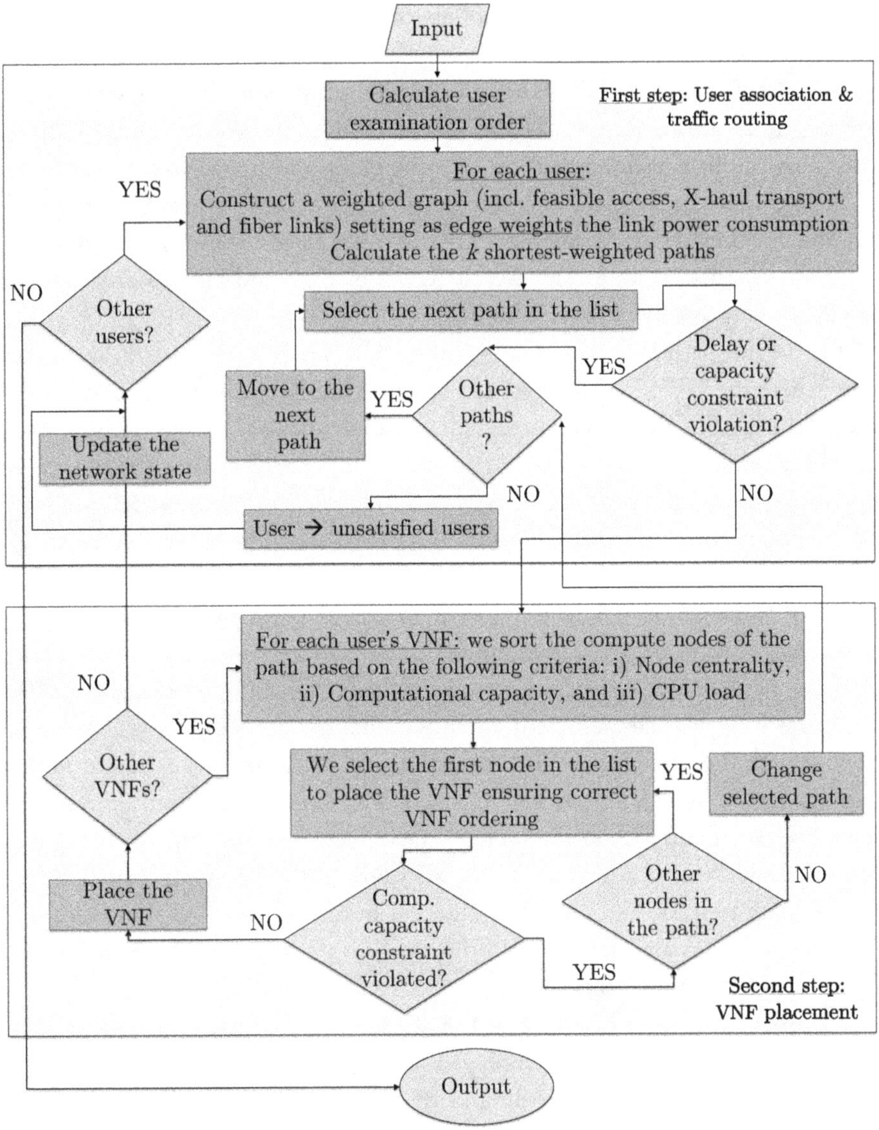

Figure 6.9. Flowchart of the proposed energy-efficient VNF placement, traffic routing, and user association algorithm (HERO) [40].

the network energy efficiency while ensuring low UE blocking probability. HERO consists of two steps, as shown in Figure 6.9. In the first step, the traffic path is selected (i.e., user association and routing are performed), while in the second one, VNF placement takes place, ensuring correct VNF ordering.

Initially, to ensure a high UE acceptance ratio, the UEs are sorted based on their service demands, giving priority to the UEs with the most delay-intolerant services. For UEs with the same delay requirements, priority is given to the UEs with higher rate demands. Then, for each UE, a weighted graph is constructed from the service traffic source to the UE with all feasible links and their respective power consumption acting as weights. HERO calculates the k shortest-weighted paths and starts with the first, as long as it satisfies the delay and link capacity constraints, taking into account the decisions for the already examined users. Otherwise, the next path is selected until either a path that satisfies all constraints is found or there are no other paths. In the latter case, the UE is blocked and HERO proceeds with the next as long as there is one.

After a valid path is found for the current UE, HERO proceeds to the second stage, where the UE VNFs are placed. To that end, for each VNF, following the order of the UE SFC, a list is constructed with all the available computational nodes based on a parameter denoted by H, which is equal to the sum of the normalized node centrality (closeness), the normalized node computational capabilities (cy), and the node central processing unit (CPU) utilization. The latter is equal to (a) 1 when the studied VNF can be placed in the examined node without initiating a new VNF instance, (b) 0.1 when there is enough computational capacity to host the studied VNF in the examined node, but a new VNF instance is required, and (c) 0 otherwise. Subsequently, the node with the highest H for the selected VNF is selected, as long as it has sufficient computational resources to host it. Otherwise, the node with the next highest H is selected, until either the VNF is placed or there is no other node to examine in the selected path. In the latter case, the algorithm returns to stage 1 and the next path out of the k calculated is examined. The process is repeated for the new path until either all VNFs of the UE are placed or there is no other path to study and the UE is blocked. In case all UE VNFs are placed, the network conditions are updated, and the algorithm proceeds to the next UE. The aforementioned steps are repeated until all UEs are examined.

Subsequently, the performance of HERO is compared with the optimal solution (a detailed analysis of the employed model can be found in [39]), and two state-of-the-art solutions, i.e., Holu and BCSP, are both proposed in [41]. These approaches first place the VNFs and then decide upon the traffic routing. In particular, Holu places the VNFs on the computational node with the highest closeness centrality and CPU utilization, and then it chooses the traffic route with the lowest power consumption that satisfies the E2E latency requirement of the service request, whereas the BCSP places the VNFs on the computational node with the highest betweenness centrality and selects the least delay route that satisfies the E2E latency requirement of the service request.

Provided that the state-of-the-art algorithms do not account for user association, the default selection criterion is applied in both, i.e., the users connect to the BS that provides the highest signal-to-interference-plus-noise ratio (SINR).

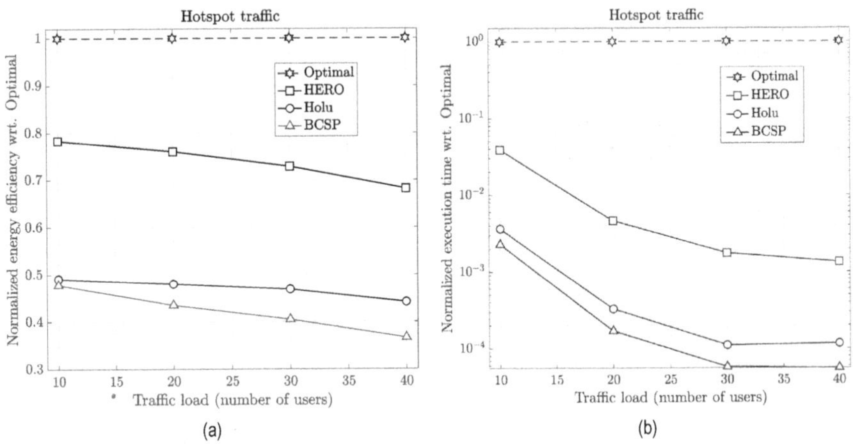

Figure 6.10. Normalized energy efficiency (bits/joule, linear scale) and execution time (s, logarithmic scale) of all studied algorithms with reference to the optimal solution for different traffic load conditions [40].

In Figure 6.10, the normalized (in comparison with the optimal solution) energy efficiency (in bits/Joule) and computational time (in logarithmic scale), respectively, are depicted for all studied algorithms versus different numbers of UEs. As can be seen, HERO provides a very good trade-off between energy efficiency and complexity compared to the other approaches, achieving up to 78% of the Optimal value, with up to 742 times lower complexity. It is worth noting that all algorithms have a 100% user acceptance ratio in all cases, except for BCSP which is a little lower for high-load traffic. This is due to the fact that, in BCSP, the CPU utilization of the computing nodes is not taken into account, resulting in less efficient VNF placement which, under higher traffic load, can lead to few UEs being blocked.

Compared to Holu and BCSP, HERO achieves up to 60% and 86% higher energy efficiency, respectively, while keeping the complexity low, as shown in Figure 6.10. This is due to the fact that HERO additionally considers user association as part of the optimization problem leading to higher flexibility at the expense of a little higher complexity. On the other hand, in both Holu and BCSP, the serving BSs are already decided (based on the best SINR criterion) and then the optimal VNF placement and traffic routing from the UE traffic source to its serving BS are performed.

It can also be observed that the power consumption of the optimal and HERO are scaling better than the state-of-the-art with increasing load (HERO still achieves 68% of the optimal energy efficiency value when N = 40).

As a final remark, the performance gains of the proposed algorithms justify the motivation of this work that user association, VNF placement, and traffic routing should be jointly considered to guarantee true optimal E2E network performance.

6.2.2.4 Energy and EMF-aware joint network and compute resource allocation

This section focuses on a typical computation offloading setting, as depicted in Figure 6.11, where, for the sake of simplicity, one device (e.g., a robotic arm) continuously generating data at a given rate is considered. According to the architecture presented in Figure 6.4, the investigated scenario relates to the interaction between end users (UE) and a gNB with MEC resources, i.e., the access network links, involving the elements located on the left-hand side of the figure. A more complex multi-user scenario, with a more detailed technical presentation, can be found in [36]. In case of (communication and computation) power and/or EMF exposure requirements, a device may be forced to proactively drop some of the incoming data traffic, in order to match the real capacity of the system, under finite E2E delay constraints. It is possible to model this through the tandem queueing system (with communication and computation buffer) qualitatively illustrated in Figure 6.11, with a fictitious control valve, used to dynamically limit the arrivals.

The idea is to proactively drop offloading requests, in order to ensure that the accepted offloading traffic is served within a finite E2E delay (i.e., the tandem queueing system is stationary stable). A joint optimization problem is cast to maximize the overall data offloading rate of the system (i.e., ratio of accepted arrivals), under constraints on (i) long-term device power consumption, (ii) long-term computing power consumption, (iii) EMF exposure in predefined zones in space, with a long-term average measure, as typically recommended by international regulation bodies [42], and (iv) E2E delay constraints.

The optimization variables generally involve wireless (bandwidth, transmit power, radio scheduling, etc.), and computing resources (CPU scheduling, workload placement, etc.). A possible solution, based on stochastic optimization, can be found in [36], with theoretical analysis inspired by [43, 44], and [45].

In Figure 6.12, a first set of results is shown, representing four metrics (normalized to their maximum value – except for the delay), all as functions of the arrivals

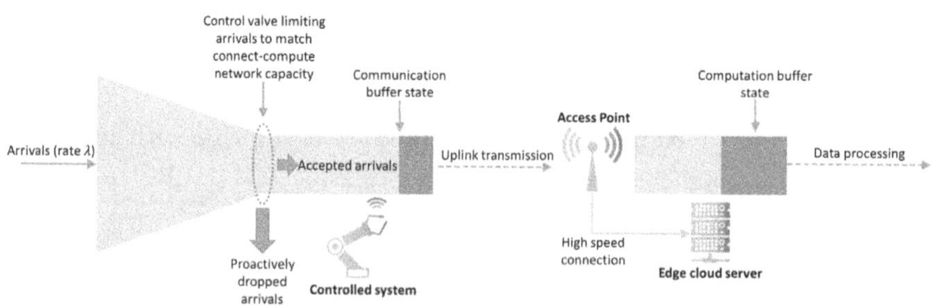

Figure 6.11. Network setting with tandem communication and computation queues.

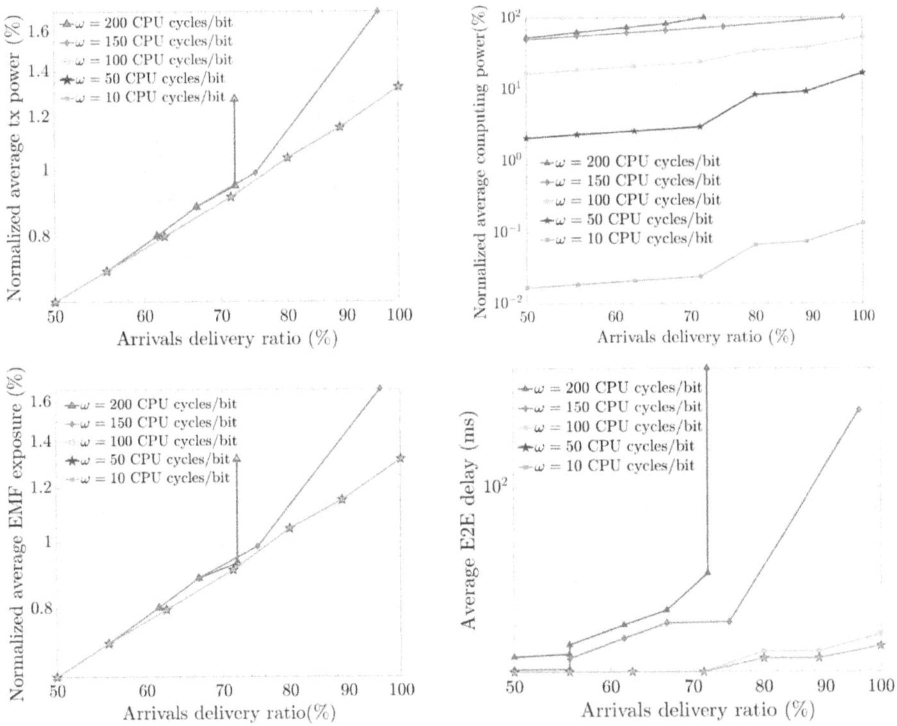

Figure 6.12. Trade-off involving communication, computation power, EMF exposure, data delivery ratio, and E2E delay.

delivery ratio (i.e., the ratio of accepted arrivals into the system, served within finite E2E delay – which is 100% in case of zero proactive drops): (i) the average device transmit power (Figure 6.12(a)), (ii) the average computing power (Figure 6.12(b)), (iii) the average EMF exposure (Figure 6.12(c)), and (iv) the average service E2E delay (Figure 6.12(d)).

Results are shown for different values of computing load (CPU cycles to be performed per offloaded bit). First, it can be noticed how, obviously, a higher delivery ratio costs higher power consumption for all network entities, as well as higher EMF exposure. Moreover, for almost all settings, the delivery ratio reaches reach 100% (left-hand points of the plot), with all entities working below their maximum power. This suggests that the connect-compute network capacity is able to afford the injected traffic arrival rate. However, for $\omega = 150$ CPU cycles/bit, the delivery ratio decreases, and, at the same time, the computing power reaches its maximum, meaning that the computing capacity is saturated by such computational load. This also reflects on the average E2E delay (Figure 6.12(d)), as the arrival rate is close to the maximum service rate and becomes even more obvious for $\omega = 200$ CPU cycles/bit, due to the higher computational load. Also, for

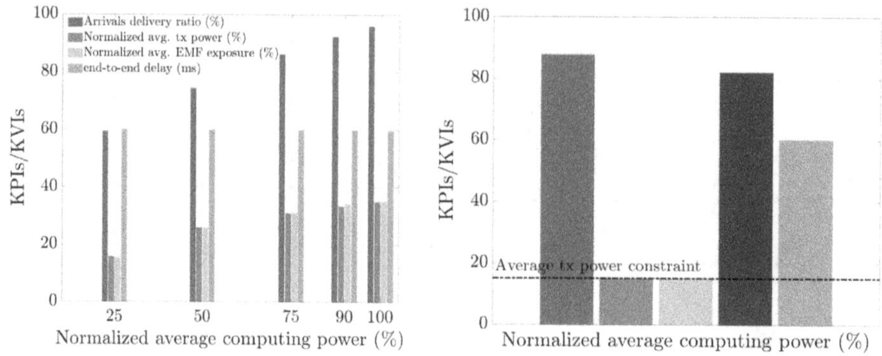

Figure 6.13. Network KPIs/KVIs versus normalized computing power.

$\omega = 200$ CPU cycles/bit, the delivery ratio drops down to 70%. On the other hand, the device's average transmitting power never exceeds 2%, suggesting that network arrivals are *limited by the computing capacity*. The latter is an example of a connect-compute wireless network setting, whose capacity is limited by computing capabilities.

Building on the previous results, different regions can be explored, namely, communication and computation capacity limited ones, from a sustainability perspective, by explicitly imposing network power consumption constraints, in order to attain the best offloading performance in terms of delivery ratio, while guaranteeing predefined network footprint bounds (such as energy and EMF exposure). The latter, under the same joint communication–computation resource allocation framework.

To this end, Figure 6.13(a) depicts the different KPIs/KVIs as functions of the normalized computing power, set as a threshold and controlled through the stochastic optimization-based algorithm, with a predefined E2E delay requirement of 60 ms in this simulation. First, in the case of full computing power, it can be noticed that, in the proposed setting, the system is not able to accommodate all requests (delivery ratio under 100% in all cases), while the device normalized transmit power is always below 40%. Again, this suggests that the system is limited by the computing capacity.

Therefore, as a final example, it is worth considering a communication capacity-limited example. As such, it is enough to take the last example, in the case of 100% computing power consumption, and set, as a requirement, a normalized average transmit power below, e.g., 15%. The result is shown in Figure 6.13(b). It can be remarked how the method is able to guarantee the predefined constraint (see the horizontal black dashed line). However, with respect to the previous example with 100% normalized computing power availability, the delivery ratio decreases, suggesting that the system has passed from a computing to a communication-limited condition. The latter observation is furtherly validated by the drop in needed computing capacity, due to the fact that the computing rate is relieved by the limiting communication resources.

This section has introduced the concept of bottleneck identification in connect-compute networks, in relation to different sustainability-oriented constraints (involving energy and EMF exposure), with performance indicators including data offloading rate and E2E delay. The idea is to let the network autonomously identify the bottleneck and proactively drop offloading requests to guarantee energy (entailing communication and computing), EMF exposure, and E2E delay requirements. Future direction towards fully Connect-Compute-Control Co-design (CoCoCoCo) will finally assess the performance of special purpose functionalities in next-generation wireless systems (6G). In particular, no evaluation of the impact of proactive dropping (e.g., packet loss rate) is presented in this section.

6.2.3 Sustainability Enablers at the Service/Application Layer

Traditionally, networks are regarded as a bit pipe for applications, sufficient in terms of QoS. This means that the stochastic behaviour of networks is not part of the application design and conversely, the application does not interplay with short-term network behaviour. However, the sustainability of the whole system can also be achieved by developing a joint understanding of the overall goal across network and application layers.

This section highlights how application-aware networks can reduce energy as well as network resource consumption to contribute to higher sustainability. This may be achieved as the network can understand the meaning of the data to be transmitted (semantic communication) and is subsequently able to optimize the network behaviour thereupon.

6.2.3.1 Sustainable application-aware networks

The joint modelling of control and (possibly fault-prone) communication provides crucial insight for a codesign that targets to ensure dependable application behaviour and at the same time reduce wireless network resource consumption for better sustainability. It is well-known that operating a closed control loop requires timely data, otherwise the control loop might become unstable. However, as control applications are also oversampled for better smoothness [46], a few packet losses can typically be tolerated as long as not *too many* occur. A classical design methodology would attempt to reduce the overall packet loss rate of the system, which significantly increases the energy for every single transmission. However, the actual goal is to guarantee a given data timeliness ("freshness"), which enables to spend less energy in cases where the most recent transmission was successful and to only

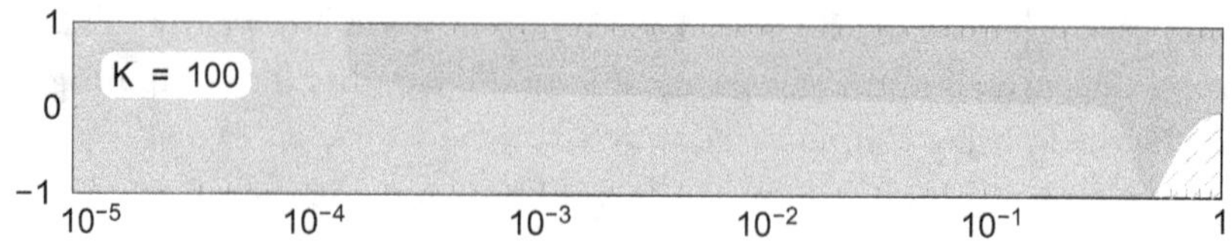

Figure 6.14. Negatively correlating packet losses (y-axis) has a significant impact on control performance. Green area = stable and red area = unstable.

gradually increase the transmission energy upon an increasing number of consecutive packet losses. What constitutes "*loosing too many*" may be investigated through a Markov jump linear system (MJLS) modelling approach and will be outlined in the following.

MJLS theory enables deriving stability boundaries for linear-time-variant (LTV) control systems that can be described as a set of switching linear-time-invariant (LTI) control systems as long as the switching behaviour can be modelled through a Markov Chain, i.e., the probability of being in a state at time $t = t_0 + 1$ only depends upon the state at time $t = t_0$ (and not any $t < t_0$) [47].

With a system model consisting of four different operating modes (uplink and downlink; only uplink; only downlink; and none), the Markov chain defines the temporal packet loss dependencies. A temporal packet loss correlation coefficient $-1 \leq \rho \leq 1$ may be defined, ranging from fully negative correlation (-1: a second packet loss cannot occur after a loss) to fully positive correlation (1: upon first packet loss, all packets until a maximum boundary K will also be lost).

As expected, the results show that a negative packet loss correlation is highly beneficial for closed-loop control applications, as demonstrated here for an example AGV use case (compare Figure 6.14). A negative temporal packet loss correlation leads to short packet loss sequences and therefore stabilizes the control application. In this context, the "meaning" of a packet equals the time that has passed since the last successful transmission, which not only constitutes a simplification over more sophisticated modelling approaches but also greatly reduces complexity and enables more general insights. Here, a more sustainable operating point would be to allow for energy-efficient, resource-efficient best-effort transmissions if the time since the last successful transmission is low and to increase the packet success probability (decrease energy efficiency) as the time since the last successful transmission increases, and therefore, packets' importance increases.

It was previously shown that negatively correlating packet losses has a beneficial impact on control stability. On the network side, this can be achieved by spending increasingly more resources, depending on the number of packets that have been recently lost. In the following, a multi-connectivity network scenario is targeted. The Markov model in Figure 6.15 is used to describe the state of the current service:

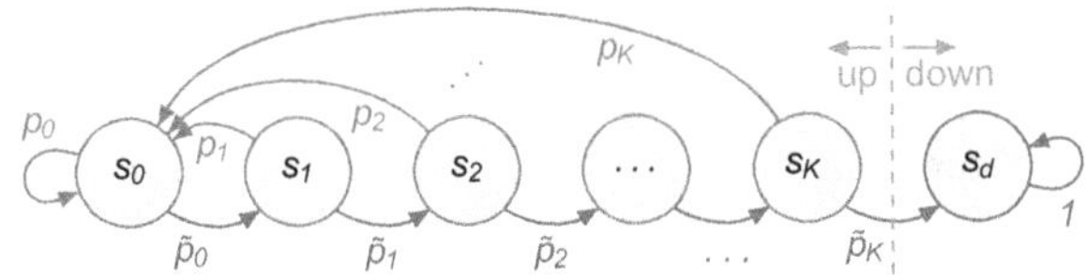

Figure 6.15. Negatively correlating packet losses (y-axis) has a significant impact on control performance. Green area = stable and red area = unstable.

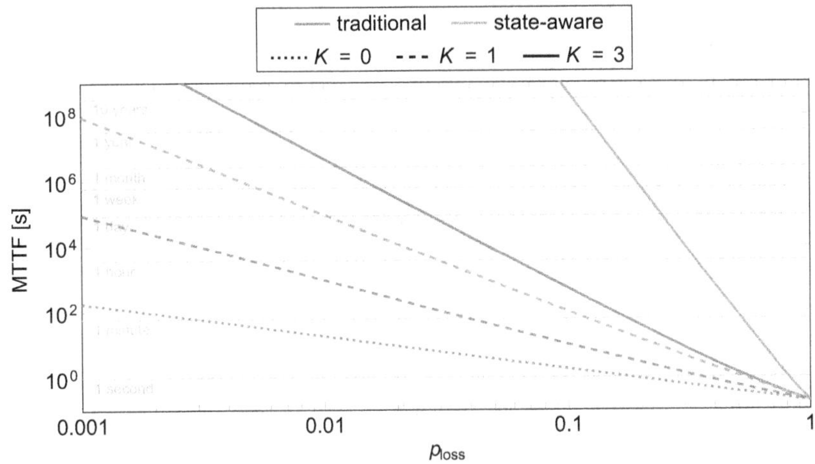

Figure 6.16. State-aware resource allocation allows increasing the mean time to failure by orders of magnitude while preserving low resource and energy consumption.

a service is considered in state s_k with $k \in \{0, \ldots, K\}$, where k denotes the number of consecutive packet losses that occurred immediately before entering s_k. For instance, in s_2, the last two transmissions failed. Consequently, when a transmission succeeds (green transitions), the state s_0 is entered. Whenever a packet is lost (red transitions) with transition probability $\tilde{p}_k$, the state index is incremented by one. The rightmost state s_d denotes the failure state and refers to $K+1$ consecutive packet losses. All other states are considered up.

This model allows determining the probability of entering the down state s_d subject to changing the state-specific transition probabilities that may be influenced by increasing the number of parallel links, increasing transmission power, and so on. Figure 6.16 shows how the mean time to failure (MTTF), i.e., the mean time until s_d is reached, can be shaped by assigning resources according to the time that has passed since the last successful transmission. p_{loss} constitutes the per-link packet loss probability. The plots for the *traditional* network design (red) show that if the application may tolerate an increasing number of consecutive losses (increasing K), the gain in terms of MTTF is only marginal. On the other hand, if the resources may be increased in a *state-aware* fashion, the MTTF increase is tremendous (green lines compared to red lines), all while not spending additional resources. For example, if three consecutively lost packets can be tolerated (but not four, see solid lines), the MTTF can be increased from ~30 min to >10 years (1,000,000x improvement) when assuming a best-effort packet loss rate $p_{\text{loss}} = 10\%$.

6.2.4 Cross-layer Sustainability Enablers

Supplementing the novel sustainability enablers of previous sections, this section focuses on *cross-layer sustainability enablers*, highlighting the 6G key technology of digital twining and how this can lead to sustainable networks. Further sustainability gains can be achieved by the combination of the aforementioned techniques with renewable sources of energy, targeting at environment-friendly network solutions.

6.2.4.1 Sustainable radio-aware digital twin

Digital Twins are an up-to-date digital/virtual representation of a physical asset or process that can simulate or predict the status of the process. They are expected to optimize operations and enable higher levels of productivity and efficiency which in turn makes the operation more sustainable.

A controlled environment is considered, such as a fully automated factory with few or no humans on the factory floor, and where positions of the terminals and access points are known and/or can be controlled. In this scenario, a digital twin of the radio propagation environment is targeted, which is a radio-aware digital twin. Given the locations of the access points and positions/trajectories of the terminals, the radio-aware digital twin can predict link conditions through ray tracing or machine learning models for pro-active, anticipatory, intent-based resource allocation, and beam management to improve network capacity and reliability. This framework can also be utilized to optimize the energy efficiency of the network through trajectory and network planning.

Therefore, in the following, a radio-aware trajectory optimization strategy of an automatic guided vehicle (AGV)/unmanned aerial vehicle (UAV) is described for a mission (for instance, video surveillance) under certain constraints, which may include:

- a time constraint within which the AGV/UAV must reach the destination
- a data rate constraint that may have to be satisfied at every point along the trajectory with high reliability (e.g., for control traffic).

Under these constraints, minimization of the total energy consumed by the AGV/UAV is targeted to improve sustainability.

In this regard, the shortest path between the start and destination points is selected as the baseline. This solution provides the lowest flight time and also

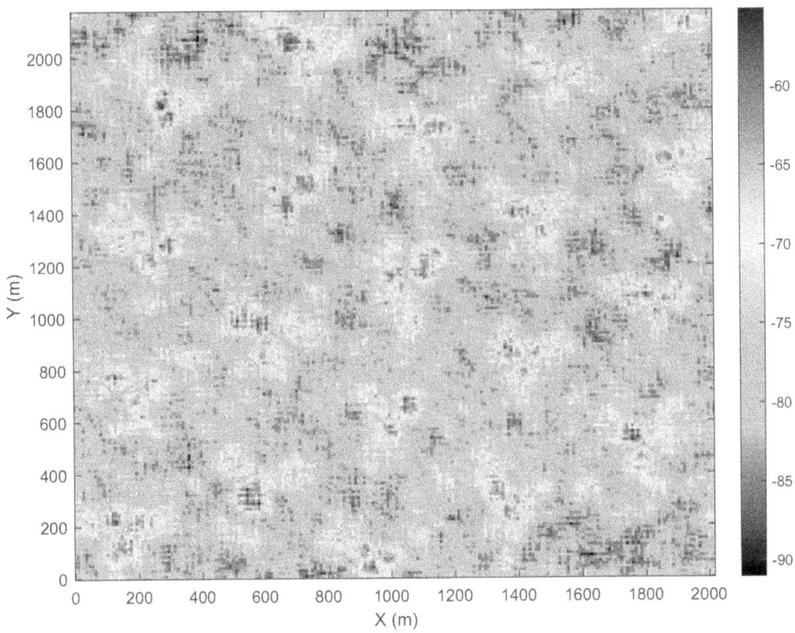

Figure 6.17. Radio environment map showing the path loss with a resolution of 5 m × 5 m.

consumes the lowest energy as the overall energy consumption is dominated by flight. However, this solution may violate the minimum data rate required for command and control and hence may be infeasible for practical use.

To solve this optimization problem, a radio-aware digital twin is adopted to generate a radio-environment map (REM) which contains estimates of the path loss as well as the interference, as shown in Figure 6.17. The REM in Figure 6.17 is the path loss as seen by a UAV flying at a height of 50 m over a 2 sq. km area (although the rest of the section focuses on UAVs, the idea can be also applied to AGVs). Such a REM can be obtained with ray tracing or training a machine-learning model with previously collected data. In addition, the energy consumption can be modelled accurately based on the size of the UAV and the payload carried. It is also straightforward to include the energy required for hovering, turning, taking off, and landing, as well as flying at a certain speed. The model also factors in wind relative to the UAV's direction.

Figure 6.18 shows the flight time and data rate for three different algorithms: the optimal solution, a greedy solution, and the shortest path. As mentioned before, the shortest path has the shortest flight time and the lowest energy consumption as the energy consumption is almost directly proportional to and dominated by the flight time. However, the shortest path also has the lowest aggregate data rate and may be infeasible as it may not satisfy the minimum throughput required for command and control traffic.

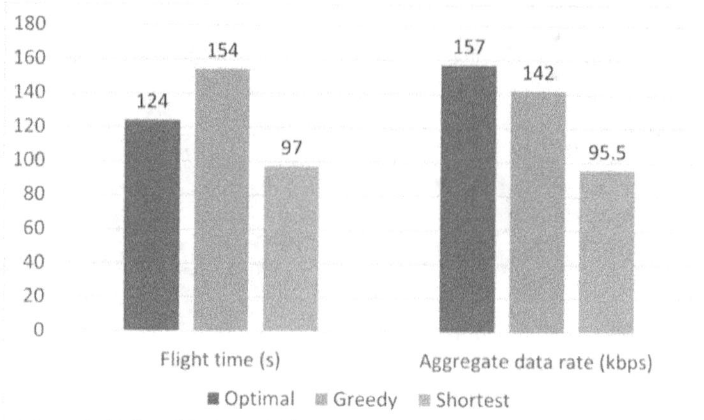

Figure 6.18. Flight time and aggregate data rate for the optimal, greedy, and shortest UAV trajectory.

A simple alternative is a greedy algorithm where the UAV trajectory and speed in the next instant are solely based on a weighted combination of the currently observed path loss (as a proxy for data rate) and energy consumption. While the greedy algorithm offers a much higher data rate than the shortest path, the flight time is shown to be considerably higher, which implies a higher energy consumption.

Finally, the optimal algorithm, which uses dynamic programming to optimize a weighted sum of the path loss and energy consumption, can be seen to offer the highest aggregate data rate but at a much lower flight time when compared with the greedy path, resulting in energy savings of about 25% with respect to the greedy algorithm. In a practical deployment, where it would be necessary to minimize the overall energy consumption to meet sustainability targets, weighting the energy consumption higher than the data rate would result in higher energy savings at the cost of lower throughput.

Note that the proposed algorithm that can optimize the flight time and minimize power consumption is possible only with a REM that is provided by a radio-aware digital twin.

6.3 Summary and Outlook

This chapter presented the European view on the key sustainability enablers of 6G networks. Sustainability was captured from two perspectives: (i) *Sustainable 6G*, referring to the need for 6G itself to be sustainable and mapped to network energy efficiency as well as material efficiency (circularity) and environmental footprint, and (ii) *6G for Sustainability*, referring to 6G as an enabler for sustainable growth in other markets and value chains. Three main targeted KPIs were identified, and a number of key sustainability enablers were detailed in order to meet them. The sustainability enablers were divided into four categories based on the way they achieve sustainability: (i) the enablers *at the deployment level* that include architectural innovations (disaggregated and virtualized RAN) or hardware innovations (energy-neutral devices); (ii) the enablers *at the management/orchestration level* (of algorithmic nature) that target at network operation efficiency maximization (sustainable resource allocation); (iii) the ones *at the service/application layer* (application-aware networks); and (iv) the *cross-layer sustainability enablers* (sustainable radio-aware digital twin) that include innovations in two or more layers, respectively. Although this chapter focuses on the sustainability enablers that target at solving specific problems, it is worth noting that there are many other problems closely coupled to sustainability that require attention, such as the network operation under a limited energy/power budget (e.g., in battery-powered devices) and consequently the optimal power allocation in such a setup. Furthermore, given that the efficient network operation is closely related to its energy efficiency, key enabling technologies presented in other chapters, such as AI, detailed in Chapter 5, are expected to play a key role, as discussed in Section 6.2.2, while allowing many more devices to connect to the network than it is possible today. All in all, it can be concluded that "*sustainability is about to be a core building block of 6G and 6G a core building block of sustainability.*"

References

[1] United Nation (UN), "Sustainable Development Goals," 2023. Accessed: April 6, 2023. [Online]. Available: https://sdgs.un.org/goals.

[2] Hexa-X, "D1.2 – Expanded 6G vision, use cases and societal values – including aspects of sustainability, security and spectrum," 2021. Accessed: April 6, 2023. [Online]. Available: https://ec.europa.eu/research/participants/documents/downloadPublic?documentIds=080166e5dc8b611b&appId=PPGMS.

[3] *Methodology for environmental life cycle assessments of information and communication technology goods, networks and services*, ITU-T Recommendation L.1410 (12/14), 2014.

[4] GSMA, "ICT Industry Agrees Landmark Science-Based Pathway to Reach Net Zero Emissions", Feb. 2020.

[5] Hexa-X, "D1.3 – Targets and requirements for 6G – initial E2E architecture", 2022. Accessed: April 6, 2023. [Online]. Available: https://ec.europa.eu/research/participants/documents/downloadPublic?documentIds=080166e5e90431aa&appId=PPGMS.

[6] 5GPPP Architecture WG, "The 6G Architecture Landscape European Perspective", White paper, Feb. 2023.
[7] *Enabling the Net Zero transition: Assessing how the use of information and communication technology solutions impact greenhouse gas emissions of other sectors*, ITU-T Recommendation L.1480, Dec 2022. Accessed: April 6, 2023. [Online]. Available: https://www.itu.int/rec/T-REC-L.1480-202212-I.
[8] Z. Ghadialy, "Understanding the TCO of a Mobile Network," The 3G4G Blog, 2020. Accessed: April 6, 2023. [Online]. Available: https://blog.3g4g.co.uk/2020/10/understanding-tco-of-mobile-network.html.
[9] *Environmental Engineering (EE); Assessment of mobile network energy efficiency*, ETSI, ES 203 228 v1.3.1, Oct. 2020.
[10] *Study on application architecture for enabling Edge Applications*, 3rd generation partnership project (3GPP), Technical Report (TR) 23.758, v17.0.0, Dec. 2019.
[11] 5G-CITY project, 2020. Accessed: April 6, 2023. [Online]. Available: https://cordis.europa.eu/project/id/761508.
[12] ETSI White paper "MEC in 5G networks", Jun. 2018, Accessed: April 6, 2023. [Online]. Available: https://www.etsi.org/images/files/ETSIWhitePapers/etsi_wp28_mec_in_5G_FINAL.pdf.
[13] *Mobile Edge Computing (MEC); Deployment of Mobile Edge Computing in an NFV environment*, ETSI MEC 017, v1.1.1, Feb. 2018.
[14] 5G-PPP Software Networking Working Group, "Cloud native and 5G verticals' services," Feb 2020. Accessed: April 6, 2023. [Online]. Available: https://5g-ppp.eu/wp-content/uploads/2020/02/5G-PPP-SN-WG-5G-and-Cloud-Native.pdf.
[15] *Network Functions Virtualisation (NFV) Release 3; Architecture; Report on the Enhancements of the NFV architecture towards "Cloud-native" and "PaaS"*, ETSI, NFV-IFA 029, v.3.3.1, Nov. 2019. Accessed: April 6, 2023. [Online]. Available: https://www.etsi.org/deliver/etsi_gr/NFV-IFA/001_099/029/03.03.01_60/.
[16] J. S. Vardakas, K. Ramantas, E. Datsika, M. Payaró, S. Pollin, E. Vinogradov, M. Varvarigos, P. Kokkinos, R. González-Sánchez, J. J. V. Olmos, I. Chochliouros, P. Chanclou, P. Samarati, A. Flizikowski, M. A. Rahman, and C. Verikoukis, "Towards Machine-Learning-Based 5G and Beyond Intelligent Networks: The MARSAL Project Vision," In *2021 IEEE International Mediterranean Conference on Communications and Networking (MeditCom)*, pp. 488–493, 2021, doi: 10.1109/MeditCom49071.2021.9647671.
[17] MARSAL, "D4.2 Initial report on elastic MEC platform design and data-driven orchestration and automation," 2022. Accessed: April 6, 2023.

[Online]. Available: https://ec.europa.eu/research/participants/documents/downloadPublic?documentIds=080166e5ee3d850d&appId=PPGMS.

[18] *Multi-access Edge Computing (MEC); Framework and Reference Architecture*, ETSI Industry Specification Group (ISG) Multi-access Edge Computing (MEC). ETSI GS MEC 003 V3.1.1, Mar. 2022. Accessed: April 6, 2023. [Online]. Available: https://www.etsi.org/deliver/etsi_gs/MEC/001_099/003/03.01.01_60/.

[19] *Network Functions Virtualisation (NFV); Architectural Framework*, ETSI GS NFV 002 V1.2.1, 2014. Accessed: April 6, 2023. [Online]. Available: https://www.etsi.org/deliver/etsi_gs/nfv/001_099/002/01.02.01_60/.

[20] B. Cox, C. Buyle, D. Delabie, L. De Strycker, and L. Van der Perre, "Positioning Energy-Neutral Devices: Technological Status and Hybrid RF-Acoustic Experiments," In *Future Internet*, vol. 14, no. 5, pp. 156, 2022. Accessed: April 6, 2023. [Online]. Available: https://doi.org/10.3390/fi14050156.

[21] C. M. Lastoskie and Q. Dai, "Comparative life cycle assessment of laminated and vacuum vapor-deposited thin film solid-state batteries," In *Journal of Cleaner Production*, vol. 91, no. 15, pp. 158–169, Mar. 2015.

[22] G. Dolci, C. Tua, M. Grosso, and L. Rigamonti, "Life cycle assessment of consumption choices: a comparison between disposable and rechargeable household batteries," In *The International Journal of Life Cycle Assessment*, vol. 21, no. 12, pp. 1691–1705, Dec. 2016.

[23] J. F. Peters, M. Baumann, B. Zimmermann, J. Braun, and M. Weil, "The environmental impact of Li-Ion batteries and the role of key parameters – A review," In *Renewable and Sustainable Energy Reviews*, vol. 67, pp. 491–506, Jan. 2017.

[24] T. Pirson and D. Bol, "Assessing the embodied carbon footprint of IoT edge devices with a bottom-up life-cycle approach," In *Journal of Cleaner Production*, vol. 322, no. 1, pp. 128966. Accessed: April 6, 2023. [Online]. Available: https://doi.org/10.1016/j.jclepro.2021.128966.

[25] REINDEER "D2.1 Initial assessment of architectures and hardware resources for a RadioWeaves infrastructure", Jan 2022. Accessed: April 6, 2023. [Online]. Available: https://ec.europa.eu/research/participants/documents/downloadPublic?documentIds=080166e5e7bcfeeb&appId=PPGMS, doi: https://doi.org/10.5281/zenodo.5938909.

[26] REINDEER, "D1.1 Use case-driven specifications and technical requirements and initial channel model", Jan 2022. Accessed: April 6, 2023. [Online]. Available: https://ec.europa.eu/research/participants/documents/downloadPublic?documentIds=080166e5e2ac056c&appId=PPGMS. doi: https://doi.org/10.5281/zenodo.5561844.

[27] G. Callebaut, W. Tärneberg, L. Van der Perre, and E. Fitzgerald, "Dynamic Federations for 6G Cell-Free Networking: Concepts and Terminology," *2022 IEEE 23rd International Workshop on Signal Processing Advances in Wireless Communication (SPAWC)*, Oulu, Finland, pp. 1–5, 2022. doi: https://doi.org/10.1109/SPAWC51304.2022.9833918.

[28] J. Van Mulders, D. Delabie, C. Lecluyse, C. Buyle, G. Callebaut, L. Van der Perre, and L. De Strycker, "Wireless Power Transfer: Systems, Circuits, Standards, and Use Cases," In *Sensors*, vol. 22, no. 15, pp. 5573, 2022. doi: 10.3390/s22155573. PMID: 35898075; PMCID: PMC9371050.

[29] B. J. B. Deutschmann, T. Wilding, E.G. Larsson, and K. Witrisal, "Location-based Initial Access for Wireless Power Transfer with Physically Large Arrays," In *2022 IEEE International Conference on Communications Workshops (ICC Workshops)*, 16–20 May, 2022.

[30] REINDEER "D4.1 System design study for energy-neutral devices interacting with the RadioWeaves infrastructure", Oct. 2022, to appear online at: https://cordis.europa.eu/project/id/101013425/results.

[31] REINDEER "D3.2 Methods for Communication and Initial Access with RadioWeaves", Oct. 2022, to appear online at: https://cordis.europa.eu/project/id/101013425/results.

[32] S. Kekki, W. Featherstone, Y. Fang, P. Kuure, A. Li, A. Ranjan, D. Purkayastha, F. Jiangping, D. Frydman, G. Verin, K.-W. Wen, K. Kim, R. Arora, A. Odgers, L. M. Contreras, and S. Scarpina, "MEC in 5G networks," The European Telecommunications Standards Institute (ETSI), Tech. Rep. ETSI White Paper No. 28, 2018.

[33] J. Oueis and E. Calvanese Strinati, "Uplink traffic in future mobile networks: Pulling the alarm," In *International Conference on Cognitive Radio Oriented Wireless Networks*, pp. 583–593, 2016.

[34] L. Chiaraviglio, S. Rossetti, S. Saida, S. Bartoletti, and N. Blefari-Melazzi, "Pencil Beamforming Increases Human Exposure to ElectroMagnetic Fields: True or False?," In *IEEE Access*, vol. 9, pp. 25158–25171, 2021, doi: https://doi.org/10.1109/ACCESS.2021.3057237.

[35] J. Galán-Jiménez and L. Chiaraviglio, "Measuring the impact of ICNIRP vs. stricter-than-ICNIRP exposure limits on QoS and EMF from cellular networks," In *Computer Networks*, vol. 187, no. 14, pp. 107824, 2021. Accessed: April 6, 2023. [Online]. Available: https://doi.org/10.1016/j.comnet.2021.107824.

[36] M. Merluzzi, S. Bories, and E. C. Strinati, "Energy-Efficient Dynamic Edge Computing with Electromagnetic Field Exposure Constraints," In *2022 Joint European Conference on Networks and Communications & 6G Summit*

(EuCNC/6G Summit), pp. 202–207, 2022, doi: https://doi.org/10.1109/EuCNC/6GSummit54941.2022.9815625.

[37] 5G-COMPLETE, "D2.3 Final report on 5G-COMPLETE network architecture and interfaces", Jul. 2022. Accessed: April 6, 2023. [Online]. Available: https://doi.org/10.13140/RG.2.2.35846.45128.

[38] L. Liu, S. Gou, G. Liu, and Y. Yang, "Joint Dynamical VNF Placement and SFC Routing in NFV-Enabled SDNs," In *IEEE Transactions on Network and Service Management*, vol. 18, no. 4, pp. 4263–4276, Dec. 2021.

[39] M. Gatzianas, A. Mesodiakaki, G. Kalfas, and N. Pleros, "Energy-efficient Joint Computational and Network Resource Planning in Beyond 5G Networks," In *2021 IEEE Global Communications Conference (GLOBECOM)*, pp. 1–6, Dec. 2021.

[40] M. Gatzianas, A. Mesodiakaki, G. Kalfas, N. Pleros, F. Moscatelli,G. Landi, N. Ciulli, and L. Lossi, "Offline Joint Network and Computational Resource Allocation for Energy-Efficient 5G and beyond Networks", In *Appl. Sci.* vol. 11, no. 22, pp. 10547, 2021. Accessed: April 6, 2023. [Online]. Available: https://doi.org/10.3390/app112210547.

[41] A. Varasteh, B. Madiwalar, A. Van Bemten, W. Kellerer, and C. Mas-Machuca, "Holu: Power-aware and delay-constrained VNF placement and chaining," In *IEEE Transactions on Network and Service Management*, vol. 18, no. 2, pp. 1524–1539, 2021.

[42] International Commission on Non-Ionizing Radiation Protection (ICNIRP), "ICNIRP guidelines for limiting exposure to electromagnetic fields (100 kHz – 300 GHz)," 2020. Accessed: March: 31, 2023. [Online]. Available: https://www.icnirp.org/cms/upload/publications/ICNIRPrfgdl2020.pdf.

[43] M. J. Neely, E. Modiano, and C.-P. Li, "Fairness and optimal stochastic control for heterogeneous networks," In *IEEE/ACM Transactions on Networking*, vol. 16, no. 2, pp. 396–409, 2008.

[44] S. Lakshminarayana and T. Q. Quek, "Throughput maximization with channel acquisition in energy harvesting systems," In *2014 IEEE Int. Conf. on Communications (ICC)*, pp. 2430–2435, 2014.

[45] M. J. Neely, *Stochastic Network Optimization with Application to Communication and Queueing Systems*. Morgan & Claypool Publishers, 2010.

[46] G. F. Franklin, M. L. Workman, and D. Powell, *Digital Control of Dynamic Systems*, 3rd ed. Boston, MA, USA: Addison-Wesley Longman Publishing Co., Inc., 1997.

[47] O. L. V. Costa, M. D. Fragoso, and R. P. Marques, *Discrete-Time Markov Jump Linear Systems*. London: Springer-Verlag, 2005. doi: https://doi.org/10.1007/b138575.

DOI: 10.1561/9781638282396.ch7

Chapter 7

Towards Continuously Programmable Networks

By Dimitris Tsolkas, et al.[1]

7.1 Introduction

While programmability has been a feature of network devices for a long time, the past decade has seen significant enhancement of programming capability for network functions and nodes, spearheaded by the ongoing trend towards softwarization and cloudification. In his context, new design principles and technology enablers are introduced (Section 7.2) which reside[2] at: (i) service/application provisioning level, (ii) network and resource management level, as well as (iii) network deployment and connectivity level.

At the service/application provisioning level, the exposure of Application Programming Interfaces (APIs) from the network core and edge creates new opportunities to third parties for interaction with the network [1]. The work of 3rd Generation Partnership Project (3GPP) SA6 towards vertical application enablers

1. The full list of chapter authors is provided in the Contributing Authors section of the book.
2. The categorization used here is from a programmability and enabler point of view and is not strictly mapped to the Chapter 2 architectural structure.

(VAEs) is a representative paradigm in this field. At the network and resource management level, there are disruptive changes at all the domains of the service provisioning chain. Key enablers can be considered the adoption of the cloud-native approach from the communication service providers (CSPs), as well as the recent work from Open RAN Alliance (ORAN) towards vendor-agnostic management of radio access components. Finally, at the network deployment and connectivity level (including edge compute capabilities exposed to third parties), the deployment of private networks is well specified already, while the concepts of the Cell-Free paradigm and the provisioning of a connectivity mesh topology to end devices are expected to support the vision of a truly flexible access network.

In this context, disruptive architectural and service concepts emerge, anticipating multi-connectivity structures (multiple coordinated point-to-point connections), potentially including edge compute and third-party service function chains. In addition, service abstractions, programmability features, and new application with capabilities to interact and negotiate with the network, are being developed accordingly. The notion of third party is re-contextualised, including two main views: (i) one is within the 6th Generation (6G) system (platform), for network applications that are developed by third parties and (ii) the other is on top of the 6G system (platform), for services provided from third parties through the system (platform) and might be specific to a vertical.

Due to the above-mentioned multilevel evolution, a total transformation of the conventional mobile networks is expected towards operating as open service provisioning platforms. The cornerstone of this transformation is the definition of common interfaces and reference points that enable interaction by third parties with the network functions and nodes at any of the above-mentioned levels and for all the communication planes (control, management, and data plane).

The realization of such interaction is facilitated through various types of interaction-enabling frameworks (representative frameworks can be found in Sections 7.3, 7.4, 7.5, and 7.6) that on their southbound can securely consume native APIs and access network nodes (e.g., switches, gNBs, and NFs of network core) to on-board new applications or enforcing new policies, while on their northbound they can expose (e.g., vertical-oriented) services to support any kind of network-aware application and services (see Figure 7.1). Those frameworks shall take advantage of programming languages, such as the protocol-independent packet processors – P4 [2] (further explained in Section 7.5), as well as common data models, such as the OPC UA[3] information models which refer to vertical-specific companion specifications for the industrial/manufacturing nodes.

3. https://opcfoundation.org/developer-tools/specifications-opc-ua-information-models

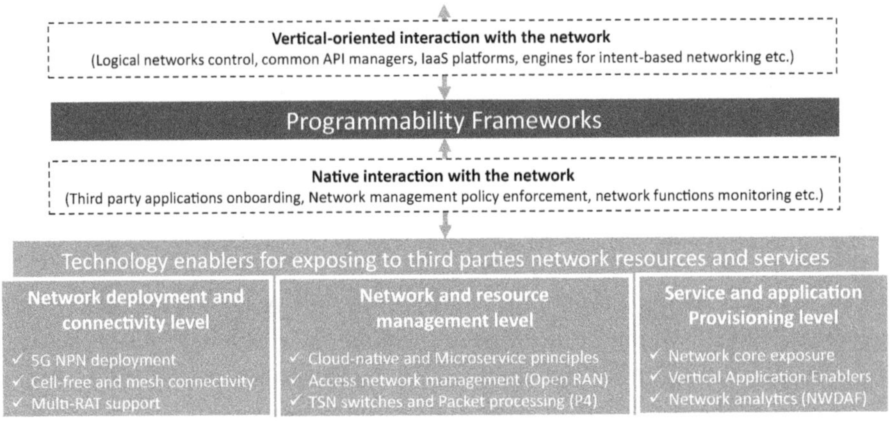

Figure 7.1. Enabling interaction of verticals with the underlay network.

Overall, the implementation of frameworks that exploit common/standardized languages, APIs, and data models provides the means for the enforcement of programmability in the next-generation networks, a concept that incorporates the capability of a device or a network to accept a new set of instructions that may alter the device or network behaviour [3]. A representative example of the programmability potential is the concept of intent-based network (IBN) which has emerged to introduce a layer of artificial intelligence (AI) in the 6G networks. It promises to solve problems of traditional networks in terms of efficiency, flexibility, and security [4, 5]. This technology has revolutionized the way that interaction is performed with systems by starting to communicate through intents. This paradigm is further studied in Section 7.7.

From the business perspective, the above-mentioned programmability frameworks create a new potential around the development of the so-called network applications. Network applications (or Network Apps) are third-party applications that interact (through standardized APIs) with the network to provide network- or vertical-oriented services. Network applications provide network- or vertical- oriented services, meaning that they can assist/enhance either the network operation/management[4] or the vertical application.[5] As third-party apps, the Network Apps should interact with network functions/nodes though open and

4. For instance, in the EVOLVED-5G project, a related contribution to 3GPP SA6 work has emerged (3GPP/TSG SA2/eNA_Ph2 Rel.17: Contribution "Support of DN performance analytics by NWDAF" – S2-2101388) under the scope of extending the NWDAF analytics APIs so that network applications can retrieve data from vertical apps, and the NWDAF build performance analytics and predictions by using inputs from network applications.

5. The ICT-41 projects (5GPPP, phase 3, part 6 projects), work towards providing network applications that fulfil needs and requests from various vertical industries, e.g., automotive (5GIANA, 5GASP), Industry 4.0/manufacturing (5GINDUCE, EVOLVED5G, 5GERA), transport & logistics (VITAL5G, 5GERA),

standardized interfaces/APIs that can reside at any plane (user, control, management) or any domain (core, radio, transport). To support the Network Apps lifecycle, experimentation facilities emerge, which share common principles and functionalities as summarized in Section 7.8 (for more information, see [14]).

7.2 Technology Enablers for Network Programmability

Towards the 6G era, a continuous evolution of mobile systems is foreseen, by the introduction of new design principles and technology enablers. Considering network programmability, those enablers, could reside at: the *service/application provisioning level*, the *network and resource management level*, as well as the *network deployment and connectivity level.*

7.2.1 Enablers at Deployment and Connectivity Level

At the deployment level, programmability can be facilitated by the introduction of on-demand deployment capabilities as well as through extensibility at the access and transport domain. On the one hand, the concept of on-demand deployment refers to the effort towards creating the technologies and the business potential for establishing mobile networks in a similar way that local area networks are currently deployed. This refers to the concept of 5th Generation (5G) private networks [4]. On the other hand, extendible deployment refers to the interfacing capabilities that mobile networks provide for engaging other network technologies at the transport and access domain, such as the N3WIF – Non-3GPP Interworking Function or the Time-Sensitive Networking (TSN) gateway. All these concepts *transform the monolithic communication infrastructure of a mobile network (conventionally dedicated to only mobile network customers) to a platform that can engage different access and transport technologies (i.e., support different types of devices and protocols) and can be fully accessible by the vertical provider when it comes to private and ad hoc developments.*

7.2.1.1 Non-public networks

Already, 5G has brought the concept of the so-called 5G non-public networks (5G NPN) [4]. 5G NPN refers to a 5G system deployment that is dedicated to providing 5G network services for private use (i.e., to a specific organization, e.g., inside a factory). Such a network is deployed on the organization's premises, for instance, inside a factory. The introduction of such networks is beneficial for the verticals for a series of reasons, but, mainly due to the potential to isolate from other (public)

media (5GMediaHub), public protection and disaster relief (5G-EPICENTRE, 5GERA, 5GGASP), healthcare (5GERA).

networks and the capability to provide to the vertical users administrative and management services on the mobile network. This isolation is desirable for reasons such as performance, security, and privacy. For those reasons, the actual implementation and utilization of NPN are expected to be part of 6G networks as well. Already there are a few deployment options for the 5G NPN that can be primarily clustered into two groups.

- Standalone non-public network: an isolated network operated by an NPN operator. The NPN operator can be either the vertical itself, by using locally available 5G spectrum, or by an operator (that potentially deploys also the NPN at vertical's premises), by using licensed 5G frequencies. The key aspect in this category is that deployments do not rely on network functions provided by a public network.
- Public network integrated NPN: a non-public network deployed with the support of a public network. The above-mentioned support can include different levels of interaction with the public network at any domain of the service provisioning chain (RAN, edge, or core). The key enabler for this deployment is the concept of network slicing.

7.2.1.2 Non-3GPP interworking function

From the previous generation of mobile networks, has emerged the potential to unify heterogeneous access networks to mobile cellular technologies (e.g., [6]). 5G materializes this potential, by exploiting the concepts of untrusted non-3GPP access (introduced in Release 15), and trusted non-3GPP access (introduced in Release 16 [7]). This is realized by the addition of non-3GPP access network and wireline access support via the Non-3GPP Interworking Function (N3IWF) component, i.e., the same 5G core network (CN) is used to provide services to a wide range of wireless and wireline access technologies, enabling integration and convergence between new and legacy networks. The way that this concept will be exploited in 6G networks is to be defined; however, the combination of multiple access technologies under a common control and management system is a key feature that has not yet revealed its full potential.

7.2.1.3 Support of time-sensitive networking

Today, the vast majority of communication technologies used in manufacturing are various Ethernet-based technologies (e.g., Sercos®, PROFINET®, and EtherCAT®) [8] and field buses (e.g., PROFIBUS®, CAN®, etc.).[6] To overcome this

6. "Industrial communication technology handbook," 2nd edition, Richard Zurawski, CRC Press, August 2014.

heterogeneity, the objective of the IEEE TSN task group[7] is to provide deterministic services through IEEE 802 networks, i.e., guaranteed packet transport with bounded latency, low packet delay variation, and low packet loss. There have been defined several standards covering aspects, such as synchronization, stream reservation, pre-emption, scheduling, and *Frame Replication and Elimination for Reliability*. Thus, TSN is an important functionality of industrial communication networks. Such industrial communication networks are usually IEEE 802.1-based networks with Ethernet links (non-3GPP networks). The IEC/IEEE 60802 profile[8] specifies the application of TSN for industrial automation and describes what a mobile network (5G or beyond 5G) needs to support. The challenge in future networks is the proper integration of mobile and TSN networks as it requires the acquisition of service requirements that are not yet in the scope of the current 5G QoS framework. This is also related to the capability to achieve end-to-end URLL (ultra-reliable and low latency) communications.

7.2.2 Enablers at the Management Level

At the management level, programmability can be materialized by the cloud toolbox (which has recently supported the network slicing concept) as well as by the design of a flexible and scalable interface with the access and transport domains of the network.

The first concept is based on the way that has been followed for the design of the service-based architecture (SBA) [9]. 3GPP defines the SBA as a set of functional components, known as interconnected network functions (NFs), where each one can use standardized interfaces, or service-based interfaces, to access and consume services of other NFs through an API-based internal communication. Towards 6G, the software industry investigates the capability to improve the modularity of services that are offered through the current SBA. In this context, a service can be broken down into fundamental service components, allowing third-party developers to mix and match components from different vendors into a single service chain.

For the second concept, programmability in the radio access domain is studied by the O-RAN Alliance, through the so-called functional split concept where the functionality of the RAN is softwarized and migrated to enable monitoring and automation services. In a similar way, for the transport domain, programming

7. Avnu Alliance® White Paper, "Industrial Wireless Time-Sensitive Networking: RFC on the Path Forward," Jan 2018.

8. IEC/IEEE 60802 TSN Profile for Industrial Automation, IEC CD/IEEE 802.1 TSN TG ballot.

languages like protocol-independent packet processors – P4 [2] can be exploited for end-to-end forwarding management and performance monitoring.

With the above-mentioned concepts, *the management capabilities at the service provisioning chain are expanded due to the usage of the cloud toolbox, while a use case-specific parading is born where third-party solutions can provide network automation and control solutions on top of flexible access and transport features (e.g., O-RAN and P4-programmability).*

7.2.2.1 Cloud-based services

The softwarization/cloudification of the NF (i.e., their implementation as virtual network functions (VNFs) or containerized network functions (CNFs)) is an evolution that brought cloud-based management services to the telecommunication sector. Indeed, extensive capabilities and tools primarily designed for cloud-native applications are provided based on the following aspects:

- Realization of Core NFs as Micro-Services: The realization of the 5GC NFs as micro-services (on containerization engines) provides capabilities, such as agile 5GC creation and flexible deployment, while it enables automated and lightweight lifecycle management. The main benefits are (i) the stateless implementation of the NF that facilitates the on-the-fly migration of the 5GC functions in different domains (e.g., to the edge) and (ii) the highly efficient sharing of the infrastructure resources. The cost for those benefits is the complexity of network-based chaining of the containers, as well as the additional computation cost required for the APIs to expose and consume processes.
- Adoption of Agile software development methods: This is a requirement of the new era in service provisioning, where the services should be able to continuously and easily update with improvements that reflect changing market demand. Such an approach is the DevOps methodology that integrates software development and IT operations.

In the same context, network slicing has emerged as a cutting-edge technology that allows for the creation of multiple virtual networks on top of shared physical infrastructure, allowing operators to provide portions of their networks that meet the needs of various vertical industries. A network slice is a collection of multiple sub-slices from various domains, such as the Core Datacentre, the transport network, and one or more edge locations. A key tool for the realization of network slices is the exploitation of ETSI VNF principles and existing frameworks, such as the Open-Source MANO[9] (an ETSI-hosted project to develop an Open Source

9. https://osm.etsi.org/

NFV Management and Orchestration software stack aligned with ETSI NFV). By providing specialized virtualized network slices for each vertical, network slicing will play a crucial role in addressing a variety of vertical applications. This is because new interactions and uses are anticipated to be brought about by the future ecosystem of smart connectivity. Section 7.3 provides a related solution, specified to the management of interconnected network slices though a logical network-as-a-service (LNaaS) approach.

Already network slicing tools have been released, such as the Katana Slice Manager[10] and the Open Slice framework.[11] In general, a network slice manager gets network slice template (NEST) for generating network slices through the north bound interface and offers the API for controlling and monitoring them. From Release 16 and onwards, 3GPP is working on providing higher flexibility and better modularization of the 5G System for the easier definition of different network slices and to enable better re-use of the defined services. Towards the enhanced service-based architecture (eSBA), as defined in Release 16, a service communication proxy (SCP) is introduced that can have a role in service selection, load balancing, and other common functions. This is an effort that supports the transformation of 5GC architecture to be as much compatible with cloud native as possible.

7.2.2.2 Procedure-based service structure – organic architecture

All the telecommunication network architectures were based on the concept of network functions as representing the basic functional element of a telecom network defined and standardized by its input interfaces, output interfaces, transfer function, and subscriber state. Ultimately, a network function is a system by itself that functions independently of other network functions which enables their parallel development and testing in isolation preparing them for system interoperability tests.

However, the complete independence of the network functions as part of the same subscriber connectivity service has significant limitations, which increase the complexity of the overall system, as illustrated in Figure 7.2:

- Maintaining a logically independent state in each of the network functions is creating an information multiplication as well as it requires many messages to be exchanged between the components, especially the acknowledgment that a specific step of a procedure was executed correctly. The additional steps create additional synchronization and scheduling functionality in the components themselves as well as prolong the end-to-end procedure duration.

---

10. https://github.com/medianetlab/katana-slice_manager

11. https://openslice.readthedocs.io/en/stable/

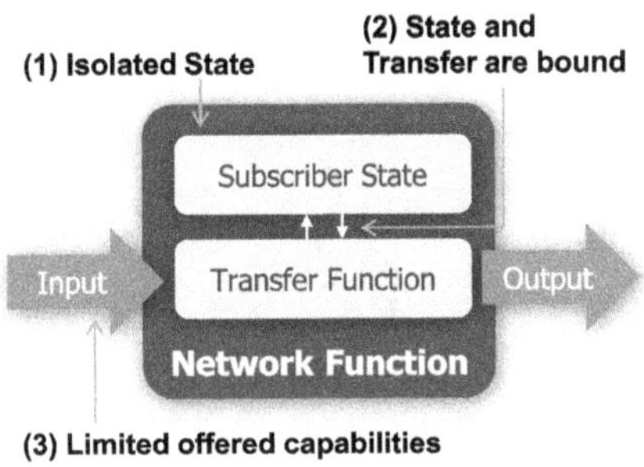

Figure 7.2. Network function concept elements.

- The state and the transfer functions are directly bound to each other. No matter if the subscriber state is grouped in a separate *Unstructured Data Storage Function* as in the 5G core network, it still must be fetched for at the beginning of the execution of each procedure step, modified, and pushed back at the end. These steps are executed in each of the network functions which receive the messages during the procedures inducing an additional execution delay.
- For the system to easily interoperate, the input interfaces must be clearly defined. This complete definition is also defining which type of messages can be received and processed. When aiming to introduce a new network function into the system, it must either use the existing input interfaces, thus the existing limited exposed functionality, or would require the modification of the other network functions. Considering that the 5G core network already has a very high split of functionality, when aiming to introduce a new service, represented by one or more new network functions, many interfaces would need to be modified. Due to this large entanglement, the system behaves highly monolithically in regard to adding new functionality.

However, as the core network functionality is expected to remain software only, the current network function concept does not have to be maintained. Instead, other concepts from Information Technologies (IT), specifically from web services can be adopted for a more flexible [10], reliable [11], and less complex network system [12]. Instead of splitting the functionality into network functions to describe different functionality in the core network, the network is split directly into services, as illustrated in Figure 7.3.

First, the core network is seen as a single large web service able to offer multiple services directly to subscribers. The user equipment (UE) is communicating with a single front-end component. The front end has very similar functionality to the proxy-call state control function (P-CSCF) in the IP multimedia system (IMS).

Instead of splitting the processing of the workers based on network functions, a new split is considered based on the procedure that has to be executed by the

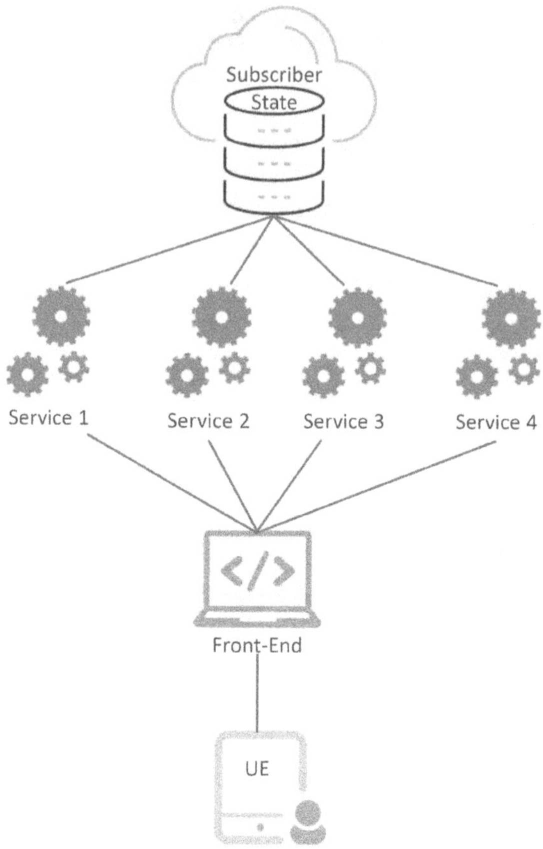

Figure 7.3. Services concept adapted for organic core networks.

system. Depending on the type of request transmitted by the subscriber, the front end will send the requests to a specific worker which will execute all the steps of a specific procedure. After receiving a request, the worker is fetching the subscriber state from the data base. Using the request and the subscriber state information, the worker will process all the steps of the request and generate new state information to be pushed to the database and the response to be transmitted to the subscriber.

Adapted to the current 5G network functionality, the concept will translate into the following functional split, illustrated in Figure 7.4.

Front end – a network component that receives all the requests from the UE. A security association is created with the FE during the initial authentication and authorization. This secure session is used to exchange all the messages of the UE. For this, a connection state is maintained with the UE. However, this state is independent of the connectivity service state as provided by the packet core. It represents only the secure association with the core network and not the subscriber connectivity session. And this could be skipped, although not recommended, if secure

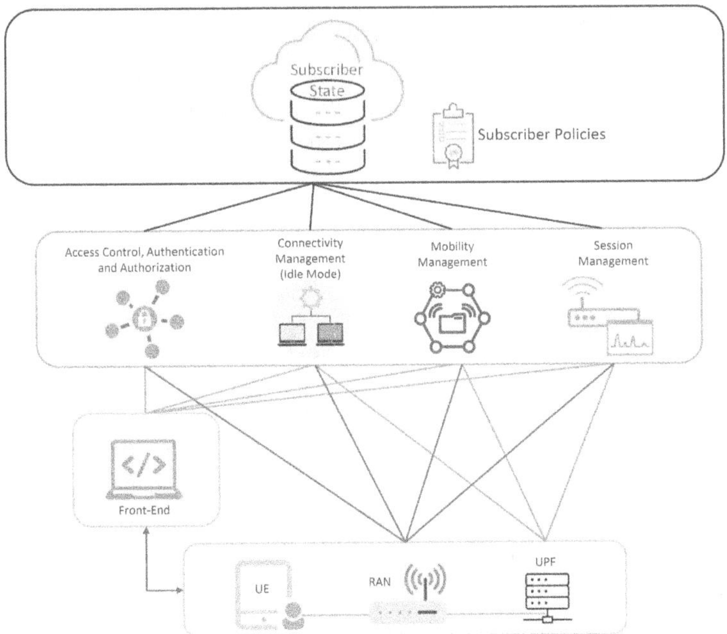

Figure 7.4. Organic core networks concept applied to 5G networks.

communication of the UE is not a requirement. The front-end role is like the access and mobility function (AMF) role in the 5G architecture in receiving the non-access stratum (NAS) messages of the UE. Also, it schedules the messages to specific workers, like the AMF for session management.

Workers – the workers are stateless components that execute the specific procedures of the core network and respond to the specific services. As a minimal example, the following basic services are considered:

- Access control, authentication, and authorization – this worker is executing the registration procedures with all their steps being equivalent to the 5G core AMF and authentication service function (AUSF) and to the 4G mobility management entity (MME).
- Connectivity management (idle mode) – a single service handling all the functionality related to the entering idle mode state, the subscriber-related tracking, service activations, and paging.
- Mobility management – facilitates the UE handover procedures similarly to the 5G NG Application Protocol (NGAP) UE. Compared to the 5G system, it is in charge of the execution of the handover procedures as well as the execution of the resource reservations during handovers, previously executed by the session management function (SMF).

- Session management – enables the resource reservations for the UE including dedicated resources. This worker is equivalent to the SMF main functionality.

Subscriber state – the subscriber state includes all the information that the core network maintains for a subscriber. With all the information maintained in the same place, there is no need for additional procedures of synchronizing state. This would significantly reduce the number of messages required for the communication.

7.2.2.3 Flexible and scalable management of access network

The potential to spit the radio access domain has emerged as an effort to softwarize as much as possible from the access functionality. This allows for (i) hosting centrally management functionality, a choice that can lead to an optimal (at the system level and not as conventionally at the base station level) management of multiple radio access nodes, (ii) supporting the networks slicing concept to the far edge of the service provisioning chain (access part of the network), and (iii) enable RAN management applications based on automatic control loops. Such an approach requires vendor-agnostic interoperability among different components, such as the central units, the distributed units, and the radio heads. In this direction, common interfaces are developed and open challenges are addressed by the O-RAN Alliance.

O-RAN Alliance was founded as a group of vendors, operators, and academic contributors towards producing standards to allow an inter-operable Open RAN deployment. It was formed by a merger between two different organizations, the C-RAN Alliance and XRAN forum. Three main control loops are defined in O-RAN architecture, as presented in [13]. They run in parallel and they interact with each other depending on the use case.

- Non-real-time RIC control loops have a timing of 1 second or more. rApps are used in this control loop as an independent software plug-in to provide extra functionality.
- Near-real-time RIC control loops have executing time between 10 ms and 1 second. xApps are used in this control loop to provide extra functionality.
- O-DU Control loops have a timing of less than 10 ms, but it should be noted that it is still not standardized as of yet by O-RAN, and the main focus is on the first two control loops.

Section 7.4 provides a representative solution for hardware and software programmability at the RAN domain, based on FlexRIC [14], which is in line in line with the reference O-RAN RIC approach, and provides better performance compared to the O-RAN's reference implementation.

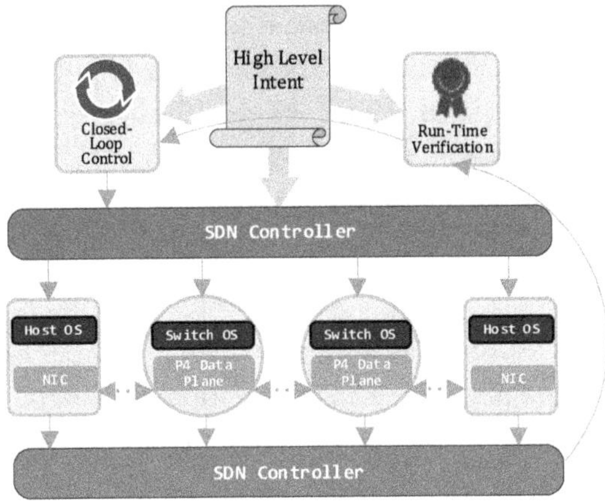

Figure 7.5. Reference architecture supporting closed-loop control with E2E programmability.

7.2.2.4 Flexible and scalable management of transport network

In an IP-based transport network, programmability could be applicable at the control and management planes, as well as at the data plane. For example, the function running on a programmable device can be changed from performing IPv4 routing to any other function simply by loading a new NF programme on the programmable networking device. This enables agile updating of functionalities executed on the UE and network side when programmability on these two ends is enabled.

Network deployments supporting closed-loop control with E2E programmability [15] can enable more effective resource management and operational robustness. P4-based E2E network programmability addresses packet forwarding policies as well as fine-grained telemetry. Fine-grained telemetry is supported by altering packets to contain information about the packet provenance (e.g., the path they took, the rules they followed, the delays they encountered, etc.). In network deployments supporting closed-loop control with E2E programmability, it is feasible to identify mistakes and make necessary repairs within milliseconds by keeping an eye on the network state and validating against packet telemetry.

A reference network architecture supporting closed-loop control with E2E programmability and fine-grained telemetry [15] is illustrated in Figure 7.5.

The elements of the architecture shown in Figure 7.5 and described in [15] are as follows:

- The data plane is made up of programmable switches and/or SmartNICss, which are usually realized using FPGA providing packet processing at line rate while being able to instantly update the forwarding behaviour in response to changes in the local state.

- The switch and host operating system includes the supporting IP network routing protocols (such as Open Shortest Path First (OSPF), P4 Runtime, and others), the associated algorithms, and the aggregation of measurement information received from the data plane.
- The SDN controller executes complex algorithms that would be challenging to represent as distributed computations while maintaining a network-wide perspective of the current topology and operational conditions.

The 5G System (5GS) of today already enables a number of closed-loop control use cases in the data plane. The above-mentioned fine-grained telemetry and UP programmability could further boost the effectiveness and performance of the network. These use cases are of interest:

- Traffic load-balancing control when multi-access protocol data unit (PDU) sessions are used supporting different access networks, including untrusted and trusted non-3GPP access networks, wireline 5G access networks, and so on.
- Traffic load-balancing control when redundant UP paths with dual connectivity (DC) are used.
- Traffic load-balancing control when the redundant transmission on N3/N9 interfaces is used.
- Control of ultra-reliable low-latency communication (URLLC) services.
- Decision for relocation of the UPF when acting as the PDU Session Anchor.

We believe that 6G will require full E2E programmability of the data plane with all the involved user plane nodes and the UE (controlled via NAS protocol). Currently, there are various technologies and programming abstractions that allow P4 E2E programming networks and end-hosts as explained in Section 7.5. Complementary to this, the open-source initiative by ETSI to further develop and evolve the TeraFlow SDN controller is presented in Section 7.3.

7.2.3 Enablers at the Service/Application Level

The openness at the service/application level is reflected in the wide adoption of APIs-based interaction in the telecommunication sector. Already, during the last decades, the use of APIs has served as a bridge between mobile operators and start-ups in emerging markets.[12] Operators have begun to consider whether to open their APIs, starting form APIs related to mobile messaging, operator billing, and so on. Irrefutably, this openness creates a powerful cycle of innovation as start-ups/third parties can combine several APIs to create new services. For example, a

12. https://www.gsma.com/mobilefordevelopment/wp-content/uploads/2016/07/GSMA_Mobile-operators-start-ups-in-emerging-markets.pdf

start-up can offer SMS-based localized content to its users depending on their city or area and then charge them by deducting the amount from their mobile airtime. In the same direction, the TM Forum's 60+ REST-based Open APIs are developed collaboratively by TM Forum members working on the Open API project [16].

Overall, the availability/exposure of APIs from the network is the major enabler for network programmability, since it is a key necessity for any other technology enabler at management and connectivity layer described in previous sections.

At the network core domain, the network exposure function (NEF) of the SBA, exposes capabilities, provided by the other 5GC NFs, to external systems, i.e., NEF offers the appropriate APIs for the usage of SBIs from externals. More precisely, NEF enables external applications to communicate with the 5G core's SBA via APIs (i.e., Secure provision of information from an external application to the 3GPP network). Practically, it adapts and transforms telecom network interfaces to RESTful APIs. Considering the SBA, the main consumer of NEF APIs is the Application Function (AF). AF may or may not reside in the domain of the infrastructure owner (operator's domain), and its main functionality includes the provision of Packet-Flow Descriptors (PFDs) to NEF, and the consumption of RESTful APIs to utilize services and capabilities securely exposed by NEF. This openness of the 5GC provides third-party innovators (vertical industries) with the required toolset to (i) control the service deployment process, (ii) monitor the performance of their services, and (iii) feed 5GC NFs NF with information from the application layer.

The same scope of 5G-core openness to third parties is also served by the APIs that are developed on top of NEF, namely, the VAEs [17], the service enabler architecture layer (SEAL) services, as well as edge services through the multi-access edge computing (MEC) APIs. Verticals and operators can develop their own APIs and expose them securely by utilizing the 3GPP Common API Framework (CAPIF) [18].

Section 7.6 provides a representative development where an API programmability framework for interaction with the network core is provided as an open-source project.

7.3 Programmability Through ETSI TeraFlow SDN

This section includes anticipated service capabilities and concepts enabled by interconnected public and private networks in the context of 6G networks and slicing. The *Logical Networks as a Service* (*LNaaS*) concept and service model will become far more extensive, flexible, and pervasive as compared with 5G. While 5G is basically offering advanced connectivity service with the support of QoS from the UE/device to a data network (identified by a data network name), *with 6G, it is expected that a richer topology of connectivity can be supported, controlled and managed as a service, as an effort to support good or better customer/user experience, while achieving improved overall resource utilization and network performance.* We consider even connections from/to multiple UEs/devices to/from multiple data network gateways and/or edge compute nodes/facilities. Complementary to the UE/device connection services, a mesh connectivity is expected among the edge compute facilities (cf. edge continuum) to support distributed application functions and services to the UEs/devices.

To support and complement this edge- and device-oriented connectivity, complementary connectivity capabilities are needed that will manage to reach remote end-points or vertical enterprise sites, in order to enable roaming services or third-party application services not located on the edge. For instance, there can be reached end-points located in the local operator network, or in a medium or long distance away across public interconnected networks. The above-mentioned enhanced connectivity can be based on basic Internet access services, or specialized connectivity services, whether enabled by a multi-mode extended Internet or some dedicated single or multi-stakeholder network complementing the current Internet.

We anticipate that the above will build from and extend the LNaaS concepts and offerings of 4G (based on the access point name, APN) and 5G (data network name, DNN, extending APN-based capabilities into a 5G slicing context) towards 6G, where the logical networks can be dedicated for an online application provider, a vertical enterprise customer (VEC) or established along with specific specialized application services offered by a CSP. Those logical networks should consider and enable a broad variety of advanced 5G and emerging 6G use cases, including integrated connectivity and compute service models and wireless end-points, where combined wireless communication and sensing is part of the 6G landscape, as presented in [19].

7.3.1 Transport Network Slice as a Service

Considering the LNaaS context, in the TeraFlow project,[13] logical networks are oriented towards VEC or OAP and they are supported and enabled by the concept of transport network slice as a service (TNSaaS). The TNSaaS concept must cater to elements and capabilities such as:

- Managed specialized connectivity services on-demand, with richer topologies and flexibility to accommodate the 6G-driven edge-cloud continuum.
- Supporting interconnection between private and public networks in various contexts and scenarios.

---

13. https://www.teraflow-h2020.eu/

- The UEs/devices can be attached to public networks, fully private networks, or private networks based on slicing in public networks; also, extending L2/L3 VPNs, in combination with the above.
- Including a variety of business model enablers to support existing and future business models.
- Supporting a variety of tenant networks, considering multi-stakeholder partnering, and security support.

All the above are enabled through further standardization and open-source initiatives, having as cornerstone SDN control and management, namely, the ETSI TeraFlow SDN (TFS) [20] controller, as described below. We consider the concept below as representing a baseline from which additional requirements driven by 6G can emerge while identifying gaps in today's standards and open-source initiatives.

ETSI TeraFlow SDN controller

TFS controller and the service concepts and information models developed in the Teraflow project are targeted to enable and facilitate logical network as a service to (vertical) enterprise customers as well as operator internal and inter-NSP connectivity and networking (cf. TNSaaS). TFS is constructed as an open-source software project within the ETSI community and is developed under an Apache2 license.

Figure 7.6 outlines the architecture and functional components of the TFS controller that will facilitate the deployment and operation of TNSaaS for Beyond 5G (B5G) 6G slices.

Several data models for services and devices were required to facilitate the request and tear-down of end-to-end slices for future applications and services. Also, industry standards and specifications are considered in the definition of the architecture of the TFS controller, crating in that way the potential for tangible impact in standardization and open-source communities. Figure 7.7 depicts related standard defining organizations (SDOs) and open-source software communities. It also highlights potential overlap among them.

Ongoing work is analysing and proposing service concepts and enablers to align ongoing work in different standardization camps, leveraging the core TeraFlow concept of transport network slice as a service, and developing the enabling functional components, APIs, and data models.

The TeraFlow SDO development activity is a parallel effort managed by technology experts from the project team. This activity has specific contributions in mind and participation in supporting TFS Controller development goals and interoperability plans; this SDO effort is categorized below with the technical area and objectives summarized:

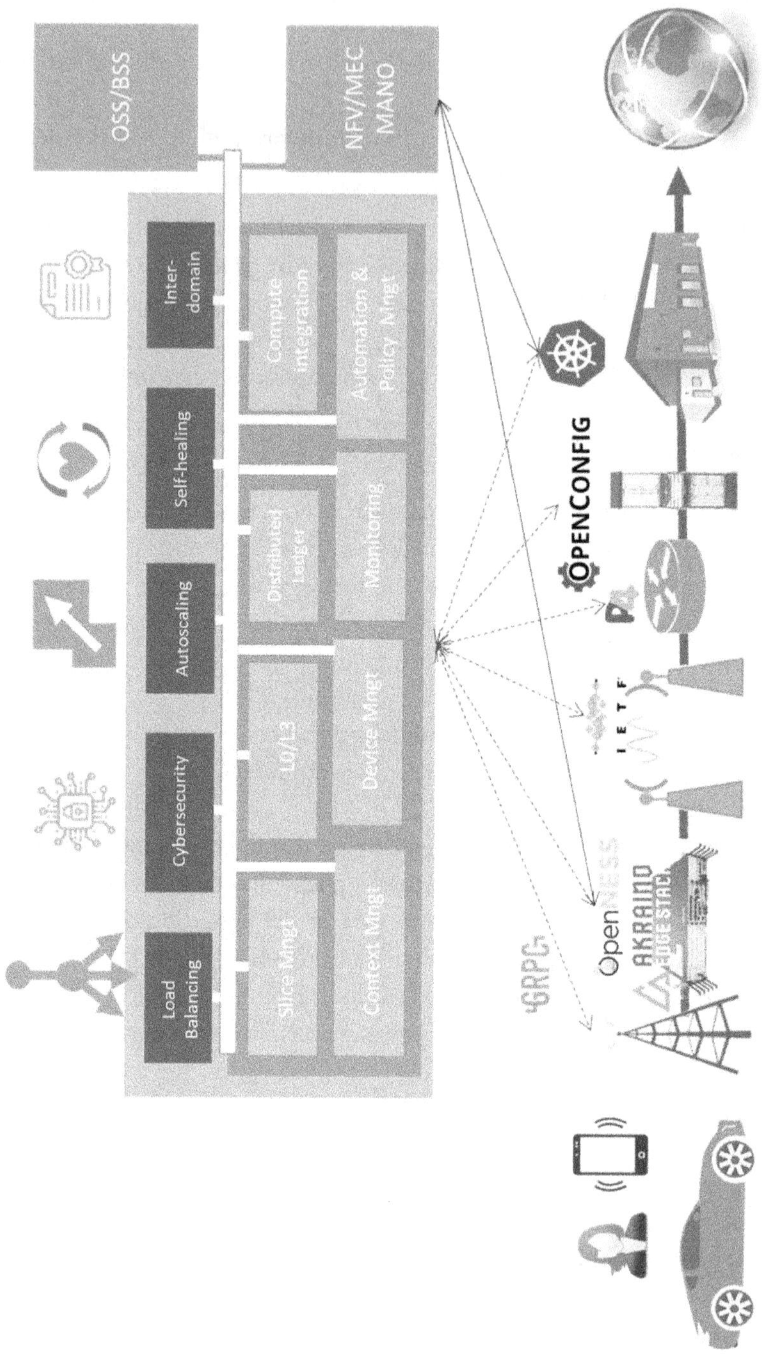

Figure 7.6 TFS controller architecture.

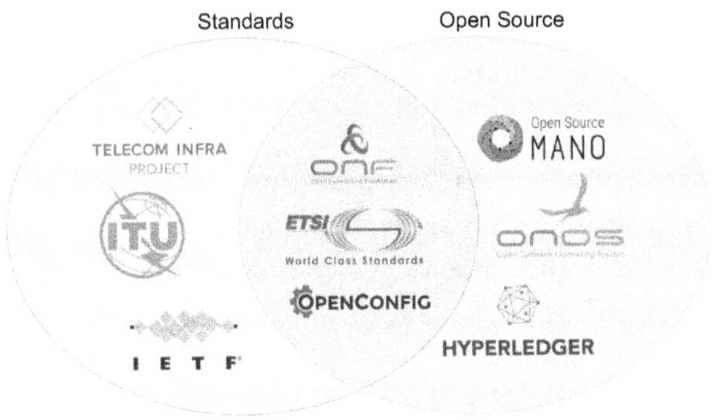

Figure 7.7. SDOs and open-source software communities related to the TFS controller.

- **TFS service and slicing models:** This IETF-based activity develops industry-recognized service models that will allow for end-to-end layer-3 and layer-2 network services to be requested. Additionally, the IETF has developed a network slicing framework. It establishes the general principles of network slicing in the IETF context and how it relates to non-IETF transport networks such as 3GPP network slices. TeraFlow project partners are authors of crucial service models [21], the project leads the editing and development of the IETF network slicing framework document [22].
- 3GPP Network Slices
  - Network operator internal slice to enable 5G operator services.
  - 5G slice is offered as a service.
- **Open networking foundation (ONF):** With a key objective of developing a cloud-native and scalable SDN controller (TFS), standardization efforts related to the development of data models and interfaces enable a hierarchy of controllers. In this sense, the ONF Open Transport Configuration and Control (OTCC) project aims to promote common configuration and control interfaces for transport networks in SDN. One of the project work items is the specification of the transport application programming interfaces (TAPI) data models, publishing open standard interfaces, whose main application domain is the controllers north bound interfaces. Lately, TeraFlow community has participated in the development of the TAPI Reference Implementation Agreement.
- **Telecom infra project:** The Open Optical and Packet Transport (OOPT) project of TIP works on the definition of open technologies, architectures, and interfaces in transport networks. Telefonica and SIAE are key members of TIP and active contributors in multiple OOPT subgroups.
- **Distributed ledger technologies:** The ETSI ISG PDL (Industry Specification Group Permissioned Distributed Ledger) facilitates the utilization of

blockchain technologies for the creation of open and trustworthy ecosystems of industrial digital solutions. TeraFlow project partners lead several activities in this ISG.

These open standards will need further development for B5G, 6G, and beyond. In addition, as new use cases, requirements, and emerging applications and services are identified, new slice models and techniques must be considered. Initial considerations are provided in the fourth version of the 5GPPP White Paper on the 5G and beyond architecture.[14]

As 6G research continues, several emerging applications and services have been identified. These new types of network applications include:

- Tactile Internet of senses.
- Embedded intelligence and connected machines.
- Cognitive networks.
- Device and network twinning.

Investigation and discussion have already started, impacting how logical network slices are requested, designed, and deployed to support the previously mentioned 6G applications and services. There will be specific standards and open-source software required that would provide building blocks, including:

- Highly distributed Cloud-native infrastructure and orchestration, supporting massive scale rapid turn-up of container-based virtual functions.
- Location awareness and trust zones, as users and applications are attached to multiple network locations.
- Green and sustainable network technologies, especially the increased use of photonic systems (using up to 70% less power than electron-based communication).

Entirely new routing and addressing architectures may also be required to meet the demand of emerging 6G applications and services. For example, within the TeraFlow project, a study has been initiated on a new proposal called semantic routing and addressing [23], where packet forwarding and path selection decisions are based on contextual information carried in the packet header.

7.4 Programmability Through O-RAN-Compliant SDK

Software- and/or hardware-based solutions operating in the RAN, edge, transport, and core segments of E2E 5G/6G infrastructure provide programmable data path

14. https://5g-ppp.eu/wp-content/uploads/2021/11/Architecture-WP-V4.0-final.pdf

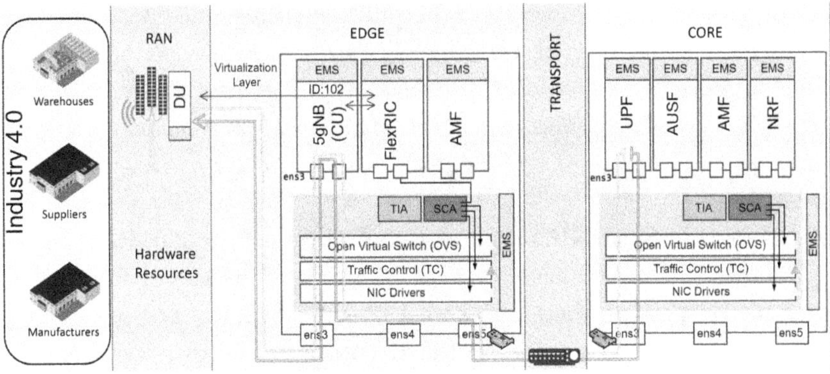

Figure 7.8. E2E programmable data path and data path network slicing.

and advanced network slicing capabilities, which allow user-definable flexible classification and prioritization of the traffic in complex networking environments featuring 5G/6G encapsulation overlays and industrial protocols such as EtherCAT and so on due to vertical business use cases and guarantee the QoS of the prioritized traffic accordingly. In this context, software solutions based on Flexible RAN Intelligent Controller (FlexRIC), Traffic Control (TC), Open vSwitch (OVS), and so on can offer cost-efficient programmable data path capabilities in the kernel of the system, while Smart Network Interface Cards (SmartNICs) such as NetFPGA and Netronome can provide performance boost benefiting from hardware acceleration. Furthermore, software- and hardware-based solutions can be exploited in a hybrid manner wherever appropriate along the E2E path. Figure 7.8 shows an architectural view of an E2E programmable data path and corresponding data path network slicing in 5G and beyond networks for Industry 4.0 and other use cases.

In this architecture, FlexRIC, TC, OVS, and SmartNICs operating in the data plane achieve E2E network slicing by leveraging the programmability of these data path components.

FlexRIC [24] represents a software-defined RAN controller, built through a server library, controller-internal Applications, and optionally a communication interface, all offered by the FlexRIC SDK. The FlexRIC SDK consists of server and agent libraries that are used to build *controller-internal applications*, which, on the one hand, help specialize the RAN controller towards specific use cases and, on the other hand, enable *external applications* (xApps) to conveniently control different RAN functions. In the case of network slicing, the fast control loop of FlexRIC can enable reinforcement learning-based xApps to correctly follow their Markov decision process modelling of the RAN slice scheduling problem, as if they are running in a "zero delay state-to-action" simulation. Furthermore, FlexRIC convenient scalability in supporting multiple xApps of different languages also allows developers to conduct complex orchestration of multiple components, such as machine learning operations. Together, developers can expect to have their simulation-trained xApps ready for production. Overall, *the controller-internal applications allow modularly building specialized RAN controllers for specific use cases, and optionally, they can expose information to external applications (xApps) via a northbound communication interface to allow the xApps to control the RAN in line with the reference O-RAN RIC approach.*

TC, OVS, and SmartNICs represent controllers for the non-RAN segments (edge, transport, and core networks). TC [25] is a Linux user-space utility programme for configuring the Linux kernel packet scheduler. The scheduler with advanced queuing disciplines and dedicated filtering expressions (covering 6G network traffic) can optimize and guarantee performance, reduce latency, and increase usable bandwidth for services with specific network requirements. TC is recommended in systems that are not equipped with higher-performance programmable data-plane network devices such as those based on SmartNICs or kernel bypass techniques, or in those scenarios where the link (L2) and network (L3) layers of a resource are required to work together. OVS [26] is a software switch in virtualized network environments, and it can be extended to allow 5G/6G-compatible data path network slicing with customized network slicing extension profiles and network slicing extension features. Experimental results show that this approach is able to provide connectivity for up to 1 million IoT devices in mMTC traffic, achieve over 19 Gbps bandwidth in congested eMBB scenarios, or ensure delays in the order of μs for critical-missions communications, providing high reliability in all tested scenarios (0% packet loss ratio). As to SmartNIC-based controllers, NetFPGA [27] and Netronome [28] are promising platforms, among others. For instance, NetFPGA can be explored to achieve a data path network slicing enabler based on the SimpleSumeSwitch model and by leveraging the P4-NetFPGA project. Performance enhancements can be expected in certain use cases thanks to hardware-based acceleration, as reported in [29].

In the control plane, the *Slice Controller Agent* coordinates the various data plane controllers in collaboration with the network slice management plane (not shown in the figure), and it is network topology aware with the help of the *Topology Inventory Agent* to initiate and control the network segment specific controllers. More details of controller-specific implementation architectures can be found in [30].

7.5 P4-Based Framework for E2E Programmability

7.5.1 Network Programmability with P4

We choose the P4 language [2] as an example of programming abstraction to show the benefits that can be achieved when programmability is employed in networks.

The P4 language is one initiative to abstract the packet processing pipeline of network devices and flexibly define its behaviour. P4 is platform-independent, meaning that it can be used to programme various classes of packet processors such as software switches, SmartNICs (NIC stands for network interface card), field-programmable gate arrays (FPGAs), and application-specific integrated circuits (ASICs). These different classes of packet processors can be used to build the infrastructure of the next cellular generation clouds.

7.5.1.1 Performance-aware management of programmable networks

When using programmable networking devices, it is important to understand their forwarding performance and to guarantee certain performance levels. This is achievable when utilizing models that can predict the packet forwarding latency when running any P4 programme on arbitrary P4 devices as it has been proposed in [31]. This model is based on an extensive measurement campaign that identifies the base processing delay of different P4 devices, in addition to the marginal delay of executing atomic P4 operations on these devices. First, performance profiles for different types of packet processors are created using the collected measurement results. Then, any given P4 programme is decomposed to its atomic operations whose processing latency is already quantified and recorded in the pre-developed performance profiles. For example, in the case of IPv4 Forwarding NF, these atomic operations are extracting and manipulating Ethernet and IPv4 headers. Finally, the performance model estimates the packet forwarding latency when running arbitrary NFs on any P4 device as the sum of the base processing delay of that device and the marginal latency cost of executing the NF's constituent atomic P4 constructs. The model showed sub-microsecond prediction accuracy when tested for different NFs and P4 devices. To optimize the management of programmable networks, the optimal placement of NF workloads on heterogeneous programmable substrates should be investigated. We propose in [32] a mathematical formulation for an optimization problem that aims to optimally place different NFs into P4-based cloud environments. The network orchestrator can use the performance model described above in [31] to perform the placement in a performance-aware manner. This allows for achieving the highest levels of performance when managing these programmable networks and meeting QoS requirements. The problem formulation takes into account various capabilities and characteristics of the hosting P4 device. These device characteristics include the supported P4 architecture, supported extern functions (i.e., non-native P4 functions such as cryptography), basic processing latency, marginal latency for executing atomic P4 constructs, latency for executing P4 extern functions, throughput capacity, processing resources availability, and cost.

In addition, the problem formulation considers the NF workload's requirements in terms of desired throughput, desired P4 architecture, desired P4 extern functions, and constituent P4 constructs.

For a given NF workload, the objective function is to find the optimal set of P4 devices and the optimal embedding of NFs into these devices while minimizing both the total forwarding latency in the system and the capital cost for building the system. The performance model described before (as that in [31]) is used to calculate *a priori* the delay that results from different placement options as shown in the following equation:

$$\Delta_d^f = \delta_d^{BP} + \sum_{c \in C_f} \delta_d^c + \sum_{e \in E_f} \delta_d^e$$

where Δ_d^f is the forwarding latency when running NF f on P4 device d. It is equal to the sum of three components: (i) the base processing latency on P4 device d, denoted as δ_d^{BP}; (ii) the sum of the marginal latency when running the atomic P4 constructs $c \in C_f$ that constitute NF f on P4 device d denoted by δ_d^c; and (iii) the sum of the latency of extern functions $e \in E_f$ required by NF f denoted by δ_d^e.

Several constraints should be satisfied. These include the following: (i) an NF should be placed on a P4 device only if the device supports a compatible P4 architecture and includes all the extern functions required by the NF; (ii) the sum of throughput required by different NFs to be placed on a device shall not surpass the limited throughput of that device; (iii) the processing resources capacity of a P4 device shall not be surpassed; and (iv) some NFs like the IPv4 forwarding NF must be placed on every used device to guarantee proper packet forwarding between devices, and so on.

To evaluate the proposed workflow, the target use case includes a combination of different types of P4 devices to build a programmable substrate for a cloud environment. These devices belong to different classes of processing platforms such as CPU, FPGA, NPU (network processing unit), and ASIC. The capabilities and performance of these devices are already studied in the literature, and a summary of these findings can be found in [32]. Figure 7.9 depicts a web chart that summarizes the comparative advantages of different P4 device types based on different criteria. A set of common NFs such as IPv4 forwarding load balancer, and so on are selected as the workload to be handled by the network. More details about the evaluation in terms of the requirements of the used NFs and the capabilities of the selected P4 devices can be found in [32].

Four scenarios are defined and evaluated wherein the weights of the two selected objective functions vary (i.e., the forwarding delay in the system and the cost of building the system). Scenario 1 (S1) targets achieving the best performance without worrying about costs, while Scenario 2 (S2) targets finding the cheapest solution where best-effort performance is sufficient. Scenario 3 (S3) targets achieving a balanced solution in terms of the best performance and minimum costs. Finally, Scenario 4 (S4) targets achieving the best performance under the limited budget of $100k.

The placement results for Scenarios 1, 2, and 3 are trivial where the optimal solution included an increasing number of a homogeneous set of devices as the workload increases. The ASIC devices with the highest performance were selected in S1, while the CPU-based devices were selected in S2 as they are the cheapest solution. In S3, the NPU-based devices were selected since they achieve the best trade-off between cost and performance.

The placement results of S4, wherein a cost limit of $100k is defined, are more interesting to analyse. Figure 7.10 depicts the results corresponding to this scenario showing the number of instances of each device type required for an increasing number of NFs to be placed. When the workload is low, a single ASIC device is sufficient to handle the load while providing the best performance. When up to 22 NFs must be placed, another ASIC device is needed to handle the load since the first device's processing resources are exhausted. As the load increases further, no more ASIC devices could be utilized since the remaining budget only allows for the second-best performing device, which is in this case FPGA-based. At this stage, up to four FPGA devices are required to process the increased load until reaching a total of 90 NFs. Afterward, one of the two ASIC devices is discarded to afford to employ a bigger number of cheaper FPGA devices to handle the increased workload. When the number of NFs to be deployed reaches 174, the second ASIC device is also sacrificed and replaced with additional FPGAs whose number increases to 20 FPGAs when the number of NFs workload reaches 200. After this point, the optimal solution tends to further sacrifice performance through replacing FPGA devices with the next best performant device, i.e., NPUs, to enable handling the increased number of NFs within the available budget.

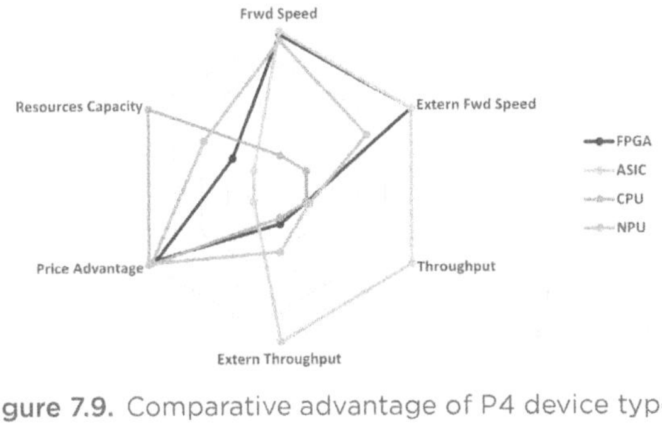

Figure 7.9. Comparative advantage of P4 device types.

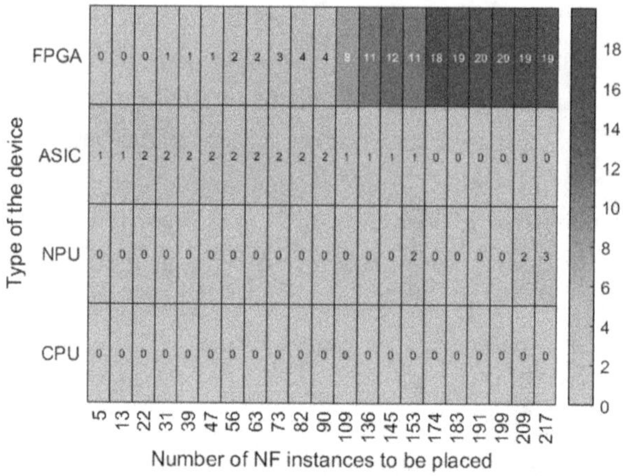

Figure 7.10. Used P4 devices in Scenario 4.

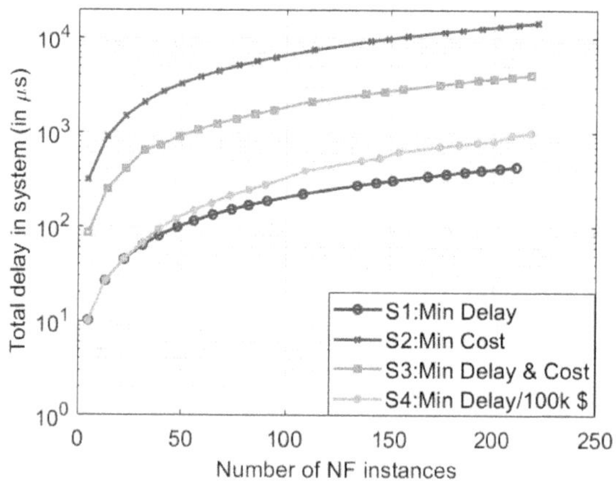

Figure 7.11. Total forwarding delay in the system.

The delay and cost functions of the optimal solution are depicted in Figures 7.11 and 7.12, respectively, for various scenarios as a function of the number of NFs to be placed. Following the defined objective, the overall delay in S1 is minimal, while the overall cost in S2 is the lowest. The results for S3 reveal the trade-off between the two objective functions, where the overall delay and cost of the system are both minimized. The results of S4 show that the delay in the system is as low as that of S1 (when only the delay is minimized) until the limit on the budget is reached after 22 NFs. After this point, the system's delay begins to increase diverging from the delay in S1, while the cost stays always below the limited budget of $100k.

In summary, utilizing precise performance models for programmable network devices and intelligently managing these devices enables the creation of flexible networks without sacrificing performance. Additionally, proper consideration of

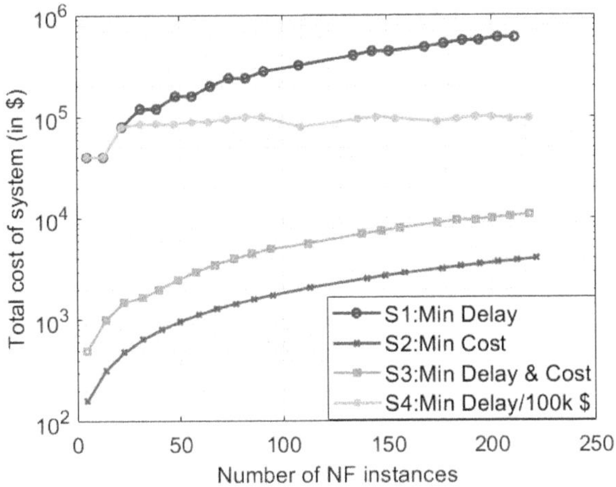

Figure 7.12. Total cost of the system.

the network substrate's capabilities and costs enables cost-effective infrastructure planning to reduce the system's total cost of ownership.

7.5.2 Extensions Towards UE Programmability

The standardization process in 3GPP time and time has proven its paramount value for numerous successful generations of cellular networks. There are, however, some hiccups: the process is time-consuming; some features are niche and do not interest all members; it takes years for a feature to be introduced and there is no guarantee it will be implemented. To some extent, this could become a barrier to innovation, especially when considering aspects, such as network flexibility for different deployments and use cases. For 6G, the UE programmability concept aims to significantly decrease the time to innovation for features having an impact on the air interface protocols. The concept refers to defining API(s) for the UE associated with actions/routines/sub-routines that can be exposed to a programmer entity so that a programmer entity is able to modify/add one or multiple behaviour(s) at the UE associated to the air interface protocols. A high-level signalling diagram and architecture are shown in Figure 7.13.

As a result, with a reduced amount of 3GPP standardization, the programmer entity should be able to define new behaviours/features for the programmed UE, such as a new message received or transmitted, new information elements and associated interpretation, new reports, new trigger for that message or report, and new/additional triggers for existing messages.

At a high level, some initial components are essential to enable UE programmability. Those components can be considered high-level solutions that are needed initially to realize the concept. Here, three of those are provided, which are API

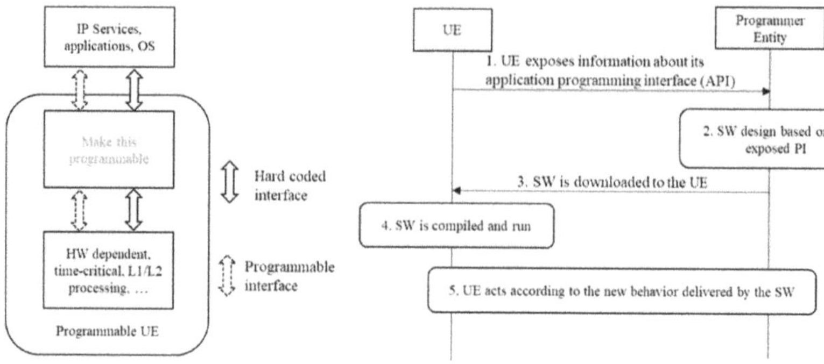

Figure 7.13. To the left, high-level architecture of a programmable UE, to the right, signalling diagram for programmability.

exposure towards the network, initial access mechanism for programmable UEs, and software version management.

The API exposure mechanism is needed so that the programmer entity knows the specific capabilities of programmable UEs in order to utilize their capability. Hence, there needs to be a signalling procedure to convey the programmable capabilities and their specifics to the programmer entity. Such capability exposure can be initiated at specific events such as UE registering with the network or can be on-demand using dedicated signalling towards specific UEs. The capabilities regarding programmability can be standardized to include specific APIs and UEs can indicate which subset of standardized APIs are supported. Therefore, the programmer entity receiving the capability can design suitable SW for specific UEs based on available APIs.

Upon initial access, programmable UEs may have an SW version for a specific behaviour that is different from the one on the network side. Hence, the UE may not be able to connect to the network because of SW incompatibility. A bootstrapping broadcast channel can potentially solve this problem by providing information on the current version of the SW on the network side and how to acquire it. A programmable UE upon initial access first checks the bootstrap channel to acquire information on the active SWs and how to acquire them. Based on such information, the UE can download the proper SW version and hence connect to the network.

Finally, SW management mechanisms are needed because for specific behaviour implemented by an SW, there could be various versions corresponding to new updates or different interests of operators/vendors. Hence, an SW management solution is required to enable synchronized operation of the UE and network with respect to SW versions. Such a solution will serve to store and manage multiple SW versions including adding, removing, and updating SW. Moreover, the solution will allow to select a specific SW version to be initialized upon request from the network.

However, there are many challenges to overcome for the successful adoption of the concept which faces manifold open research questions.

One aspect is to ensure that the programmability solution does not disrupt 3GPP. As mentioned before, the 3GPP way of working is vital to the success of cellular network evolution and is trusted by every player in the ecosystem. The concept must be developed in harmony with the 3GPP and complement it rather than disrupting an already successful framework. This can be achieved by defining an overall framework for UE programmability by defining a bare minimum for the concept in 3GPP, such as methods of downloading a software and defining and exposing APIs.

Another challenge is that there are potentially many flavours of programmability, each with advantages and disadvantages. Each flavour balances a trade-off between its capability and pragmatism. At one extreme edge, one can envision a downloadable UE stack paradigm potentially offering full programmability. However, developing this approach from concept to reality will face many difficulties ranging from technical issues, split of responsibilities and concerns among different entities/vendors, trust and privacy issues, and acceptance from 3GPP. On the other hand, the programmability could be introduced in a limited way by allowing only specific features, e.g., radio resource management measurement, to be programmed. Such an approach will be more acceptable from the point of pragmatism at the risk of being very limited and potentially leading to the introduction of multiple APIs per protocol stack.

Another challenge is related to a typical UE hardware. The UE hardware typically has a small footprint and is packed with optimized codes to achieve extreme efficiency in contrast to more flexible hardware such as a VM. Thus, any framework should make an extra effort to accommodate this constraint.

Privacy and security aspects should be considered fundamental parts of the concept. A security/privacy functionality needs to ensure that UE always receives a safe programme, the received programme does not introduce security concerns, the privacy of the UE is never compromised, and UE is implemented with a mechanism to ensure trusted computing.

Another challenge is considering the mobility aspects. The programmability framework should not restrict the UE from moving freely in the network. When a new behaviour is programmed to the UE, e.g., by an SW patch, then the framework should ensure that the UE is not interrupted when moving to another part of the network because either that specific SW is not available or have non-compatible versions. This also raises the multivendor issues and the need to properly handle trustworthiness aspects when it comes to code exposure.

Beyond the high-level solutions presented here, enabling the realization of the concept in a general framework, the next step is to develop a concrete architecture for the UE programmability and specific use cases that it can realize for a programmable configuration of the air interface. The introduction of conditional handover [33] in NR enables the UE to participate in the decision to reconfigure the air interface and this facilitates to pursue a programmable reconfiguration use case.

7.6 Programmability Through the 3GPP API Framework

API-based interaction of third parties with the network is needed for the support of openness at deployment, management, and application levels. In this context, the development of a Common API framework (CAPIF) has been coined in 3GPP as an effort to avoid duplication and inconsistency between the various existing API specifications. As such, 3GPP CAPIF includes, under a common architecture, aspects that are applicable to any service APIs at the northbound of a mobile network. More precisely, CAPIF is a complete 3GPP API framework that covers functionality related to on-board and off-board API invokers, register and release APIs that need to be exposed, discovering APIs by third entities, as well as authorization and authentication. From the market perspective, the need for such a management framework has well recognized, while the CAPIF implementations[15] and products are still under development. For instance, proprietary solutions have emerged, such as the Red Hat API management.[16] It is based on the 3scale enterprise API management product of the company, and it provides authentication, governance, security, analytics, automated documentation, developer portals, and monetization for API services.

The 3GPP CAPIF functional architecture is covered in 3GPP TS 23.222 [18, 34] "Functional architecture and information flows to support Common API Framework for 3GPP Northbound APIs" (since Release 15). Based on the architecture and the procedures defined in these reports, additional information and requirements are considered, regarding CAPIF security features and security mechanisms, as presented in 3GPP TS 33.122 [35]. The specification of the CAPIF APIs that are needed for realizing the CAPIF functionality is part of 3GPP TS 29.222 [36]. Actually, within this technical specification, the interacting protocol for the CAPIF Northbound APIs is described.

CAPIF functionality is considered a cornerstone in the realization of mobile network openness, since it allows secure exposure of core network APIs to third-party domains, and also, enables third parties to define and expose their own APIs.

---

15. https://github.com/EVOLVED-5G/CAPIF_API_Services

16. https://www.redhat.com/en/blog/api-management-3scale-service-provider-use-case

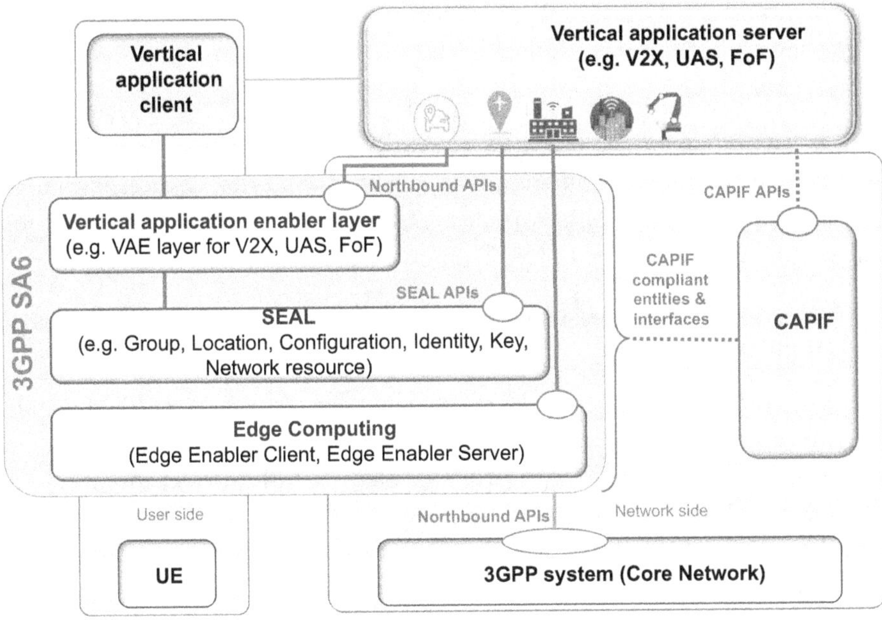

Figure 7.14. CAPIF in the context of 3GPP SA6 activities.

Indeed, CAPIF has become already a fundamental feature for the 3GPP SA6, targeting the interaction of Verticals with the 5G system. In this context, 3GPP introduced the concept of VAEs, enabling the efficient use and deployment of vertical apps over 3GPP systems. The specifications and the architecture are based on the notion of the VAE layer that interfaces with one or more Vertical apps. VAEs communicate via network-based interfaces that are well-defined and version-controlled. The focus of VAEs is to provide key capabilities, such as message distribution, service continuity, application resource management, dynamic group management, and vertical app server APIs over the 3GPP system capabilities. The importance of realizing CAPIF is reflected in the fact that CAPIF compliance is required (Figure 7.14) in (i) the development of VAEs for various vertical industries (V2X, Factories of the Future, etc.), (ii) the realization of the service enabled architecture layer (SEAL), as well as (iii) the implementation of the service side of edge computing services.

7.6.1 CAPIF Services and Implementation

CAPIF architecture is presented in 3GPP TS 23.222 [18] and includes three main entities, namely, the API invoker, the CAPIF core function, and the API provider.

The API invoker is typically provided by a third-party application that supports capabilities, such as supporting the authentication by providing the API invoker identity and other information required for authentication of the API invoker and discovering service API information.

The CAPIF core function is the main entity of the CAPIF and it consists of engines that among other capabilities authenticate the API invoker based on identity and/or other information, authorize API invokers prior to accessing service APIs, on-board/off-board API invokers, monitor service API invocations, and store policy configurations related to CAPIF and service APIs.

The API provider is an entity that provides API exposing, publishing, and management functions.

- The *API exposing function (AEF)* is the provider of the service APIs and is also the service communication entry point of the service API to the API invokers.
- The *API publishing function (APF)* enables the API provider to publish the service API information in order to enable the discovery of service APIs by the API invoker.
- The *API management function (AMF)* enables the API provider to perform administration of the service APIs.

Based on the three fundamental entities that CAPIF architecture has defined, a set of reference points (interfaces), with associated management APIs, for enabling the interaction between API Invokers and AEFs, are specified as well.

To facilitate any further contribution in the area, the currently available open-source implementation of CAPIF[17] (as it is being developed in the EVOLVED-5G project by Telefonica and Fogus innovation and services[18]) [36] follows the principles of microservice programming, and it is released together with a set of evaluation tests. The major aspects of this implementation are further described below.

Based on the CAPIF architecture, the core part of CAPIF is the CAPIF Core Function (CCF). To implement the API services of the CCF, the first thing needed is the CAPIF API definitions/signatures. Those have been specified in 3GPP TS 29.222 [37], and 3GPP has published the related YAML files[19] as well. The following services have been defined for the CCF:

- Discover Service: API to ask CCF the list of APIs published and available in CAPIF.
- Publish Service: API to publish API information from APF/AEFs.

17. https://github.com/EVOLVED-5G/CAPIF_API_Services

18. https://fogus.gr/

19. https://forge.3gpp.org/rep/all/5G_APIs
https://github.com/jdegre/5GC_APIs

- Events: API to manage Event subscriptions that enable event notification from CCF.
- API Invoker Management: API to enable the onboarding of API Invokers into CCF.
- Security: API to enable setting security profiles and retrieve security Tokens.
- Access Control Policy: API to manage access control rules in CCF.
- Logging API Invocation: API to add logs on API consumption.
- Auditing: API to query and retrieve service API invocation logs stored on the CAPIF core function.
- AEF Authentication: API for AEF security management.
- API Provider Management: API for API provider domain functions management.
- Routing Information: API to provide API routing information.

The YAML files of those services can be used by a Swagger editor to inspect all information elements in JSON. Each of these YAML files defines one or more APIs and the supported methods to use them (POST, GET, DELETE, and PUT). With the YAML files of the services described above, it is possible to generate automatic code that implements HTTP/HTTPS Endpoints that act as Servers that accept HTTP requests.

To build the CAPIF core function, several software tools have been used to develop, implement, build, and test the CAPIF API services.

- **OpenAPI Generator[20]:** This software programme allows the generation of API clients SDKs (Software Development Kit tools) or API servers given an OpenAPI specification. Moreover, it is possible to generate code in more than 20 different programming languages.
- **MongoDB:** Mongo is a non-SQL and open-source database tool used to provide storage to different CAPIF core functionalities.
- **Nginx:** Nginx is an open-source web serving technology that is used as a reverse proxy to forward requests to the different CAPIF modules.
- **Flask:** Flask is a micro-service web framework written in python used to build CAPIF. The main advantage of this framework is its modularity. This feature allows us to run different CAPIF services and mix them directly with other services as database with relative ease.
- **Robot framework:** Robot is a generic test automation framework used for acceptance testing and acceptance test-driven development.

20. https://github.com/OpenAPITools/openapi-generator

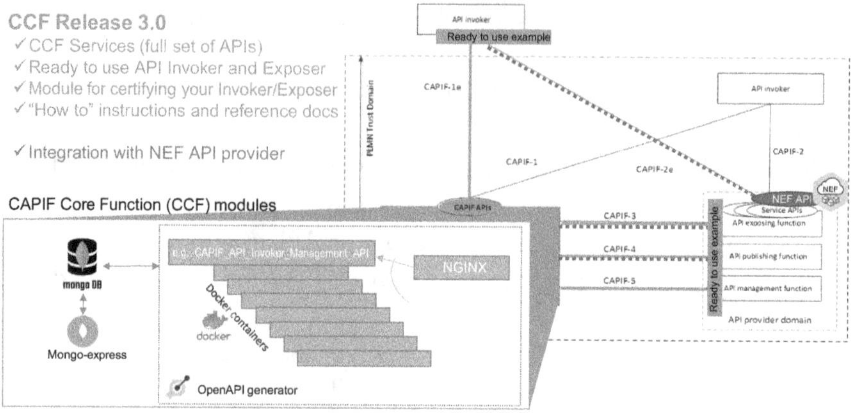

Figure 7.15. Open-source implementation of the 3GPP CCF (Release 3.0 features).

- **Docker:** Docker is an open-source containerization tool for building, running, and managing containers, where software is deployed. We use Docker to build each service of the CAPIF. In this way, each CAPIF service development is kept isolated from other services. This design pattern is known as a microservices-oriented architecture and is widely used in software development due to its innumerable advantages like better fault isolation and improved scalability, among others.

In order to guarantee the integrity of our implementation, an automated test suite is used, covering the core functionality of each CAPIF API service. Robot framework is the testing tool that ensures the quality and robustness of developed code. Moreover, it is defined as a test strategy to improve the code quality. This test strategy is composed of two steps:

- **Test plan document elaboration:** In this step, test plans are described, including various behaviour scenarios (considering both success and failure cases). The test plan structure includes clarifications on the pre-conditions, the action that takes place, and the post-conditions (response/result expected). The horizon of the test plans that can be defined, moves beyond the request–respond information that is available in the related 3GPP specifications, in a sense that behavioural scenarios beyond the basic functionality are defined to stress test the implementation integrity.
- **Test implementation and execution:** This step continues after finishing the elaboration of the test plan documentation. Each test suite is implemented and included in an automation pipeline that checks the status of the code after every deployment in the platform.

7.6.2 NEF as API Exposing Function

One first example of the exposing function is hiding all the underlying topology of the core network and playing the role of API provider, as defined in the CAPIF architecture. In future systems, this role can be played by other functions that refer to core, transport, or radio domain, and expose APIs for interaction from control, data, or management plane. With no loss of generality, here further analysis is provided for the case of network exposing function (NEF). NEF comprises several services that can be described as monitoring services, policy and charging services, application provisioning services, analytics services, Industry 4.0/IoT (Internet of Things) specific services, and security services.

The exploitation of the NEF capabilities from industry has begun already; however, before taking full advantage of NEF-based network exposure, many challenges are to be addressed. Indeed, topics regarding the exposure capabilities of the network are still to be considered in 3GPP Release 18, while telecommunication vendors are working to adapt the SBA and implement the already specified exposure functionalities.

To enable interaction with the core network, the implementation of NEF functionality as an open tool is a prerequisite since currently there are no open commercial solutions implementing the entire SBA and the southbound interfaces that NEF requires in order to expose the standardized APIs. Nevertheless, a NEF Simulator [38] has been developed by NCSR "Demokritos" aiming at surpassing this challenge by creating simulated and emulated events. The architecture of the Simulator[21] is decomposed into three distinguishing parts depicted in Figure 7.16.

The main features of the NEF architecture are described below:

- **Exposure layer (NEF APIs):** The principal idea of the simulator is the provision of the APIs that 5GC's exposure function (i.e., NEF) defines. Therefore, the available APIs have been placed in the 5G Exposure layer. Currently, the available APIs include monitoring events (i.e., location) and session establishment with QoS. As the work progresses, new APIs will be gradually added to the simulator and the existing ones will be enhanced.
- **Simulation environment:** As mentioned above, currently, communication with the southbound interface (i.e., 5GC) is a demanding task. However, the simulator can tackle this challenge by creating simulated events. To achieve this, it provides an interactive geolocated environment where users can create different network scenarios. These scenarios are designed to simulate the basic aspects of a 5G network, required for testing the available service APIs. For

21. https://github.com/medianetlab/NEF_emulator

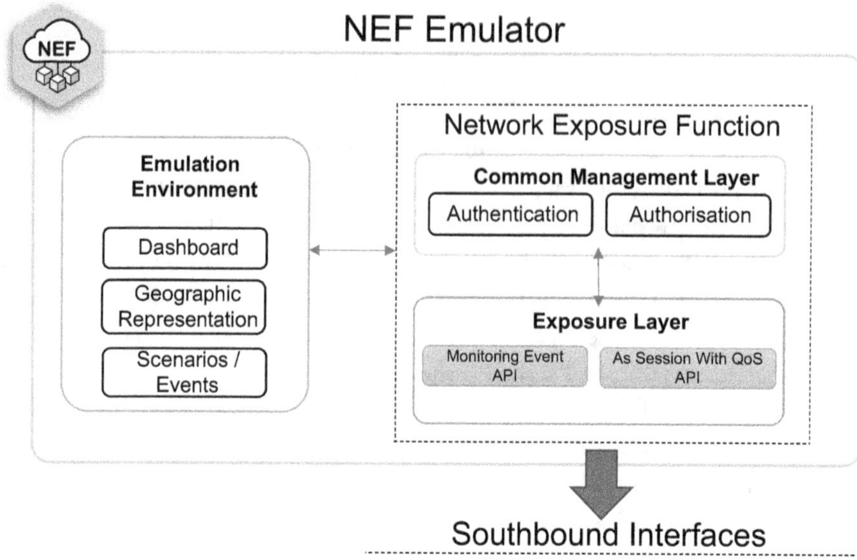

Figure 7.16. NEF simulator architecture.

example, in order to retrieve the location (i.e., cell level accuracy) of the UEs through the Monitoring Event API that NEF exposes, the simulator allows for the implementation of a scenario where UEs are moving through 5G cells. Developers are able to alter data, allowing them to define and run specific scenarios according to their needs.

- **Common management layer:** The simulator also provides common management functions such as token-based user authentication/authorization. Initially, to gain access to the simulator, there is a need for the user to create an account. After the creation of the account, an authorization step based on OAuth2.0 takes place, in order to make use of the available Northbound APIs or the Simulation Environment. Each application developer has a registered account/profile within the simulator that is considered an isolated environment; thus, NEF Simulator can store different scenarios configured by multiple users.

7.7 Programmability Enables the Network App Ecosystem

From the business perspective, programmability enables a new business potential around the development of the so-called network applications. The major challenge for the Network Apps ecosystem lies in the need for continuous development, test, and evaluation of vertical-specific network-enabled applications, on top of realistic configurable infrastructures, prior to their commercial deployment in the mobile networks (operators' infrastructures).

As a response to this challenge, a facility that could support in long term the vertical application development and provisioning over mobile networks is required. The facility should take advantage of programmability frameworks and support the entire lifecycle of the Network Apps (Figure 7.17). In the lifetime of a Network App, three main processes can be defined, namely: (i) the Network App Development process, where the actual code production is performed; (ii) the Network App Testing process, where testing at various levels and for different targets is performed (including verification tests, validation tests, and certification tests); and (iii) Network App Publication process, where the process of uploading/storing a Network App to a marketplace is performed.

From a business perspective, the facility that supports those processes is meant to serve as a collaborative platform for the infrastructure owners and the vertical industries, and thus, it should engage the creation of a vertical-specific market, analogous to the currently available ones for mobile apps (e.g., Play Store and AppStore). Next, the architectural components/environments of such a facility are provided.

7.7.1 Architectural Components of the Facility

7.7.1.1 Development environment

For the facility that will support the Network App lifecycle, the adoption of a CI/CD approach enabled by DevOps software development methodology (software development – Dev and information-technology operations – Ops) is a key approach. Already, OSM and ONAP have implemented aspects of DevOps workflows to support their respective deployments. Also, there are several related tools that can enable automated lifecycle management for the Network App development process.

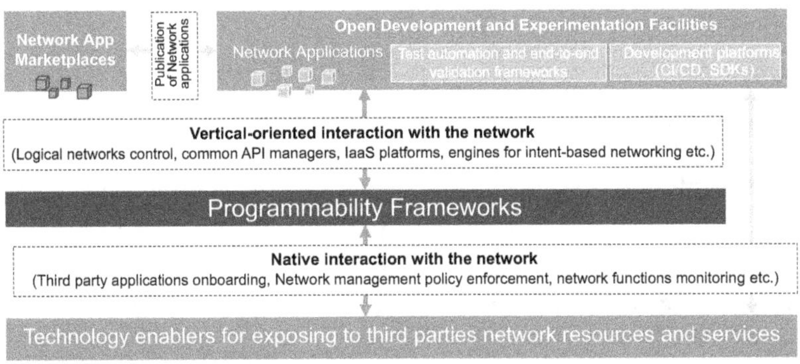

Figure 7.17. Open facility for Network App lifecycle support on top of programmability frameworks.

7.7.1.2 Validation environment

The main role of the validation environment is to provide capabilities for running automated tests under well-defined configuration/parametrization of the facility. Four main features are considered.

- **Onboarding controller:** The developments are imported into third-party servers that are part of the open experimentation facility in order to be validated and tested; thus, capabilities for controlling the onboarding process are added to the platform. To make this procedure dynamic, a set of virtual infrastructure managers and orchestrators at the facility is required.
- **Test execution manager:** It is responsible for configuring and scheduling the facility based on the target validation tests. This is performed by using the information available from the facility in conjunction with a test descriptor. The test descriptor is a structured form of information needed for conducting a test. Already, there is comprehensive work on that direction from EC Horizon 2020 5GPPP projects,[22] and 5GPPP has published a comprehensive white paper on that [39].
- **Test automation tool:** It is responsible for applying the commands from the test execution manager to the facility and to host agents or plugins required for executing a test. This is an important part of the framework since it allows the verticals to reuse a pool of tests and related plugins on demand. Thus, the design of the test and the implementation of the related agents are decoupled from the vertical service development process.
- **Results analysis and visualization:** Vertical to network-level measurement campaigns are considered in the proposed framework. The measurements can be collected from exposure APIs and directly from agents. The diversity of the available data and their volume can enable analytics. The visualization of the results is also provided, for real-time network resource monitoring and, also, for dashboarding of post-processed data.

7.7.1.3 Certification environment

As the technology evolution moves network functions to the software layer and through virtualization allows open and dynamic composition of network services extending capabilities to the business through the Network Apps concept, the established certification practice in the mobile network business needs to extend beyond the current practice and include supplementary software specification conformance and quality assessments. Network Apps, as primarily third-party software interworking with the network, shall need certification in accordance with

22. https://5g-ppp.eu/5g-ppp-phase-3-1-projects/

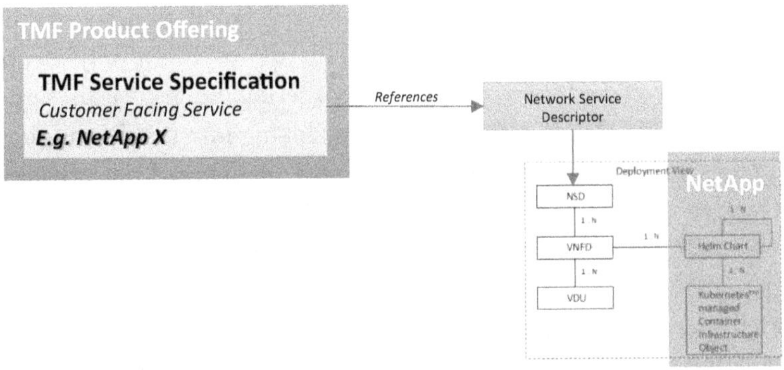

Figure 7.18. Network App exposed as TMF product offering in a marketplace.

the equipment paradigm. In that sense, the certification environment refers to the programming environment that provides all the tools and processed required for certifying Network Apps. The certification process includes testing against standards and regulations, so as to guarantee that a Network App functions properly under any scenario and data load. In contrast to the verification and validation processes where the testing is application- or scenario-oriented where the focus is on the efficiency of the Network App functionality, the certification process focuses on the correctness of the interfacing, based on the related specifications, so as to maximize the interoperability of the Network App.

7.7.1.4 Publication environment – marketplace

From the business perspective, the Network App ecosystem requires a collaborative platform for the infrastructure owners and the vertical industries, and as such, it engages the creation of a vertical-specific market, analogous to the currently available ones for mobile apps (e.g., Play Store and AppStore). Thus, the developments (Network Apps) become available to any interested party either for purchase/utilization and/or for reuse/enhance towards a new development. The role of that collaborative platform can be played by a Network App publication environment, with functionality analogous to a mobile app marketplace. Already, related 5GPPP projects develop such marketplaces, for instance, the EVOLVED-5G market place[23]; however, the usage of existing (of more general purpose) platforms (AWS marketplace, Google, etc.) as a basis is not excluded as long as the related Network App certification is guaranteed.

Another approach from the 5GASP project proposes to use the TMF's Product resource model[24] for publishing Network Apps to a Marketplace and therefore

23. https://github.com/EVOLVED-5G/marketplace

24. TMF620 – Product Catalog Management API REST Specification.

making it publicly available after the DevOps experimentation and certification readiness lifecycle is successfully completed. This approach not only incorporates adequate resources to describe a Network App offering but also ensures interoperability among other industry implementations. Taking into consideration the various properties of the aforementioned model, the highlighted ones (Figure 7.18) will be leveraged to describe a Marketplace asset. Briefly, a Marketplace asset might contain an attachment (e.g., logo, images, certification links, or files), topological information about the offered deployment, pricing, asset's specific characteristics, service level agreement (SLA) reference, and lastly, a reference to the actual services ordered and employed, i.e., hosting network slice, Network App, and test descriptor. Notable mention should be made of the latter entity, namely, Service Candidate Ref of the Product Offering resource model. This entity associates the product offering with the Network Apps Service Specifications constituting the onboarding and deployment model.

To end up, utilizing TMF's product aims at:

- Consistency between the ordering and deployment model.
- Introduction of business aspects, such as pricing, product options, and market segment.
- Imposing an abstraction layer between customers and service providers.
- Effortlessly interacting with other production systems.

7.8 Programmability Enables Intent-Based Networking

An intent is an expression of the desired state that you want to be realized and can be considered [6, 40] as portable and abstract. Portable, in the sense that it can be moved between the different controller and network implementations and remain valid; and abstract since it must not contain any details of a specific network. The advantage of an Intent is flexibility, as it allows users to express policies using concepts and terminology that are familiar to the user without having specific knowledge in the field. There are several applications where intents can be applied, including service model and orchestration ([41–44]), network orchestration ([45, 46]), monitoring and resource exposure ([47, 48]), and intent deployment and configuration [49].

To make possible the implementation of intents, without compromising the system, it is necessary to study the life cycle of an intent from its creation to its installation (see Section 7.8.1). On top of the principles that are defined from the lifecycle of the intents, programmable frameworks and middleware can be created to enable the development of related (IBN-based) Network Apps (see Section 7.8.2).

7.8.1 State Machine for IBN-enabled Industrial Networks

A reference state machine for IBN-enabled Industrial networks is depicted in Figure 7.19. This state machine is divided into six sections. After the user requests in the first phase, the validation phase verifies that the request contains all the necessary information to implement the desired action. If the user has forgotten to mention necessary information for the request to be implemented or if the information, what he has provided, is wrong, this phase assigns an invalid status, which will require user interaction. On the other hand, if the user provides all the necessary information for the request to proceed, it assigns a valid status and continues to the next phase.

The conflict phase is the second checkpoint of this state machine. Here, the already validated attempts are subject to a comparison with the requests already implemented and stored in the database to conclude conflicts. If the request is redundant, or if the new request consists of information contrary to that already implemented, the conflict phase assigns a conflict state which will be resolved with the help of the user. If the information does not conflict, the compilation section is enabled.

When the request reaches the compilation phase, the system tries to convert into rules the policies that the user wants to implement, i.e., the system converts the high-level language, from the intent, into a language that can be interpreted by the controller so that it is then possible to install them. Sometimes, the desired information may not be supported by the controller, or the desired operations may not be operational, and in this case, the compilation phase gives a compilation error that can be solved with the help of the user. If the rules to be installed are retrieved, the installation phase is enabled.

Once the rules are obtained, the next step is to install them on the controller. If the objectives are supported by the controller and installed, the monitoring phase is enabled. If the goals are not achievable because they may be offline or non-existent, the user is alerted that the requested application was not installed.

In the final phase, monitoring consists of a cycle that constantly checks if the existing requests in the database are being fulfilled or if any violation has occurred. If any non-compliance occurs, the system, through intelligent algorithms, tries to identify how to solve the problem without human intervention automatically. To make this possible, the area of ML enters this phase, whereby capturing data from the network, the system tries to identify patterns in order to validate existing policies and intervene if necessary. If the intelligent algorithms do not identify a solution that corrects the problem, the user is alerted.

This cycle of intents installation avoids problems of redundancy and inconsistency in the infrastructure, allowing the user to orchestrate networks by expressing

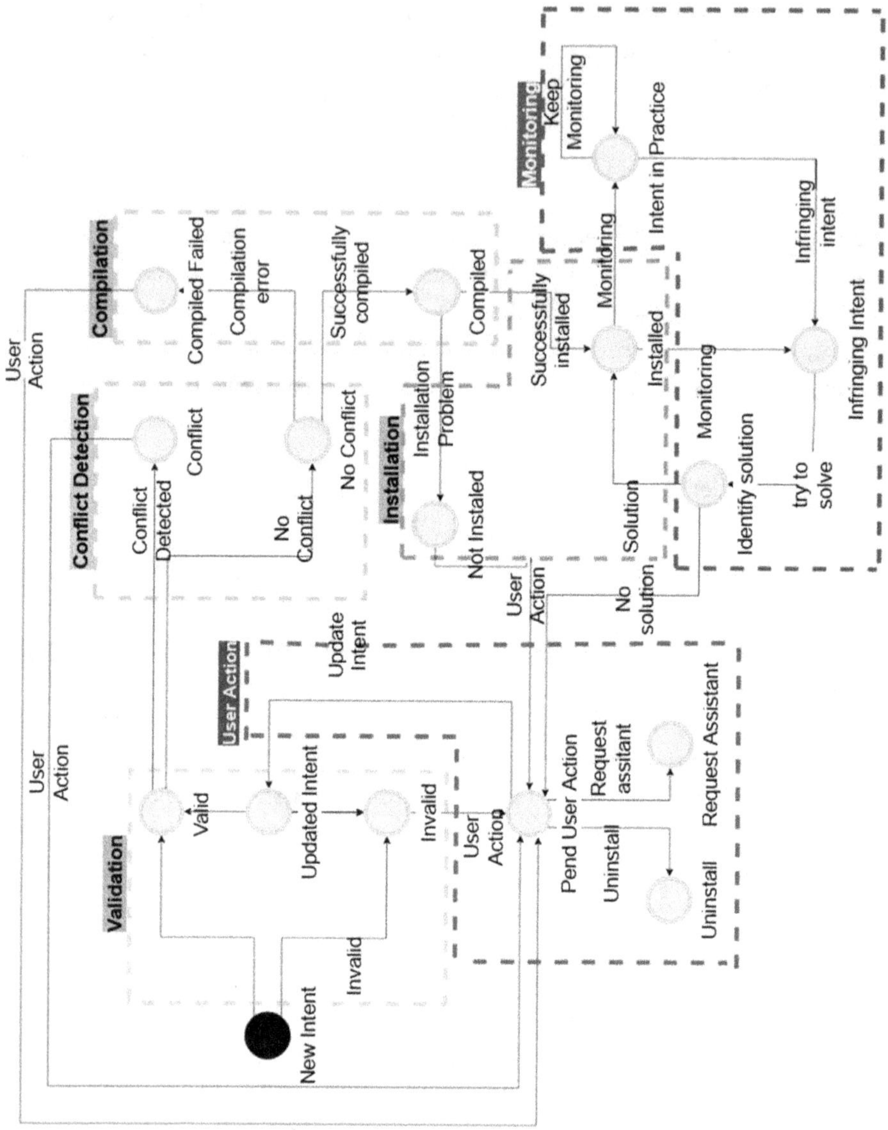

Figure 7.19 State machine for enabling IBN.

only what they want to happen, without having to worry about how it will be implemented.

Perhaps the most important application of intent-based networking in 5G industry use is automation [50], automation that is readily derived from IBN has helped to overcome the traditional massive device-to-device connection, thus ensuring great efficiency, turnaround, and scalability. With the use of a well-designed artificial model, operators are not only able to automate several processes but also provide service-level assurances.

7.8.2 Middleware for Intent-based Networking

Adopting the concept of intent-based networking, an "intent engine" is envisioned, which acts as an OSS/BSS leveraging AI/ML algorithms to automate the network application lifecycle management depending on the end users' intent, providing it to a middleware, which allows the intent engine to control the full operation of network services and network slices under its control.

The domain layer generates IBN policy based on specific domain knowledge pre-stored in a semantic model and an information model. The policy will be used by the infrastructure enablement layer to manage virtual resources.

A predefined sequence may not be compatible with the workflow of an autonomous operation. Semantic models and intent-based networking is supposed to improve service time. During the verification process, the amount of time taken to launch network service under three typical scenarios will be tested to verify the improvement of the domain knowledge on the phase.

The intent-based networking predicts the need for robots and specifies policy towards OSM and RAN controllers to deliver management, topology, placement, and resource optimization within 5G Cloud environments. An automation mechanism to align 5G orchestrators such as performance management, VNFs placement, life cycle management, and event monitoring will be implemented to reflect the intents in the optimized way. By identifying intentions and parameters autonomously specifically for the vertical domain of autonomous robots, they automate the 5G testing process accordingly.

OSM uses information model management procedure, resource, and topology. The current models do not understand the vertical users' logic behind the behaviours. A semantic model can then be used to fill the gap in between. It contains QoE models and a cookbook:

QoE models are application-specific models for translating user requirements into KPIs of QoS. The models will be derived from relevant use case patterns per application. It is essential for OSM to understand the workflow, workloads, and topology and enables the middleware to optimize service provisions for individual applications.

The cookbook contains many recipes as templates. They are prepared for vertical with multiple concurrent applications. A semantic model will create a type system to describe possible building blocks (such as a "Compute" node type, a "Network" node type, or a generic "Database" node type) of applications. They will be used in the model for constructing a behaviour template together with QoE models. The type system will then be used to define service templates (robot service, edge service, Cloud service, and collective service) which consist of lifecycle operations and behaviours of orchestration engines.

Combining QoE models and Template, user intents will be applied to derive the order of component instantiation, manage lifecycle operations, and instantiate single components at runtime with strong interpretability and interoperability.

The intent-based networking is realized by middleware with

- Semantic interpretation engine.
- Lifecycle management engine.
- Performance management engine.
- Fault management engine.
- Package management engine.
- Security manager.

An IB policy generator is the centre of the middleware, and it derives the context of the requests and then translates the request into policy through corresponding engines. The intent-based networking is supported by pre-defined receipts from the semantic database, and ML toolboxes help the engine to derive the current states of the system using events gathered. Fault, package, lifecycle, and performance management engine helps the orchestrator to define management procedures. Their policies are also specified as receipts within the rich domain model. The focus is on translating behaviours derived by the semantic interpretation engine to procedures within the information model, which is subsequently understandable by OSM and RAN Controller.

7.8.2.1 Enabling network applications for robot autonomy

In order to achieve robot autonomy, many advancements are needed beyond just another provider-centric 5G architecture or framework solely to improve quality of service (QoS). The ambition must be on the user-centric paradigm of integrating vertical knowledge into the existing 5G solutions. Under the umbrella, the software architecture proposed focuses on bridging OSM and ROS/network applications development.

The architecture is divided into four spiral layers as shown in Figure 7.20.

- The domain layer generates IBN policy from specific domain knowledge pre-stored in semantic models and information models.

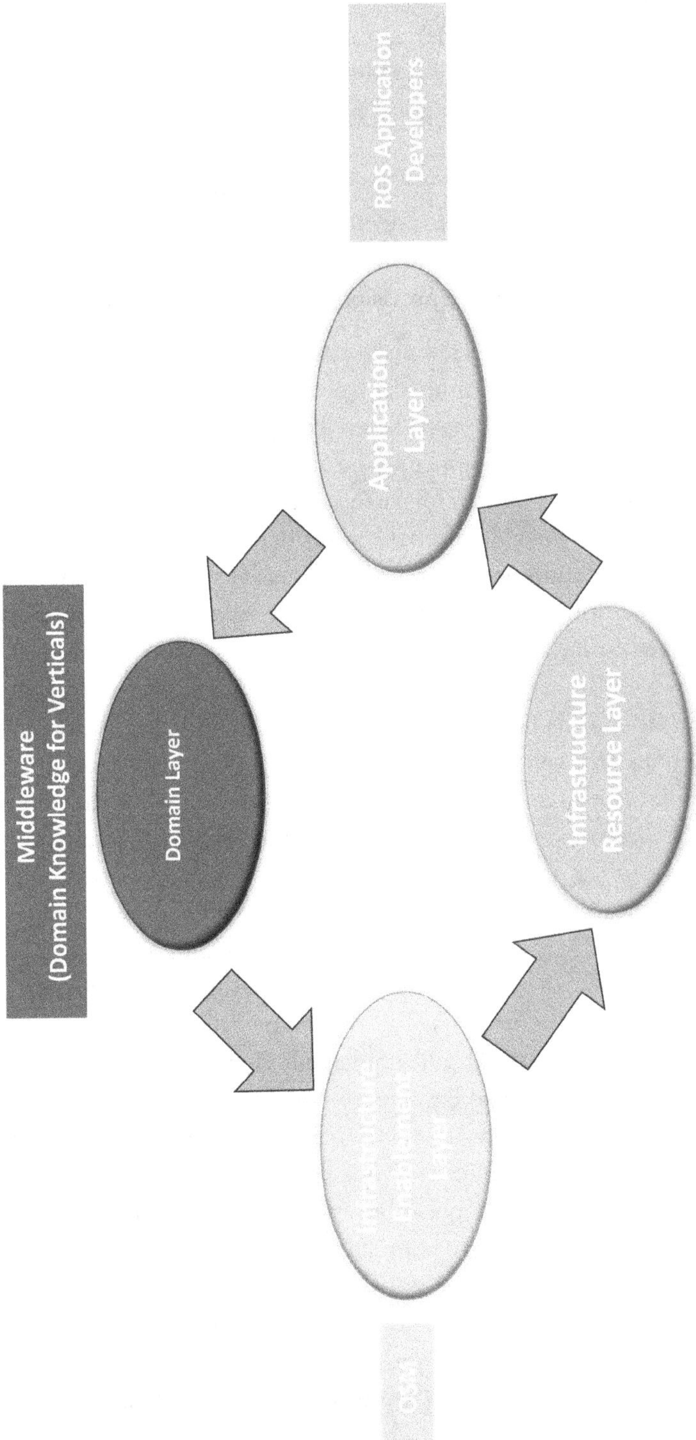

Figure 7.20 Interconnected layers.

- The infrastructure enablement layer contains ETSI OSM and RAN controller to control the core network and RAN.
- The infrastructure resource layer is closely linked to the 5G testbeds, it provides physical resources and enables network services to be deployed on Robots, Edge, and Cloud.
- The application layer delivers the network application using VNFs and KNFs. The network application is deployed to enable the ROS network for fundamental robot capabilities such as perception and user interaction. It also enables the Cloud-native service provision together with the network services.

The four layers are interconnected in a spiral shape and linked closely to the ongoing development of 5G testbeds, open-source MANO (OSM), robot operating systems (ROS), and domain-driven design. It reflects the multidisciplinary nature of the project development. The architecture enables patterns to be tangible in their specific sub-domains for further verification. The innovation leads to an automated and interpretable mechanism for deriving the placement of network functions, order of component instantiation, and effective lifecycle management. This is essential for application-driven approaches towards automatic configuration on testbeds using ML and AI and for enhanced robot autonomy in the vertical sectors.

A middleware layer of ROS-based network applications will implement common robot functions that can be invoked by the respective over-the-top (OTT) robotic applications. In this case, network applications are defined as disaggregated application enablement services, which can span across technology domains (i.e., Core and Edge).

The network applications will be implemented as VNF chains within network slice subnet instances (NSSIs), as per ETSI NFV EVE012 specifications. Thus, individual VNF instances can be optimally placed depending on resource availability, and a number of various constraints (e.g., maximum delay, maximum throughput, etc.) network applications that will implement common robotic operations such as mapping can then be shared between several robotics applications.

Thus, OTT service creation entails the instantiation of a network slice instance (NSI), which shares the network slice sub-slice instance (NSSI) resources already reserved by one or more network applications (e.g., in terms of processing power, storage, etc.), thus avoiding the costly resource reservation and VNF instantiation step and significantly reducing service creation time. Furthermore, the network applications will deliver open, standards-compliant Northbound APIs for robotics vertical applications that facilitate rapid prototyping.

The workflow that exploits NFV/SDN infrastructures for enhanced autonomy requires computing and storage to be shifted dynamically and repeatedly among robots, edges, and the central cloud. Partial information will be replicated among network services (NSs) deployed in different locations. To tailor NSs, different configurations of VNFs and KNFs are required to achieve 5G network services. Additionally, due to limited resources, robots and edges would prefer fine-grained network functions to preserve their efficiency.

A library of generic vertical services is the centre of the reference Network Applications. They are linked to ROS simulation environments, dense learning, and model-based RL learning toolbox and optimized by specific deployment requirements of robots alone, edge along, Cloud along, or collective. The network functions of the generic library will be implemented using KNFs and VNFs and controlled by VCA of OSM under the orchestrator to the VNFs network. Network slicing of the testbeds will be customized and integrated for deploying the KNFs and VNFs in "terminal," "edge," and "remote" environments optimized deployment. The topology, placement, and life cycle management of the network functions in the reference network application will be derived based on general patterns of the 5G enhanced autonomy. To realize the Cloud-native design, generic vertical services will be implemented using Micro-services. The service definition can be obtained using the reference catalogue service. The service can be ordered and replicated using the reference order service. Internally, data consistency is ensured by CQRS and ES. Authentication is traced in the reference identity service. Overall security is controlled by the security manager (which is also part of generic vertical service). A UI can be provided for high-level monitoring of the system status and result analysis.

As an example, a new open-source library can be implemented to realize distributed map services within "terminal," "edge," and "remote" environments for a shared environment representation. Topology, placement, and the life cycle management of the networked mapping functions will be implemented in the reference network application to realize the collective intelligence required by enhanced autonomy.

Finally, the reference network application demonstrates the standardization of APIs on testing facilities. The applications within the library of generic vertical services can be developed using ROS directly. Low-level events obtained from testbeds will be propagated on the event bus and translated by a semantic interpretation engine for high-level meanings. This capability ensures interpretability. Third-party vertical developers can reuse VNFs and KNFs of the generic vertical services which have validated their compatibility from the testbeds. Therefore, the experimental facilities are exposed to the developer. They can be expanded for use case-specific functions such as 5G enhanced perception, detection, and planning in vertical sectors such as PPDR and healthcare, transport, and industrial 4.0. A key innovation, namely, Cloud-native applications for NSs and standard APIs can be realized by reference network applications and ROS.

7.9 Conclusions

The next generation of mobile networks will take full advantage of the convergence between the IT and telecom sectors. Through this convergence, 5G has already transformed the mobile network infrastructure to a flexible service provisioning platform, and 6G is expected to provide mobile networks as fully programmable platforms, with native cloud capabilities at any network domain and communication plane. This chapter describes current technology enablers that contribute towards a fully programmable mobile network, by clustering them to those at the deployment and connectivity level, at the network and resource management level, and at the service and application provisioning level. To exploit those enablers, ongoing research and standardization work is being conducted, with the main target being the provisioning of programmability frameworks, i.e., frameworks, that abstract the network underlay infrastructure and its capabilities so that they are dynamically controlled and configurable. Some indicative approaches are described in this chapter, including the deployment of common API managers, the exploitation of P4-programmable switches, the usage of open interfaces of O-RAN, and the design of SDKs for providing network slices as a service. The potential of programmable networks is high and yet to be investigated in detail. However, as it has been indicated already, concepts such as intent-based networking take advantage of network programmability features. Overall, new business models emerge since, through programmability, third parties can develop and integrate their solutions (e.g., their network applications) into the underlay connectivity and compute infrastructure.

References

[1] D. Tsolkas and H. Koumaras, "On the Development and Provisioning of Vertical Applications in the Beyond 5G Era," in *IEEE Networking Letters*, vol. 4, no. 1, pp. 43–47, March 2022, doi: 10.1109/LNET.2022.3142088.

[2] P. Bosshart, D. Daly, G. Gibb, M. Izzard, N. McKeown, J. Rexford, C. Schlesinger, D. Talayco, A. Vahdat, G. Varghese, and. Walker, "P4: Programming protocol-independent packet processors," In *ACM SIGCOMM Computer Communication Review*, vol. 44, no. 3, pp. 87–95, 2014.

[3] M. Boucadair and C. Jacquenet, "Introducing Automation in Service Delivery Procedures: An Overview.," in *Handbook of Research on redesigning the future of internet architectures*, Hershey, PA, USA: Information Science Reference, an imprint of IGI Global, 2015.

[4] A. Aijaz, "Private 5G: The Future of Industrial Wireless," In *IEEE Industrial Electronics Magazine*, vol. 14, no. 4, pp. 136–145, Dec. 2020, doi: 10.1109/MIE.2020.3004975.

[5] Y. Wei, M. Peng, Y. Liu, "Intent-based networks for 6G: Insights and challenges," In *Digit. Commun. Networks*, vol. 6, pp. 270–280, 2020.

[6] A. K. Salkintzis, "Interworking techniques and architectures for WLAN/3G integration toward 4G mobile data networks," In *IEEE Wireless Communications*, vol. 11, no. 3, pp. 50–61, 2004.

[7] *Summary of rel-16 work items*, Technical Report (TR) 21.916, v16.0.0, 3GPP, June 2021. Accessed: April 6, 2021. [Online] Available: https://www.3gpp.org/ftp/Specs/archive/21_series/21.916/.

[8] *Study on Communication for Automation in Vertical Domains (Release 16)*, technical Report (TR) 22.804, v16.2.0 3GPP, Dec 2018.

[9] *System architecture for the 5G System (5GS); Stage 2 (Release 17),* Technical Specification (TS) 23.501 v17.5.0, 3GPP, June 2022.

[10] M. Corici, E. Troudt, P. Chakraborty, and T. Magedanz, "An Ultra-Flexible Software Architecture Concept for 6G Core Networks," In *2021 IEEE 4th 5G World Forum (5GWF)*, pp. 400–405, 2021, doi: 10.1109/5GWF52925.2021.00077.

[11] M. Corici, E. Troudt, and T. Magedanz, "An Organic 6G Core Network Architecture," In *2022 25th Conference on Innovation in Clouds, Internet and Networks (ICIN)*, pp. 1–7, 2022, doi: 10.1109/ICIN53892.2022.9758088.

[12] M. Corici, E. Troudt, T. Magedanz, and H. Schotten, "Organic 6G Networks: Decomplexification of Software-based Core Networks," In *2022 Joint European Conference on Networks and Communications (EUCNC) & 6G Summit*, pp. 541–546, 2022, doi: 10.1109/EuCNC/6GSummit54941.2022.9815730.

[13] *O-RAN Use Cases Analysis Report*, Technical Report v10.00 Rel. 003, O-RAN WG1, March 2023.

[14] 5G-PPP Software Network Working Group Network Applications: Opening up 5G and beyond networks 5G-PPP projects analysis September 2022, DOI: 10.5281/zenodo.7123919.

[15] N. Foster, J. McKeown, G. Rexford, L. Parulkar, Peterson, and O. Sunay, "Using deep programmability to put network owners in control" In ACM SIGCOMM Computer Communication Review, vol. 50, no. 4, pp. 82–88, Oct. 2020, doi: https://doi.org/10.1145/3431832.3431842.

[16] TM Forum, "The Open API project," 2023. Accessed: April 6, 2023. [Online]. Available: https://www.tmforum.org/collaboration/open-api-project/.

[17] D. Fragkos, G. Makropoulos, P. Sarantos, H. Koumaras, A. -S. Charismiadis, and D. Tsolkas, "5G Vertical Application Enablers Implementation Challenges and Perspectives," In *2021 IEEE International Mediterranean Conference on Communications and Networking (MeditCom)*, pp. 117–122, 2021, doi: 10.1109/MeditCom49071.2021.9647460.

[18] *Functional architecture and information flows to support Common API Framework for 3GPP Northbound APIs; Stage 2, (Release 17)*, Technical Specification (TS) 23.222, v17.6.0, June 2022.

[19] 5G-IA, "EU vision on 6G," Whitepaper, June 2021. Accessed: April 6, 2023. [Online]. Available: https://5g-ppp.eu/european-vision-for-the-6g-network-ecosystem/.

[20] ETSI "ETSI TeraFlow SDN (TFS)," 2023. Accessed: April 6, 2023. [Online]. Available: https://tfs.etsi.org/.

[21] H. Lønsethagen, S. Lange, T. Zinner, H. Øverby, L. M. Contreras, N. Ciulli, E. Dotaro, "Towards Smart Public Interconnected Networks and Services – Approaching the Stumbling Blocks," Preprint in *TechRxiv*, 2022. Accessed: April 6, 2023. [Online]. Available: https://doi.org/10.36227/techrxiv.19690570.v1.

[22] IETF, "Framework for IETF Network Slices," 2023. Accessed: April 6, 2023. [Online]. Available: https://datatracker.ietf.org/doc/draft-ietf-teas-ietf-network-slices/.

[23] IETF, "Challenges for the Internet Routing Infrastructure Introduced by Changes in Address Semantics," 2023. Accessed: April 6, 2023. [Online]. Available: https://datatracker.ietf.org/doc/draft-king-irtf-challenges-in-routing/.

[24] R. Schmidt, M. Irazabal, and N. Nikaein, "FlexRIC: an SDK for next-generation SD-RANS," In *Proc. 17th International Conference on Emerging Networking EXperiments and Technologies (CONEXT 2021)*, 7–10 December 2021, Munich, Germany (Virtual Conference).

[25] Linux Foundation, "Traffic Control tc (8), Linux man page," 2023. Accessed: April 6, 2023. [Online]. Available: https://linux.die.net/man/8/tc.

[26] Linux Foundation, "Open vSwitch," 2023. Accessed: April 6, 2023. [Online]. Available: https://www.openvswitch.org/.

[27] NetFPGA, "About NetFPGA," 2023. Accessed: April 6, 2023. [Online]. Available: https://netfpga.org/About.html.

[28] Netronome, "About Agilio SmartNICs," 2023. Accessed: April 6, 2023. [Online]. Available: https://www.netronome.com/products/smartnic/overview/.

[29] R. Ricart-Sanchez, P. Malagón, A. M. Escolar, J. M. Alcaraz Calero, and Q. Wang "Toward hardware-accelerated QoS-aware 5G network slicing based

on data plane programmability," In *Transactions on Emerging Telecommunications Technologies*, vol. 31, no. 4, January 2020.

[30] 6G BRAINS, "D5.1 E2E network slicing control enablers," December 2021. Accessed: April 6, 2023. [Online]. Available: https://ec.europa.eu/research/participants/documents/downloadPublic?documentIds=080166e5f560ab13&appId=PPGMS.

[31] H. Harkous, M. Jarschel, M. He, R. Pries, and W. Kellerer, "P8: P4 with predictable packet processing performance," In *IEEE Transactions on Network and Service Management*, vol. 18, no. 3, pp. 2846–2859, 2021.

[32] H. Harkous, B. A. Hosn, M. He, M. Jarschel, R. Pries, and W. Kellerer, "Towards performance-aware management of p4-based cloud environments," In *2021 IEEE Conference on Network Function Virtualization and Software Defined Networks (NFV- SDN)*, pp. 87–90, 2021.

[33] *NR and NG-RAN Overall description, Release 17*, Technical Specification (TS) 38.300, v17.2.0, 3GPP, September 2022.

[34] ETSI, Common API Framework for 3GPP Northbound APIs (3GPP TS 23.222 version 16.8.0 Release 16) TS 123 222 V16.8.0 (2020-10).

[35] *Security aspects of Common API Framework (CAPIF) for 3GPP northbound APIs (Release 17)*, Technical Specification (TS) 33.12, v17.0.0, 3GPP, March 2022.

[36] A. M. Sanchez, A.-S. Charismiadis, D. Tsolkas, D. Artuñedo Guillen, and J. G. Rodrigo" Offering the 3GPP Common API Framework as Microservice to Vertical Industries," In *EuCNC & 6G Summit* 2022, June 2022.

[37] *Common API Framework for 3GPP Northbound APIs; (Release 17)*, Technical Specification (TS) 29.222, v18.0.0, December 2022.

[38] D. Fragkos, G. Makropoulos, A. Gogos, H. Koumaras, and A. Kaloxylos, "NEFSim: An open experimentation framework utilizing 3GPP's exposure services," In *2022 Joint European Conference on Networks and Communications & 6G Summit (EuCNC/6G Summit)*, June 2022.

[39] L. Nielsen, A. F. Cattoni, Z. Diaz, G. Almudena, B. García, A. Gavras, M. Dieudonné, and E. Kosmatos, "Basic Testing Guide - A Starter Kit for Basic 5G KPIs Verification," 5G PPP Whitepaper, 2021. Accessed: April 6, 2023. [Online]. Available: https://doi.org/10.5281/zenodo.5704519.

[40] *Experiential Networked Intelligence (ENI); Context-Aware Policy Management Gap Analysis*, ETSI GR ENI 003, v1.1.1, March 2018. Accessed: April 6, 2023. [Online]. Available: https://www.etsi.org/deliver/etsi_gr/ENI/001_099/003/01.01.01_60/.

[41] A. Rafiq, A. Mehmood, and W. C. Song, "Intent-based slicing between containers in sdn overlay network," In *J. Commun.* Vol. 15, no. 3, March 2020.

[42] F. Paganelli, F. Paradiso, M. Gherardelli, and G. Galletti, "Network service description model for vnf orchestration leveraging intent-based sdn interfaces," In *2017 IEEE Conference on Network Softwarization (NetSoft)*, 2017.

[43] W. Cerroni, C. Buratti, S. Cerboni, G. Davoli, C. Contoli, F. Foresta, F. Callegati, and R. Verdone, "Intent-based management and orchestration of heterogeneous open-flow/iot sdn domains," In *2017 IEEE Conference on Network Softwarization (NetSoft)*, July 2017.

[44] A. Rafiq, A. Mehmood, T. A. Khan, K. Abbas, M. Afaq, and W. C. Song, "Intent-based end-to-end network service orchestration system for multi-platforms," In *Sustainability*, vol. 12, no. 7, 2020.

[45] K. Abbas, M. Afaq, T. A. Khan, A. Rafiq, and W. C. Song, "Slicing the core network and radio access network domains through intent-based networking for 5G networks," In *Electronics*, vol. 9 no. 10, pp. 1710, 2020.

[46] R. A. Addad, D. L. C. Dutra, M. Bagaa, T. Taleb, H. Flinck, and M. Namane, "Benchmarking the ONOS intent interfaces to ease 5g service management," In *2018 IEEE Global Communications Conference (GLOBECOM)*, Dec. 2018.

[47] Y. Tsuzaki and Y. Okabe, "Reactive configuration updating for intent-based networking," In *2017 International Conference on Information Networking (ICOIN)*, 2017.

[48] A. S. Jacobs, R. J. Pfitscher, R. A. Ferreira, and L. Z. Granville, "Refining network intents for self-driving networks," In *Proceedings of the Afternoon Workshop on Self-Driving Networks, SelfDN*, pp. 15–21, 2018.

[49] F. Aklamanu, S. Randriamasy, E. Renault, I. Latif, and A. Hebbar, "Intent-based real-time 5G cloud service provisioning," In *2018 IEEE Global Communications Conference (GLOBECOM)*, Dec. 2018.

[50] S. Garg, M. Guizani, Y. -C. Liang, F. Granelli, N. Prasad, and R. R. V. Prasad, "Guest Editorial Special Issue on Intent-Based Networking for 5G-Envisioned Internet of Connected Vehicles," In *IEEE Transactions on Intelligent Transportation Systems*, vol. 22, no. 8, pp. 5009–5017, Aug. 2021, doi: 10.1109/TITS.2021.3101259.

DOI: 10.1561/9781638282396.ch8

Chapter 8

Secure, Privacy-Preserving, and Trustworthy Networks

By Alexandros Kostopoulos, et al.[1]

The ability to support data-intensive applications that require the analysis of data under the control of different parties is a priority for 5th Generation/Beyond 5th Generation (5G/B5G) mobile networks [1]. Several scenarios may be considered, typically characterized by a diversity of data sources stored on different nodes, possibly under the control of different parties. Data analysis may then require data exchanges and cooperation between these different parties.

A concern related to the storage and collaborative processing of data is the lack of control over the computation and hence the uncertainty about the correctness of the result. This is a well-known problem, and the research and industrial communities have devoted many efforts to the development of techniques to assess the integrity of the result of computations outsourced to external parties (e.g., [2] and [3]). However, the problem of how to use such techniques and of assessing their effectiveness in different application scenarios still needs to be further investigated [4].

6th Generation (6G) telecommunication networks need to have a fast and efficient way of exchanging information and resources among different parties, with minimum risk of failure, to enhance the Quality of Service (QoS) they provide towards a new level of Quality of Experience (QoE) for the users. The research

1. The full list of chapter authors is provided in the Contributing Authors section of the book.

attention has turned towards a new approach to the decentralized framework. Such a framework is the blockchain platform, which can be used in several domains (e.g., network slicing, industrial Internet of Things (IoT) networks, etc.).

Another similar research issue is the protection of the users against malware and other network attacks. While various security measures have been designed to protect the availability, privacy, and integrity of user data in 6G networks from malicious and unauthorized accesses, most of them are not effective against a few specific threats that abuse trust [5]. Therefore, a third-party entity/platform could be adopted to enable "Trust as a Service" (TaaS). Moreover, the use of Machine Learning/Artifical Intelligence (ML/AI) will also impact the delivered trustworthiness as it has the potential of providing intelligence for threat detection and mitigation.

Section 2.3 presents the overall architecture, visualizing the applicable security and privacy components in all areas, intending to provide a holistic 6G security architecture, comprising today's well-proven security mechanisms. In this chapter, we investigate aspects related to network privacy and security for information sharing among different tenants and cloud-stored data, as well as end users' network security. Moreover, we present the application of blockchain-based platforms for network slicing by using smart contracts, as well as for industrial IoT networks. Finally, we focus on trusted execution, Trust as a Service (TaaS), as well as trustworthy ML/AI.

8.1 Network Privacy and Security

This section investigates network privacy and security aspects at different levels. First, we focus on security and privacy for information sharing among tenants. As a next step, we consider security and privacy for cloud-stored data. Finally, we explore aspects related to end users' network security.

8.1.1 Security and Privacy for Information Sharing Among Tenants

The number of AI applications that are used in production real-world environments has rocketed in the past years, following the amazing advances obtained in different areas. These ML-based applications range from the personalization of services or the improved healthcare offered to final users to the automatic management of networks by telecom operators in the new 6G architecture. However, these applications rely on input data coming from possibly heterogeneous sources (either human or other machines) and spread through platforms owned by different actors which may not be fully trusted. Clearly, this poses different privacy and confidentiality issues.

The study of the privacy implications for individuals has attracted the focus of different research communities in the past decades, proposing different solutions to allow data sharing without identifying specific users [6–8]. However, all these solutions are designed to keep the dataset in a human-readable format, without thinking about the implications of the modifications on downstream ML tasks. Other solutions, such as the differential privacy [9] paradigm, can ensure the privacy of ML tasks with strong theoretical guarantees, but they are difficult to apply in practice, mostly when sharing the entire datasets is required to complete a task. Moreover, they are created to avoid the identification of a single user, but they may not apply when the data are generated by machines. As an example, if different network tenants want to share data of their resource consumption to jointly train a better orchestration algorithm, the aforementioned solutions would disclose the dataset as it is (as most performance reads are similar or equal) disclosing that way the original values that may include business confidential data.

An interesting research domain focuses on the development of data transformation methods, that is, methods that are designed to improve the accuracy of a downstream ML task, without the requirement of keeping the data in a human-readable format. While this kind of transformation prevents human analysis, it could be applied in an industrial environment such as an automated 6G network, where the data collection, transformation, analysis, and action are all done in an automatic fashion, typically without human intervention. There are several methods designed to protect the raw data, allowing the sharing of training data among different actors without the privacy risks associated (i.e., the Privacy Preserving Data Publishing, PPDP [10]). Thus, these technologies open the possibility of novel applications only possible before among trusted parties in (complex) federated learning scenarios. For example, they allow different tenants in a network to share data with the operator to improve the whole network's performance.

In a typical scenario where the main goal is to ensure secure information sharing, the user (or data owner) applies a transformation to the raw data. Then, the transformed data can be shared with other actors that may use it to train an ML model (possibly by adding the transformed data to other transformed data she owns). In the optimal scenario, the accuracy of the ML model would remain similar to the accuracy obtained using the raw data. Moreover, the transformed data should not allow the reconstruction of the raw data, and ideally, should not allow the recipient of the data to infer any exact information from the original data.

However, all these solutions impose some limitations on the scenario as they either (i) *bound the kinds of attacks to a subset* or (ii) *impose an a priori knowledge on the kind of machine learning task that will be performed.* While these assumptions are valid in several scenarios, there are others in which they could be a cap that is not acceptable for the workflow in the process. For instance, if the ML task is not known or it changes over time, the ML algorithm cannot be made resilient to privacy attacks, and limiting them to a small subset may be too restrictive. This is the usual case for many network applications which may expose data to third parties without knowing the final task that is going to be performed.

8.1.2 Security and Privacy for Cloud-stored Data

5G and beyond networks are a key enabler of today's digital society, supporting new and better applications in a variety of sectors because of the availability of a powerful hyper-connected infrastructure offering unprecedented network capacity and speed. Distributed sensors, mobile and pervasive devices, and cloud/edge/fog computational and storage nodes can be involved in providing advanced services and applications [11]. At the centre of such novel applications are data, gathered, generated, shared, processed, and communicated among the different components of the infrastructure at an incredible pace. Such data can be private, sensitive, or company confidential. At the same time, different components of the infrastructure might be under different administrative domains and subject to different trust assumptions.

The full realization of the power brought by the hyper-connected infrastructure can happen only with the availability of solutions to ensure proper protection (confidentiality and integrity) of the data across their whole life cycle in the infrastructure. Many aspects have been investigated, though many are the problems still open.

With respect to confidentiality, attention has first been devoted to data in storage (to maintain data private to the storage provider itself). Data can be protected through an encryption layer that is applied before storing them at a storage provider. As encryption affects the possibility of performing computations over the data, researchers have investigated different techniques for effectively and efficiently supporting computations over encrypted data (e.g., indexes [12], data fragmentation [13], property-preserving encryption [14], searchable encryption [15], and trusted hardware [16]). While promising, such solutions still present open problems, such as the possible information leakage caused by indexes or by the encryption supporting queries [17]. The availability of high-performance networks can help in the development of efficient data anonymization solutions for the massive amount of data through the distribution of data anonymization tasks to different nodes in the network [18, 19]. A challenge here is how to perform data fragmentation and distribution among different nodes in the network, each then operating with partial knowledge of the data collection, while ensuring the utility of the overall anonymization result.

More and more emerging scenarios also require different parties to cooperate for sharing data and to perform distributed computations. This distribution and sharing process should, however, take into consideration the fact that different parties may be subject to different access restrictions. The relevance of this problem is confirmed by the existence of several existing approaches that address the problem of protecting data confidentiality in distributed computations (e.g., [20–23]). None of these proposals consider the possibility of protecting data with encryption, which has been introduced in [24]. A challenge here is how to find an allocation of the different operations of a distributed computation to subjects to minimize a parameter of interest (e.g., cost or performance) [25].

With respect to integrity, some approaches have investigated the problem of providing guarantees that data are correctly stored at the storage provider (e.g., digital signatures, provable data possession, and proof of retrievability [26]). The problem becomes more complicated when different nodes of the network infrastructure are used for performing computations, and integrity then needs to be guaranteed on such computations, meaning that it is necessary to verify whether the result of a computation is: *correct* (i.e., computed on genuine data), *complete* (i.e., computed over the whole data collection), and *fresh* (i.e., computed on the most recent version of the data). Existing solutions addressing this issue can be classified as deterministic or probabilistic [27]. Deterministic techniques are based on the definition of authenticated data structures that provide integrity of the stored data and full completeness guarantees but only for the queries operating on the attribute on which the structures have been defined. On the contrary, probabilistic techniques provide integrity assurance for any query result, at the price of offering only a probabilistic guarantee.

Interesting aspects that still need to be investigated are related to the definition of probabilistic solutions for assessing the integrity of computations distributed to different nodes in the system. In particular, the definition of a model capturing the main characteristics of probabilistic techniques (e.g., pre-computed and replicated tasks) so as to enable their controlled generation and injection in such a way that provides the best effectiveness in achieving integrity guarantees is still missing.

8.1.3 End Users' Network Security

In the past years, we have witnessed a continuous digitalization of all aspects of society, ranging from the development of e-governance solutions to the step increase in remote working and even the starting of an envisioned metaverse. However, as in many other aspects of real life, criminal organizations may try to take control of them. So, these advances came together with an increase in the number and sophistication of cyberattacks [28].

While the most complex attacks are targeted at governments or companies, they typically start with the infection of one of the end users' computers. Diverse solutions exist to prevent these attacks, ranging from simple antivirus based on malware signatures to complex endpoint detection and response (EDR) systems [29]. However, all of them depend on the action of the final user.

The protection of the users against malware and other attacks from the network is possible by monitoring the network traffic. A detection engine using ML can identify harmful sessions allowing the end of the connection before the malware is downloaded. This is due to the way malware distribution and malware command and control are done by using redirection chains. Malware is a common term used to describe malicious software (e.g., spyware, ransomware, viruses, and worms), for example, an infection scenario can be described where it is used to enrol an infected system in a botnet and then perform an attack. The attack can be divided into two different phases:

- **Delivery phase:** the malware is delivered to the infected system. It is an opportunistic phase where a large mass of systems is targeted.
- **Botnet creation and attack:** the botnet is created, and bots connect to the botnet and are used to perform an attack (i.e., distributed denial of service (DDoS), intelligence gathering, and intrusion). This is usually a targeted phase that targets specific users/services.
- **Delivery phase:**
    1. Users perform regular web activities, such as browsing, using apps, and so on.
    2. A destination website, app, or service may have a vulnerability that could be exploited (e.g., a 0-day vulnerability).
    3. A Botnet owner compromises the website, e.g., using an injected malicious advertise or a malicious link.
    4. A user visiting the compromised website is redirected into a malware delivery chain. Each site of the chain can be a known malicious website, an unknown one, or also a legit but compromised site.
    5. The user accesses the last website of the chain, which is a malware delivery website usually employing drive-by download vulnerabilities.
    6. Malware is downloaded into the user's machine.

Note that the delivery chain is very dynamic, with domain names changing often (domain-flux), and the delivery websites usually apply cloaking techniques that make them difficult to discover.

The corresponding steps for botnet creation and attack are the following:

1. The botnet owner registers DNS records to enable the discovery of the Command and Control (C&C) server.
2. The infected machine discovers the location of the C&C server.
3. The C&C server is contacted by the infected host, which is added to the botnet.
4. An attacker rents the botnet to perform an attack.
5. Attack instructions are delivered to the botnet.
6. Attack is started by botnet hosts.
7. Note that the C&C server address/domain changes often (domain-flux).
8. The network is used in both phases of the attack, and attack traces are visible in the network traces.

For instance, patterns of a sequence of visited network destinations can be used to block the malware delivery phase by recognizing a redirection chain, as depicted in Figure 8.1. Such chains are characterized by a landing web page that is legit, e.g., a compromised benign web page or a page hosting a third-party advertisement, which then initiates a series of automatic redirections towards a malware distribution website that usually exploits some vulnerabilities to implement drive-by downloads. As such, the sequence of network destinations may once more help in identifying such distribution chains.

The set of network entities contained in the traffic may provide relevant information to discover and identify the host that is already part of a botnet. In fact, recent work showed that botnets' hosts have a noisy network behaviour. For example, they assess connectivity, retrieve date information, and perform scanning to detect the Command & Control channel. Such activities generate a number of network flows towards different destinations. Such communication patterns may contain important information to detect botnets. For example, dynamic analysis tools run malware to analyse their network behaviour and generate signatures that may help identify them.

Moreover, network traffic generated when a website is visited contains the fingerprints of the website itself. Indeed, modern websites include a number of external

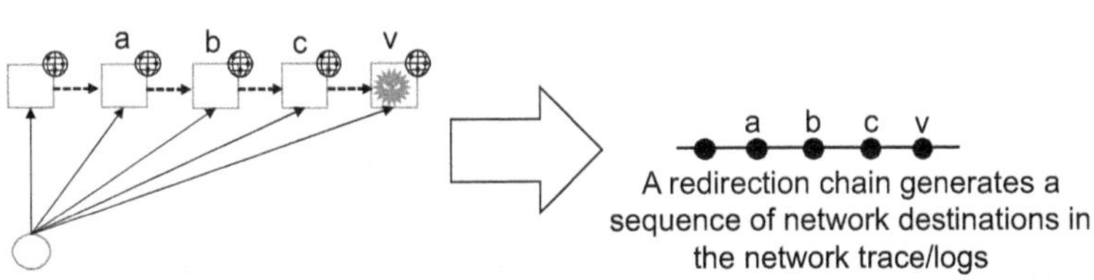

Figure 8.1. Example of redirection chain in malware distribution.

resources, hosted at domain names that may be different from the main website's domain. The nature of the different domains (e.g., news, e-commerce, advertisement, CDN, etc.) and the number of such domains contacted by a website can provide insightful information about the nature of the web page itself. In fact, even if a malicious website could change often the domain name or IP address, the external resources present on the web page are more unlikely to have changed.

This type of monitoring system (as most security solutions) requires the collection of data from the users that could be considered personal information (i.e., the browsing history of users) that could be protected by the GDPR. This situation could be especially worrisome in a work environment where users may be reluctant to have their browsing history recorded by the employer, denying that way the consent. However, the own GDPR allows the collection of data in these circumstances without the need for consent, as the security of the final user and the system is of vital interest to the own data subject.

8.2 Security and Privacy for Blockchain-Based Platforms

8.2.1 Blockchain-based Smart Contracts for Network Slicing

Traditional networks have a one-size-fits-all approach that needs to be adapted for 6G networks. The traditional rigid design can no longer be used in an environment with different requirements and diverse applications. The traditional monolithic approach gave way to an approach that is more flexible and adjustable, with the network providers using the virtualization of logical networks towards sharing their infrastructures. These logical networks are called *network slices.*

Network virtualization is not a new concept but only lately has the technology allowed for the deployment of network slices in the real world. Network slicing assumes virtualization and automated orchestration and management to provide [30]:

- *QoE*;
- *Shorter time to market and to customer*;
- *Simpler resource management*;
- *Increased automation*;
- *Flexibility and agility*;
- *Reduced risks.*

As a result of the deployment of network slices, a new market has emerged where MNO, InP, and MVNO carry out frequent transactions and exchange resources. In other words, the creation of network slices has created the need for a network broker to handle, timely execute and automate these transactions.

For decades, trusted entities were handling information exchange, money, and other asset transfers. While this was the preferred method of operation all these years, the process is time- and power-consuming. At the same time, the required trust in the central entity provides a single point of failure. Thus, 6G telecommunication networks need to enable information and resource exchanging in an efficient way.

Naturally, the attention has turned towards a new approach to the decentralized framework. Such a framework is the blockchain platform. The blockchain platform has a strong potential in 6G systems. *Blockchain* is a technology designed to store data as records, which are securely linked together using cryptographic methods, and which cannot be removed without invalidating the whole chain. It is a special type of a distributed ledger, which is regarded as a replicated, shared, and synchronized data storage where the participating nodes in a peer-to-peer network need to reach a consensus on writing the next transaction onto the ledger, i.e., agreeing upon which update is permitted, and in what order updates are done. In this way, no central trusted-by-everyone authority is needed for the data storage.

There is a plethora of application opportunities for exploiting blockchains in 6G infrastructure itself for performance gains or enabling new services such as [31]:

- *Decentralized network management structures*;
- *Pricing, charging, and billing of network services*;
- *Authentication, authorization, and accounting*;
- *Service level agreement management*;
- *Spectrum sharing*;
- *Extreme edge.*

One such application offered by the blockchain platform is network slicing through *smart contracts*. The blockchain can enable secure and automated brokerage of network slicing while providing [32]:

- *Significant savings in the operational cost*;
- *Speed up the slice negotiation process and reduce the cost of slicing agreement*;
- *Increased efficiency of operation for each network slice*;
- *Increase the security of the network slice transactions.*

Smart contracts can be used through a blockchain platform to negotiate the resources of a network and allocate them to the MVNO as needed in a timely and efficient manner. The decentralized identity of a blockchain platform removes the need for a central entity and the single point of failure of the network. Trust among the users is guaranteed through the consensus protocol, which dictates the decision mechanism of the system and prevents a single node to control the majority of the blockchain platform.

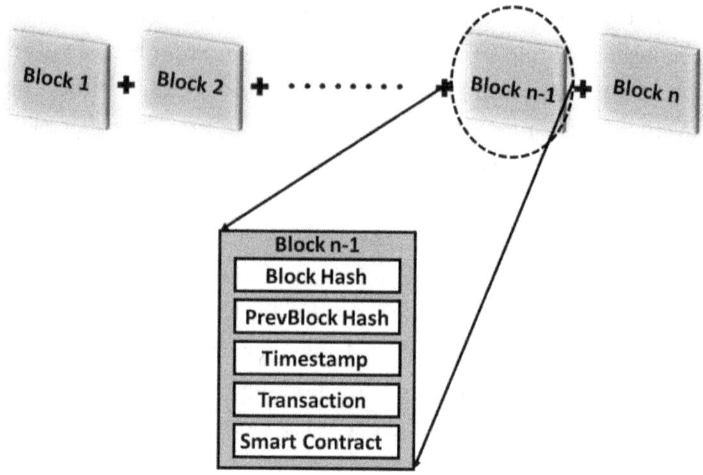

Figure 8.2. An example of a blockchain's ledger.

Furthermore, the blockchain platform allows for transparency among the users as each of them has a copy of the ledger, and it protects the privacy of its users through encryption keys and cryptographic security measures. In a centralized framework, there is the need for third parties to oversee any negotiation process among the central body and the tenants of the network, which is no longer necessary in the blockchain platform, mainly due to the chronological, time-stamped, and immutable smart contracts. Smart contracts can be programmed once and used as templates and reused repeatedly, reducing the need for a continuous overseeing entity. This transparency and immutability limit the risk of a malicious node attacking the system.

An example of a blockchain ledger is presented in Figure 8.2. As shown, each block is added after the previous block in chronological order. Each block contains a unique Hash number that identifies this block, as well as the Hash key of the previous block. Hash is a short data digest generated from the data of the block, which varies as the data of the block changes [33]. Thus, tampering with any information in a block breaks the hash pointers, ensuring the security and immutability of the platform.

Further to the hash keys, the block can contain other information, including but not limited to a timestamp and information for the transaction the block was initially created for. Additionally, the block includes the private keys and certificates of the endorsing and evoking peers to ensure the legitimacy of the transaction.

The focus of this work was on using a blockchain platform as a *Network Slicing Broker (NSB)*, with the blockchain platform of choice being the Hyperledger Fabric. The main process to verify and implement a transaction in the Hyperledger network is described below.

Initially, the network will select a leader among the participating organizations, a leader that changes to ensure trust and security in the network. When the network has elected a leader, the process to validate a transaction and adding it to the ledger is divided into four steps as follows:

(a) The leader sends a transaction request to all the nodes in the form of a smart contract.
(b) The nodes, which at this point are called endorsing peers, collect the request, simulate its execution state, sign it, and send it back to the leader.
(c) The leader collects all the endorsements from the peers and validates their authentication keys and signatures. If all information matches, it sends the endorsement peers' suggestion with the original transaction proposal to the ordering service of the network. The ordering service is the final decision-making system of the network, part of the consensus protocol.
(d) Upon receiving the transaction request, the endorsement peers' suggestion, and their authorization key, the ordering service puts the transaction into voting. If the transaction is approved, a block is created and added to the ledger before the transaction is executed. If the transaction is not approved by the majority, a block is still created and added to the ledger, but the transaction is not executed. At this point, the transaction is committed, and its results cannot be altered. The ordering service will inform the whole network of the transaction.

In the example above, the transaction can be any request originating from a peer to the leader. It can be a matter of reading the information in the ledger for audit purposes or a new request for borrowing or lending network resources. Either will be treated in the same way, with the difference that the first request will not change the status of the network, whereas the second will alter it. Such type of transaction can take just a few milliseconds to be completed in an optimized system. Thousands or even more transactions per second can take place, allowing this way almost real-time negotiation and allocation of network resources limited only by the speed of the action system.

8.2.2 Blockchain for Industrial IoT Networks

In the factory IoT setting that forms the basis of industrial 5G/6G use cases multiple communication interactions are assumed, often in an ad-hoc way. Increasingly closer collaboration between the factory operator and devices and vehicles introduced into the factory by multiple external suppliers requires establishing secure communication channels between these different actors on the factory floor. Depending on the application, a breach in an industrial IoT environment could result in risks ranging from leakage of important business information to exposures of production processes or damage to the industrial controls. One building block to address this issue is *mutual device identification and authentication.* Blockchain-based solutions can be used for implementing the Verifiable Credentials model specified by W3C [34] and mapping it to the factory IoT setting.

Blockchain implementations vary regarding multiple properties. The most distinguishing ones are a blockchain's visibility (public/private) and accessibility (permissioned/permissionless). For the multi-company verifiable credentials' scenario, a private permissioned blockchain will be implemented.

Verifiable credentials (VCs) are the digitally signed, cryptographically secure representation of physical world credentials, such as a driver's license, a birth certificate, or an inventory number. A credential has the following properties:

- It *identifies* the subject of the credential (i.e., photo, name, and identification number).
- It relates it to the *issuing authority* (i.e., government, agency, etc.).
- It is of a specific *type* (i.e., English passport, German driving license, and health insurance card).
- It proves the *assertion of specific properties* by the issuing authority about the subject.
- It informs about *constraints* on the credential (i.e., expiration date and terms of use).

VCs add digital signatures to identity information, making them more tamper-evident and more trustworthy than their physical counterparts. The rationale to use this concept for the IoT device identification lies in the cryptographically signed attestation of the device's origin and "factory compatibility."

The core roles in a VC ecosystem are (see also Figure 8.3):

- **Holder:** This role possesses one or more verifiable credentials. It receives them by the issuer role, and it presents verifiable representations, or proofs, to the verifier role.
- **Issuer:** This role asserts claims about one or more subjects, creating a VC from the claim, and transfers the credential to a holder.
- **Subject:** This is the role about which claims are made. Most of the time, a holder is identical to a subject, but sometimes they differ, e.g., a parent (holder) holding the VC of a child (subject), or a human holding the VC of its pet.
- **Verifier:** This role processes VCs to verify some property about a subject.
- **Verifiable Data Registry:** This role mediates the creation and verification of identities, keys, VC schemas and definitions, revocation registries, public

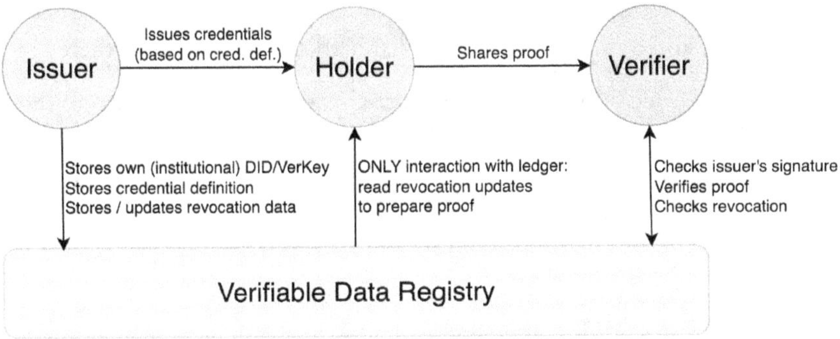

Figure 8.3. Roles that interact in a verifiable credential ecosystem.

keys of issuers, and so on. The verifiable data registry might be implemented as a distributed ledger (see above).

In the mutual device identification use case, the secure identification of automated guided vehicles (AGVs) passing the factory entrance can be modelled: An identification verification device at the factory entrance checks all incoming AGVs whether they belong to a company being part of the campus consortium that is allowed to be part of the factory's ecosystem. A further check could be the compatibility of the AGV with the factory (e.g., its type and measures).

This scenario can be mapped to the VC roles mentioned above:

- The AGV itself is both the VC *holder* and the *subject.* It holds VCs about the company that owns it and about its type and measures. In this use case, there is no need to differentiate holder and subject, because an AGV's resources should be sufficient to fulfil the requirements of the holder role.
- All companies provide their own issuing "authority," that supplies all their devices with the above-mentioned properties in the form of verifiable credentials. It thus serves as the *issuer* role which puts certain information on the data registry, such as its own identifier and verification keys, the credential definition, and revocation information.
- At the factory entrance, a verification device serves as the *verifier* role which checks the credentials (or proofs thereof) the AGVs provide. It makes use of the information on the data registry that is necessary for the verification process.
- A blockchain is used as the *verifiable data registry* to store the credential schema, as well as publicly visible necessary information by the issuers. Blockchain is well suited for this purpose: The system does not have to rely on a single trusted-by-everyone authority but stores the information in a distributed, cryptographically secure way, thus *increasing both integrity and*

availability. Furthermore, due to the blockchain's construction, once information is stored on it, it cannot be disputed to have been put there.

For human-centred applications, there are already implementations of this concept available, such as Sovrin [35] and IDunion [36] which are based on projects driven by the Hyperledger Foundation [37]. There is also some research focused on IoT scenarios [38, 39], which needs to be considered.

8.3 Trusted Execution

The 6G trust coverage must encompass hardware-based trust anchors and built-in security that work with the cloud and accelerator-based architectures expected in the 2030s. In the highly decentralized, open, and virtualized telecom architectures of the future, 6G must also utilize a secure and tamper-resistant hardware component, known as the "root of trust," to guarantee the security of data and code in untrusted environments. The 6G era is expected to see an increase in non-public networks and specialized sub-networks, many of which may be operated on-premises or dedicated cloud stacks outside of the trusted wide area network. Hardware technologies like trust anchors and execution environments will improve from the current Trusted Platform Module (TPM) developed by the Trusted Computing Group (TCG) and also incorporate Secure Boot, Trusted Execution Environments (TEE) [40], and Enclaving [41]. With 6G, further advancements in TPM and TEE are anticipated, along with new and hybrid processing units, hardware acceleration, and an accompanying acceleration abstraction layer [42].

8.3.1 Workload Isolation

Edge computing brings memory and computing power closer to the location where it is needed. In edge computing systems, computation is rather offloaded to nearby resources than to the cloud, due to latency reasons. However, the performance demand in the edge grows steadily, which makes nearby resources insufficient for many applications. Additionally, the number of parallel tasks at the edge increases, based on trends such as ML, IoT, and AI. This introduces a trade-off between the performance of the cloud and the communication latency at the edge. Furthermore, the need to be energy efficient in a mobile environment is high. The edge computing paradigm, nowadays, employs single-tenant tasks, scheduled in resource-constrained devices, needing specialized Operating Systems (OSes) to host these tasks.

To be able to move to a multi-tenancy execution model at the edge, the system must ensure non-interference and controlled data access among different and possibly concurrently running applications. To this end, virtualization plays a significant role in adding secure multi-tenancy execution at the edge. Adding another layer of abstraction facilitates unified execution frameworks but complicates the execution stack and consumes resources for ensuring correct management, isolation, and resource sharing among tenant workloads.

To alleviate the significant overhead of adding a full virtualization stack (hypervisor and VMs) at the edge, the research community has proposed the use of container technology. However, this execution model increases the attack surface, as it exposes the full OS and runtime layer to any malicious or compromised application running in a container. Related works and critical security advisories [43] have pointed out that containers are far too insecure for multi-tenancy. Recent works [44] have introduced a new type of resource virtualization, bringing the benefits of isolation and secure execution, without the burden of a full virtualization stack to support generic OSes.

***Attack Scenarios*:** A software supply chain attack [45] occurs when malicious code is purposefully added to a component that is sent to target users. The code may be introduced to the component in several different ways, such as via compromise of the source code repository, theft of signing keys, or penetration of distribution sites and channels. As a part of an authorized and normal distribution channel, customers unknowingly acquire and deploy these compromised components onto their systems and networks. Advanced malicious code typically does not disrupt normal operations and may not activate for several days or weeks, thereby remaining hidden from typical application and software testing practices.

For instance, a telecommunications company buys core network systems management software from a trusted provider; however, unbeknownst to the trusted provider, one of the components it uses in the product has been compromised and now contains malicious code. This is a threat that results from inheriting risk decisions made by a supplier within the supply chain that impacts the end user of the final product or service. The deeper into the supply chain it occurs, the more difficult it is to identify in advance. This inserted vulnerability may be used by the malicious actor as a part of a larger attack chain that uses the malicious code to gain access within the core network of the telecom and pivot towards other attack vectors.

Additionally, unauthorized access to software or network components provides a malicious actor with the opportunity to modify configurations to reduce security controls, install malware on the system, or identify weaknesses in the product. These vulnerabilities could be exploited for increased persistent and privileged access within a system or network. Malicious actors may also embed code in SDN controller applications to constrict bandwidth and negatively affect operations.

For instance, a potential attack vector stems from "persistent threats," where malware is inserted into a system in a way that the platform always boots in a compromised state, even after legitimate software is re-installed. To combat this attack, system vendors are turning to two technologies: Secure Boot and Measured Boot, to provide assurance that when a platform boots, it is running code that has not been compromised.

8.3.2 Systems Software Stack

This subsection first focuses on the reduction of the attack surface by minimizing the exposure of privileged operations. This part addresses attack vectors related to the exploitation of bugs in software, as well as component misconfiguration. Then, the hardware and software mechanisms used to enable trusted execution are described.

***Reduced Attack Surface*:** To minimize the exposed privileged operations when executing workloads at Edge environments, the use of a sandbox mechanism is introduced. Specifically, the host system is protected from any workload running on it, by (i) *executing it in a contained environment* and (ii) *reducing the exposed privileged operations to the absolute minimum required for the workload to run.*

Recent works have introduced a new type of resource virtualization [46–50]. In the context of serverless computing [51], a sandbox mechanism is one of the ways to allow multi-tenancy execution, increasing the workload consolidation factor and reducing idle resources in a cloud environment. Kata Containers [52] provide such a sandbox mechanism, allowing an OCI-compatible container to run inside a traditional VM.

To achieve (i), sandboxing mechanisms, mainly containerization and virtualization techniques, are used. In order to facilitate workload deployment, the container concept is kept, but instead of running workloads as containers on the host, the container execution is isolated using VMs. Additionally, hardware extensions are utilized when available.

To achieve (ii), unikernels [47] are used. In the last years, a new approach in lightweight virtualization aims to bridge the best from both containers and virtual machines. Unikernels are specialized single-address space machine images constructed by using library operating systems. Some of their advantages include fast boot times, low memory footprint, and increased performance while at the same time, they provide stronger security and hardware isolation. However, unikernels come with a lot of limitations, and running existing applications on top of them is not straightforward. While some frameworks try to provide a POSIX-like environment, others prefer a clean state approach, requiring the complete refactoring of an application to be able to execute on them.

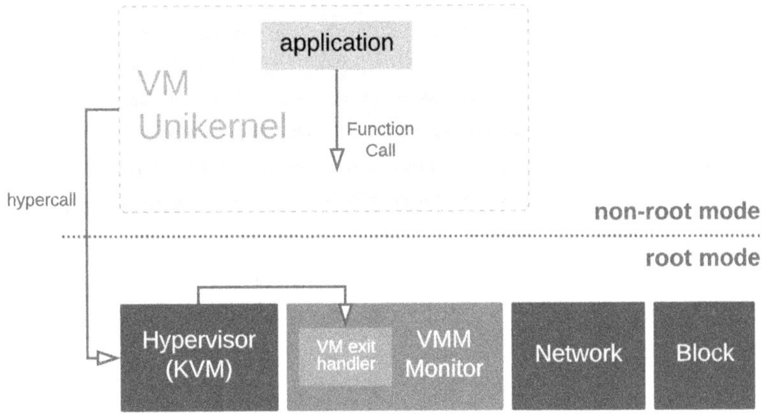

Figure 8.4. A unikernel running as a VM on HEDGE.

***Virtual Machine Monitor*:** A minimal and simplistic virtual machine monitor (VMM) that resides inside the Linux kernel interacting directly with KVM without any intervention from the user space is introduced to facilitate sandboxing. This VMM is called EDGE, Hypervisor for the Edge (Figure 8.4). HEDGE is essentially a simple dispatch handler in the kernel that services the needs of a guest. It provides an interface to the KVM API, a virtual machine execution environment for each of the VMs spawned, generic device handling (network and block), and a management layer to perform basic VM operations (i.e., create, destroy, dump console, etc.).

An important aspect of HEDGE's design is reducing the noise that VMMs enforce to handle I/O requests. In order to achieve performance, the guest needs to run uninterrupted as much as possible. Apart from removing the mode switch overhead, HEDGE handles I/O requests with the minimum possible overhead.

In the context of B5G/6G deployments, specific application code can be extracted from the relevant use cases and port it to unikernel frameworks: Unikraft [53] and rumprun [48].

8.3.3 Hardware Trust

A mobile device is susceptible to several attacks by malicious users, physical or otherwise. For instance, one could gain access to the storage backend of a mobile device (e.g., an SD card, eMMC memory, etc.) and tweak the operating systems binary that drives the device to relay data to a malicious third party. This type of attack is relevant to many use cases of B5G/6G infrastructure actors and end users. To prevent this, mechanisms such as *Secure Boot* and *Measured Boot* are introduced.

The terms Secure Boot and Measured Boot are often seen together, and they can be complementary, but they are not at all the same. Both technologies rely on a

"Root of Trust," that is, some piece of code or hardware that has been hardened well enough that it is not likely to be compromised, and either cannot be modified at all or else cannot be modified without cryptographic credentials.

For many systems, that "Root of Trust" is provided by the Unified Extended Firmware Interface (UEFI) BIOS – code that takes the place of the ad-hoc "legacy" BIOS that has been in use for years. The UEFI BIOS works with platform hardware to ensure that the flash memory that contains the BIOS cannot be modified without cryptographic authority, thus forming the "Root of Trust."

A UEFI BIOS depends on several elements to ensure the Root of Trust is not compromised:

- The BIOS contains a public key that is controlled by the equipment manufacturer. Any authorized change to the BIOS must be signed with the corresponding private key.
- The BIOS itself is required to check the validity of the signature on a proposed update, using the public key stored in a protected part of the BIOS flash.
- The BIOS must configure processor hardware features to block any unauthorized writes to the flash. In an x86 design, Protected Range Registers are one line of defence, with other mechanisms also available.

Both Secure Boot and Measured Boot start with the Root of Trust and extend a "chain of trust," starting in the root, through each component, to the operating system (and in embedded systems, often to the application itself). But once a Root of Trust is established, Secure Boot and Measured Boot do different things.

Modern platforms of all sorts often use a multi-stage boot, where firmware in flash launches an OS Loader (such as Grub2 or u-boot), which then loads and launches a series of OS components.

- In a Secure Boot chain (Figure 8.5), each step in the process checks a cryptographic signature on the executable of the next step before it is launched. Thus, the BIOS will check a signature on the loader, and the loader will check signatures on all the kernel objects that it loads. The objects in the chain are usually signed by the software manufacturer, using private keys that match

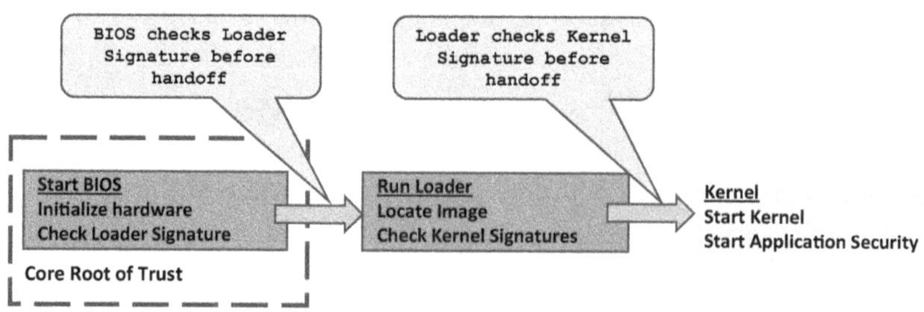

Figure 8.5. Secure boot execution flow.

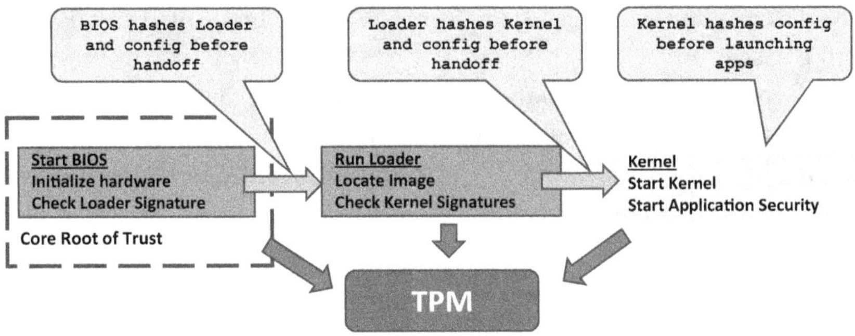

Figure 8.6. Measured boot execution flow.

up with public keys already in the BIOS. If any of the software modules in the boot chain have been hacked, then the signatures will not match, and the device will not boot the image. Because the images must be signed by the manufacturer, it is generally impractical to sign any files generated by the platform user (such as config files).

- In a Measured Boot chain (Figure 8.6), we still depend on a Root of Trust as the starting point for a chain of trust. But in this case, prior to launching the next object, the currently running object "measures" or computes the hash of the next object(s) in the chain and stores the hashes in a way that they can be securely retrieved later to find out what objects were encountered. Measured Boot does not make an implicit value judgement as to good or bad, and it does not stop the platform from running, so Measured Boot can be much more liberal about what it checks. This can include all kinds of platform configuration information such as which was the boot device, what was in the loader config file, or anything else that might be of interest.

***Secure Boot*:** Secure Boot is relatively self-contained. If the handful of signed objects have not been tampered with, the platform boots, and secure boot is done. If objects have been changed so the signature is no longer valid, the platform does not boot, and a re-installation is indicated.

***Measured Boot*:** Measured Boot is not only more flexible but also requires an important step. All those hashes must be stored in a way that there is very little chance that they can be manipulated and a very high likelihood that they can be reliably reported to a management station, using a process called attestation. As Measured Boot does not stop the platform from booting, the host OS cannot be relied upon to report the hashes.

In the case of Measure Boot, the Trusted Platform Module (TPM) is used to record these hashes. The TPM is a small self-contained security processor that can be attached to a system bus as a simple peripheral. Of the many functions a TPM can provide, one is the facility called Platform Configuration Registers (PCRs), used for storing hashes.

Secure Boot and Measured Boot can both be used at the same time. Secure Boot ensures that the system runs only authentic software, and Measured Boot gives a much more detailed picture of how the platform is configured.

8.3.4 Confidential Computing

Security has long been one of the key goals of systems design [54]. Cryptography has enabled the safe storage (at rest) and transmission (in flight) of important data. However, there is still a situation, when data can be vulnerable. The applications decrypt the data in order to save them. Therefore, the decrypted version of data is stored in RAM, CPU caches, and registers. In recent years, there has been reported a high number of memory scraping and CPU side-channel attacks. Under these circumstances, the wide adoption of cloud and edge computing, where users cannot control the underlying infrastructure, raises significant concerns regarding the security of data in use. In that context, the user cannot trust any parts of the system stack that cannot control such as the host operating system and the hypervisor.

Confidential computing aims to address the data in-use security concerns. Due to the reasons explained previously, confidential computing cannot be a solution at a software level. Accordingly, it is based on hardware extensions that modern CPUs include and provide Trusted Execution Environments (TEEs). A TEE is an enclave that isolates the code and the data of a workload from any other system component.

Depending on the implementation, a TEE might use fencing and locking mechanisms to ensure the isolation of the trusted code. As soon as a trusted code is loaded in a TEE, only specific cores and memory cases are used, aiming to avert side-channel attacks. Furthermore, TEEs can also use encryption for the data that are stored outside the TEE resources. The communication with a TEE happens through a well-defined interface, and all I/O operations are encrypted. As a result, TEEs manage to isolate the code and data running inside a TEE from any other process, user, or system component. Only the trusted code is able to view or modify the encrypted data.

The encryption and signing keys that are used from a TEE should be saved in a hardware module. That module can be the starting point (Root of Trust) and should be trustworthy. Except for encryption and signing keys, the RoT might contain other root secrets and a set of functions, needed for the encryption or validation of data. The code and data (keys) of an RoT are usually stored in a read-only memory (ROM), restricting any modifications. Trusted platform modules (TPMs) described in previous sections are examples of RoT that can generate cryptographic keys and protect important information (i.e., cryptographic and signing keys, passwords, etc.).

Using the RoT platforms can secure the underlying firmware and extend the trust to higher levels of the software stack. A verified firmware can verify the OS boot loader, which can verify the operating system, which can extend the trust to the hypervisor and/or container engine. The process of extending the trust from an RoT to higher levels of software stack is called chain of trust (CoT).

Apart from the isolation, a TEE should be able to verify the integrity of an application code. Even if the code inside a TEE is isolated and cannot be changed, there is still the danger of someone tweaking that code before it is launched inside a TEE. To be able to verify that the workload running on the hardware node is indeed the one intended by the system, attestation is used: through attestation, the workload tenant can verify that the workload is running on a genuine, authenticated platform and that the initial software stack is the expected one.

The goal is to support as many TEEs as possible. Vendors provide a wide range of security mechanisms for TEEs, from memory isolation (e.g., Intel MKTME, Arm External Memory (DRAM) Encryption and Integrity with CCA), application isolation (e.g., Intel SGX, Arm TrustZone, and IBM Application isolation technology), and virtual machine isolation (e.g., Intel Trust Domain Extensions (TDX), AMD Secure Encrypted Virtualization (SEV), or IBM Protected Execution Facility (PEF)).

8.3.5 Orchestration

To facilitate the deployment of applications in Cloud-Edge environments, and, at the same time, ensure string security guarantees for applications and data, existing Trusted Execution Environments (TEE) infrastructure support and technologies can be combined with the cloud-native word (Figure 8.7).

The key concepts to consider are the following:

- Allow cloud-native application owners to enforce application security requirements.
- Transparent deployment of unmodified containers.
- Support multiple TEE and hardware platforms.

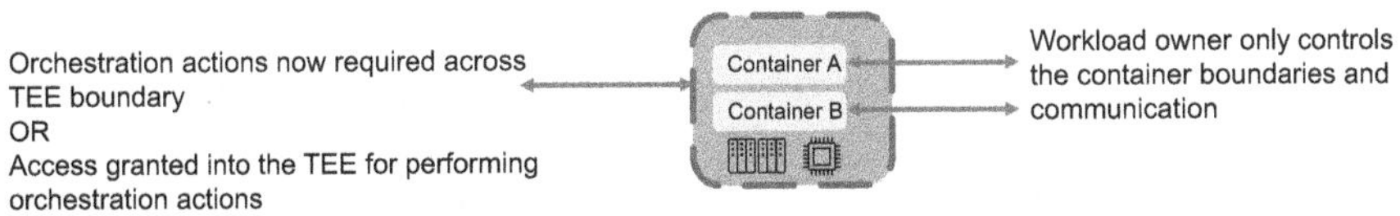

Figure 8.7. Cloud-native execution environments.

- Introduce a trust model that separates Cloud Service Providers (CSPs) from guest applications.
- Apply least privilege principles to the platform administration capabilities that impact delivering Confidential Computing for guest applications or data inside the TEE.

TEEs can be used to encapsulate different levels of the architecture stack with three key levels being *node* vs *pod* vs *container*. Container isolation was initially provided with hardware virtualization solutions, such as hypervisors. Now, pod-level support for confidential computing is addressed. Node level introduces significant challenges around the least privilege for Kubernetes cluster administration. With respect to the combination of pod and container level isolation, it is expected that the challenges explored will have relevance to future understanding of the use of TEEs at the node level.

The TEE seeks to protect the application and data from outside threats, with the application owner having complete control of all communication across the TEE boundary. The application is considered a single complete entity, and once supplied with the resources it requires, the TEE protects those resources (memory and CPU) from the infrastructure, and all communication across the TEE boundary is under the control of the Application Owner.

However, moving to a more cloud-native approach, the application is now considered a group of one or more containers, with shared storage and network resources (pod). This pod is also subject to an orchestration layer that needs to dynamically interact with the pods and containers with respect to provisioning, deployment, networking, scaling, availability, and lifecycle management.

8.4 Trust-as-a-Service

Trust is a high-level and complex value that is based on multiple basic KPIs, which are including but not limited to availability, reliability, security, privacy, integrity, and authenticity. Generally, there are three kinds of trust relation in modern mobile services:

1. A user trusts a service, sharing its confidential information with the service provider to be able to use the service. It is worth remarking that the provided service can be a data service (e.g., cloud storage), an intelligent application (e.g., cloud-based speech recognition), or even the networking service itself (e.g., radio access).
2. A service provider trusts a user. This kind of trust is not only reflected in the service provider guaranteeing certain services to the user but also allows

the service provider to exploit feedback data from the user to adapt/optimize its service.

3. A user trusts another user. This kind of trust allows the trusting user not only to exchange information with the trusted but also to exploit data shared by the latter for its own decision-making.

Though there have been numerous advanced transmission techniques developed to enhance the availability and reliability of user data in 6G networks from channel fading, interference, and hostile jamming, in addition to the various measures that are designed to protect the security, privacy, and integrity of data in from malicious and unauthorized accesses, most of them are not effective against a few specific threats that are initiated from inside of the system, which are commonly realized by abusing the trust among users and services.

One typical example of trust-abusing threats is the DoS attack, which is typically accomplished by flooding the targeted service with superfluous requests. In this scenario, the service provider's trust in the user (or massive users in the case of DDoS) initiating the attack is abused, leading to an exhaustion of its service capacity, which damages the trust of other users in this service by means of undermining the service availability.

Another common trust-oriented attack strategy is the phishing attack, which commonly combines website spoofing and social engineering techniques to manipulate users into sharing their confidential information. In such attacks, the user's inappropriate trust in another user (the sender of phishing email/message) and/or a service (e.g., a fake cloud computing platform with a fishing log-in site) is abused, which ultimately impairs the user's data privacy, or in some cases also the data availability and security (when the user is manipulated to run ransomware that disables user information, or Trojans that grant the attackers access and control to her/his system).

In addition, there has been an emerging new threat of this type: the data-injection attack, which usually aims at poisoning or manipulating the ML models to degrade the quality of data-based services by feeding fake data to the system. Such attacks have been demonstrated effective against various wireless applications, including multi-sensor localization [55] and autonomous truck platooning [57], where the trust between different users (vehicles/sensors) is abused, as well as navigation [56] and healthcare [58], where the service providers' trust in users is abused. A data-injection attack can be either initiated with social engineering techniques to obtain the control of user devices, or directly committed by malicious users through their own devices with authorized network access. In addition, even benevolent users may unintentionally report misdirecting poor data due to device malfunctions or unexpected use scenarios [58]. In all cases, destructive data can be injected into the system without violating data privacy or integrity but compromising the data authenticity. Moreover, it shall be noted that as data-injection attacks do not rely on accurate knowledge of the intelligence engine, even model-less approaches are not immune therefrom. For example, the particle swarm optimization (PSO) algorithm, which is a typical and widely applied approach to model-less emergent intelligence, has been proven fragile to biased agent reports [59].

While 6G is expected to be an infrastructure for not only data traffic but also pervasive intelligence, the intelligent services it delivers are commonly data-driven and rely on measurements or reports as feedback from users. The quality of such intelligent services, including not only mobile cloud services running over 6G networks but also the intelligent 6G networking service itself (e.g., CSI prediction, RRM, and NFV MANO), is highly dependent on the quality of such feedback data. Therefore, the threat of data-injection attacks will be significantly higher in the 6G era than ever before.

To encounter trust-based attacks, a common strategy is to assess the trustworthiness of every involved user and service provider, so that the malicious ones can be distinguished from the normal ones and discriminated against. For example, users shall be warned about phishing attacks when they try to share information with malicious users/websites, and the CSI reports from untrustworthy user devices shall not be taken with reliance for training the RRM algorithm of the 6G network. Generally, there are four main categories of methods to evaluate trust in open and dynamic multi-agent systems such as the 6G ecosystem [60], namely:

1. Direct trust evaluation models based on past local observations
2. Indirect (reputation-based) trust evaluation models based on observations from other agents in the same environment
3. Socio-cognitive trust evaluation models based on analysis of social relationships among multiple agents
4. Organizational trust evaluation models, which rely on the organizational affiliation or certificates issued by some trusted organizations.

In the 6G environment, organizational trust evaluation is commonly available for mobile devices, network equipment, and MNOs/MVNOs. However, it cannot eliminate the commitment of phishing and data-injection attacks from organizationally certified devices. Threats of such kinds can only be detected via behavioural diagnosis with direct or indirect trust evaluation models. Moreover, as it usually takes a long observation period for one agent to accurately detect the malicious behaviour of another agent, but only very few interactions to accomplish an effective trust-abusing attack, direct trust evaluation alone is unlikely sufficient to protect agents from such attacks in an open and dynamic environment. Therefore, reputation-based trust evaluation becomes the most promising solution to these

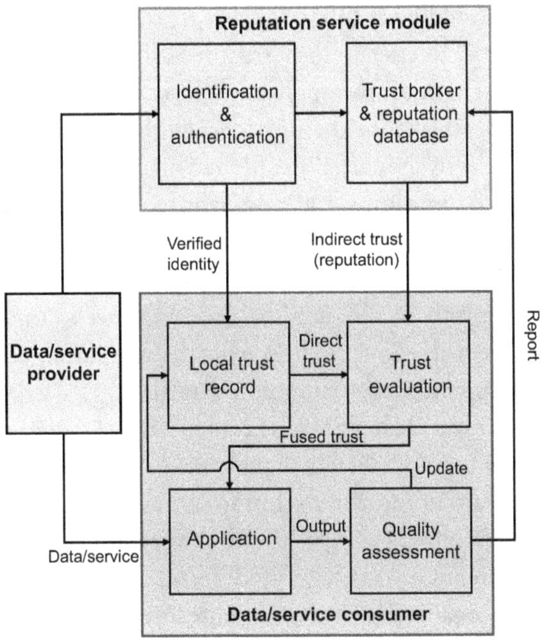

Figure 8.8. A generic TaaS framework.

threats for 6G. A pioneering effort in [60] has demonstrated its effectiveness in protecting the PSO algorithm from data-injection attacks.

To enable the indirect trust evaluation, the reputation score, i.e., the agent-assessed trust value of users or services must be sufficiently spread and shared among the agents. Therefore, a third-party entity/platform must be involved to record, maintain, update, and distribute the reputation score. This framework, as discussed by literature in various use scenarios [61–64], is known as "Trust as a Service" (TaaS).

While existing works are mostly considering cloud service and IoT applications, they generally distinguish users from service providers in an explicit manner, focusing only on generating the reputation score of cloud service providers based on user feedback and issuing references to the users. In a more generic context of 6G-enabled services of data and intelligence, we can briefly summarize the TaaS architecture in Figure 8.8. Any involved agent, depending on the type of service and the role it plays in that service, can be either a data/service provider or a data/service consumer, or in some cases both. Once an E2E data link (or service session) is established between the data/service providing and consuming agents, the reputation service module intervenes by issuing verifications to the identity and reputation score of the provider and referring them to the consumer. Upon its own interaction history with the provider, the consumer may be keeping a local trust record of the provider, which is dedicated to the latter's identity. This local record defines the direct trust, which is fused with the indirect trust (i.e., the reputation score) by a trust evaluating module. In case no local trust record is available, the indirect trust will be simply taken. Then, on the application layer, with respect to the evaluated trust score, the data/service from the provider will be either rejected, straightforwardly exploited, or selectively/partially exploited on the application layer. A quality assessing mechanism of some kind (e.g., anomaly detection) is essential on the consumer end to evaluate the quality of data/service it had obtained from the provider regarding high-layer output (on the representation and/or application layers). The assessment will be not only locally used by the consumer to update its own trust record but also reported to the trust broker in the reputation service module for further analysis and update of the reputation database.

One of the most critical challenges in the architectural design of TaaS is to allocate the reputation service module, which plays the core role in the TaaS framework, to a certain node/domain of the system. In the context of 6G mobile networks and its ecosystem, the 6G MNO/MVNO is the most appropriate stakeholder to implement and operate the reputation service module, for a three-folded reason. First, as the only original source of both direct and indirect trust, the quality assessment of data and service can only be implemented at the consumer end, because it essentially requires a semantic understanding of the message, to evaluate the quality of data and service on the representation and/or application layers, so that the malicious or poor data can be distinguished from the benevolent and reliable. Therefore, the trust broker has to be globally accessible for all agents regardless of the service domain, which can be challenging for any stakeholder but the MNO/MVNO. Second, the MNO/MVNO can leverage the Unified Data Management (UDM), or corresponding entity in the future 6G architecture, to conveniently implement the identification and authentication function. Last but not least, the TaaS itself, as the source of trust, requires the involved agents to trust the reputation data, which may create a dilemma unless the TaaS is provided by the MNO/MVNO, which is always trustworthy to a certain degree for all users and service providers that are connected through the 6G network. Therefore, we see TaaS as an essential functionality of a secure and trustworthy 6G network.

It is also worth noting that when applied solely with indirect trust evaluation, the TaaS itself can be exposed to a specific data-injection attack, i.e., reputation spoofing with a malicious fake rating. Therefore, a hybrid design that combines local direct trust evaluation by every agent and cloud-based indirect trust evaluation shall be necessary to enhance the robustness of TaaS against reputation spoofing. Furthermore, blockchain technologies may play an important role in tracing malicious agent evaluations in TaaS, which can be further combined with socio-cognitive trust evaluation models to resolve coordinated data injection attacks by multiple malicious agents.

8.5 Trustworthy ML/AI

In the 6G system, the integration and adoption of ML/AI will be significant. ML/AI-powered technology will play a pivotal role in 6G networks by automating decision-making procedures, implementing a zero-touch approach. Thus, the network and its services will be more intelligent, automated, and programmable [65]. The use of ML/AI will also impact the delivered trustworthiness as it has the potential of providing intelligence for threat detection and mitigation. On the other hand, ML/AI itself needs to assure that it is built and operates trustworthy.

The concept of trustworthy ML/AI includes both technological and societal aspects. The key requirements of trustworthy ML/AI are, as stated in EU Ethics guidelines for trustworthy AI [66], human agency and oversight, technical robustness and safety (including security aspects), privacy and data governance, transparency (including explainability aspects), accountability diversity, non-discrimination and fairness, and societal and environmental well-being.

Transparency refers to the information about the ML/AI system made available to the users interacting with the system. This encompasses the whole ML/AI pipeline and includes decisions about the pipeline. Transparency is necessary to take actions and apply rectifications against unwanted effects. Accountability reflects shared responsibility and expectations when these unwanted effects occur. Explainability means that functionalities behind the operations of the ML/AI system can be representable, and the outputs can be interpretable. This is especially needed to accomplish more reliable risk evaluations and governance.

ML/AI systems should be robust against adversarial attempts. These attempts may target to break how the ML/AI system works, its security, or privacy. ML/AI risks may stem from the data used to train the AI system, training process of ML/AI models, the operational processes, and the inference phase to the ML/AI system because of the interactions with the external users. The relevant risks should be clearly identified, and mitigations should be planned. As the 6G systems are foreseen to have a distributed and multi-stakeholder nature, the potential risk activities and their governance require considerations of this nature. To leverage the distributed nature of 6G systems, collaborative learning concepts such as federated learning and split learning are promising in addition to centralized learning concepts. They support joint optimization and data minimization and are considered privacy-aware methods as they allow training locally on end devices without requiring centralized data training. On the other hand, these methods may introduce some challenges and new threats stemming from possible malicious end-device activities during training to manipulate the overall training process [67].

Another requirement to achieve a trustworthy ML/AI system is safety from functional, operational, and human perspectives. ML/AI systems should ensure that unintentional or malicious effects do not cause any harm to the functionality, operation, or human life, health, or property. The system should be designed taking the responsible design, development, and decision practices into account. ML/AI systems should also operate to address fairness and bias concerns which refer to equality and equity considerations.

Trustworthy AI principles have also started to be considered by standardization and regulation bodies with supporting activities. The EU AI Act was proposed by the European Union proposed on April 21, 2021, and is yet to be approved. The act aims to regulate the development and use of AI in the European Union ensuring that AI is developed and used in a responsible and ethical manner that protects the rights and safety of individuals. Trustworthy AI principles brought another perspective to risk management; thus, recently ISO/IEC and NIST released their efforts to provide guidance on risk management. ISO/IEC extends the risk management framework (ISO 31000:2018) with ISO/IEC 23894:2023 [68] providing a direction to organizations involved in the ML/AI processes including development, production, or use of ML/AI-assisted products on how to manage risks that are specific to ML/AI. Similar to ISO, NIST has developed a framework in collaboration with the private and public sectors to better understand the impact of ML/AI and manage associated risk. The first version of the AI Risk Management Framework (AI RMF 1.0) [69] and its companion Playbook have been released to foster trustworthy and responsible development and use of ML/AI systems. ML/AI risk management should be adopted by organizations and should be advocated by the society to understand not only the challenges but also societal impacts. The continuation of the efforts is also important to further guide the organizations on risk evaluation and trade-offs among different characteristics of trustworthy AI like privacy and explainability.

Security, being an important key element of trustworthy AI principles, requires specific attention as it has a broader scope: ML/AI can be used to enhance the security, ML/AI can be leveraged to break the security, and ML/AI can be the target of attacks. From the standardization perspective, the ETSI Industry Specification Group on Securing Artificial Intelligence (ISG SAI) [70] focuses on these aspects and creates standards to improve ML/AI security. ETSI ISG SAI works under specific focus groups and released the following documents:

- Problem Statement and Mitigation Strategy Reports provide a state-of-the-art analysis of the threat landscape and describe mitigation strategies against those threats.
- Data Supply Chain Security focuses on the security of data pipeline, data manipulation, and its effect on ML/AI systems.
- AI Threat Ontology defines specifics of AI threats providing a view of the relationships between actors representing threats, threat agents, and assets.

- The role of hardware in the security of AI describes how to secure AI through hardware-mediated trusted execution environments and how to handle attacks on specialized AI processing hardware.
- AI Computing Platform Security Framework defines the requirements of the secure AI compute platform, its security components, and reference architecture for the platform. This secure platform aims to pave the way for an attack-resistant AI computing platform and better protection for the valuable assets of stakeholders.

Trustworthy ML/AI requirements should be addressed throughout the life cycle of design, development, deployment, and operation and should be supported with other security, privacy, and trust foundations enabled in the 6G system. The standardization efforts and advances in ML/AI risk management and regulations should be closely followed by 6G research activities.

8.6 Summary and Outlook

6G networks will transport critical and sensitive information, thus requiring security mechanisms that ensure confidentiality, integrity, and availability of the data and the networks. Superior privacy of such sensitive information that is gathered by the network is essential to make 6G networks trustworthy. Such mechanisms, either hardware- or software-based, can be applied at different layers (e.g., information sharing among tenants, cloud-stored data, and end users), because of the diversity of data sources stored on different nodes, under the control of different parties. New technologies, such as AI/ML, are adopted to cover the security requirements of the 6G architecture and the new network features and properties. Although 6G security comprises the well-proven essential security methods present in today's networks, it is crucial to think of security not only as a one-time effort to undertake when deploying a network but as a constant process over the complete lifetime of the network and its services.

References

[1] Q.V. Pham, F. Fang, V. Nguyen, Md. J. Piran, M. Le, L. B. Le, W.J. Hwang, and Z. Ding, "Survey of Multi-Access Edge Computing in 5G and Beyond: Fundamentals, Technology Integration, and State-of-the-Art", In *IEEE Access*, vol. 8, pp. 116974–117017, June 2020.

[2] S. Vimercati, S. Foresti, S. Jajodia, S. Paraboschi, and P. Samarati. "Efficient Integrity Checks for Join Queries in the Cloud," In *Journal of Computer Security*, vol. 24, no. 3, pp. 347–378, June 2016.

[3] B. Zhang, B. Dong, and. W. H. Wang, "CorrectMR: Authentication of Distributed SQL Execution on MapReduce," In *IEEE Transactions on Knowledge and Data Engineering,* vol. 33, no. 3, pp. 897–908, March 2021.

[4] S. Vimercati, S. Foresti, G. Livraga, and P. Samarati, "Data Security and Privacy in the Cloud," In *SPIE 10993, Mobile Multimedia/Image Processing, Security, and Applications,* May 2019. Doi: 10.1117/12.2523603.

[5] J. A. Ayala-Romero, A. Garcia-Saavedra, M. Gramaglia, X. Costa-Perez, and A. J. J. Alcaraz, "vrAIn: A deep learning approach tailoring computing and radio resources in virtualized RANs," In 25th Annual International Conference on Mobile Computing and Networking, pp. 1–16, Oct. 2019.

[6] P. Samarati, "Protecting Respondents' Identities in Microdata Release," In *IEEE Transactions on Knowledge and Data Engineering,* vol. 13, no. 6, pp. 1010–1027, Nov. 2001.

[7] A. Machanavajjhala, D. Kifer, J. Gehrke, and M. Venkitasubramaniam, "l-diversity: Privacy beyond k-anonymity," In *ACM Transactions on Knowledge Discovery from Data (TKDD),* vol. 1, no. 1, 3-es, Mar. 2007.

[8] N. Li, T. Li, and S. Venkatasubramanian, "t-closeness: Privacy beyond k-anonymity and l-diversity". In *IEEE 23rd International Conference on Data Engineering,* pp. 106–115, April 2007.

[9] C. Dwork, K. Talwar, A. Thakurta, and L. Zhang. "Analyze Gauss: Optimal Bounds for Privacy-Preserving Principal Component Analysis," In *Proceedings of the Forty-Sixth Annual ACM Symposium on Theory of Computing, STOC '14,* pp. 11–20. New York, NY, USA: Association for Computing Machinery, 2014. ISBN 9781450327107. Doi: 10.1145/2591796.2591883.

[10] B. C. M. Fung, K. Wang, R. Chen, and P. S. Yu. "Privacy-Preserving Data Publishing: A Survey of Recent Developments," In *ACM Comput. Surv.* Vol. 42, no. 4. ISSN 0360-0300. DOI: 10.1145/1749603.1749605, 2010.

[11] S. De Capitani di Vimercati, S. Foresti, S. Jajodia, G. Livraga, S. Paraboschi, and P. Samarati. "An Authorization Model for Query Execution in the Cloud," In *The VLDB Journal,* vol. 31, no. 3, pp. 555–579, 2022.

[12] S. De Capitani di Vimercati, D. Facchinetti, S. Foresti, G. Oldani, S. Paraboschi, M. Rossi, and P. Samarati. "Multi-Dimensional Indexes for Point and Range Queries on Outsourced Encrypted Data," In *Proc. of GLOBECOM,* Madrid, Spain, Dec. 7–11, 2021.

[13] V. Ciriani, S. De Capitani di Vimercati, S. Foresti, S. Jajodia, S. Paraboschi, and P. Samarati. "Combining Fragmentation and Encryption to Protect Privacy in Data Storage," In *ACM TISSEC,* vol. 13, no. 3, pp. 22:1–22:33, 2010.

[14] D. Li, S. Lv, Y. Huang, Y. Liu, T. Li, Z. Liu, and L. Guo. "Frequency-Hiding Order-Preserving Encryption with Small Client Storage," In *PVLDB,* vol. 14, pp. 3295–3307, 2021.

[15] G. S. Poh, J. Chin, W. Yau, K.R. Choo, and M.S. Mohamad. "Searchable Symmetric Encryption: Designs and Challenges," In *ACM CSUR*, vol. 50, no. 3, pp. 1–37, 2017.
[16] A. Arasu, K. Eguro, M. Joglekar, R. Kaushik, D. Kossmann, and R. Ramamurthy. "Transaction Processing on Confidential Data using Cipherbase," In *Proc. of ICDE*, Seoul, South Korea, April 2015.
[17] A. Arasu, R. Kaushik, D. Kossmann, and R. Ramamurthy. "Integrity-based Attacks for Encrypted Databases and Implications," In *Proc of CIDR*, virtual, Jan. 2021.
[18] S. De Capitani di Vimercati, D. Facchinetti, S. Foresti, G. Oldani, S. Paraboschi, M. Rossi, and P. Samarati. "Scalable Distributed Data Anonymization," In *Proc. of PerCom*, Kassel, Germany (virtual), March 2021.
[19] S. De Capitani di Vimercati, D. Facchinetti, S. Foresti, G. Oldani, S. Paraboschi, M. Rossi, and P. Samarati. "Artifact: Scalable Distributed Data Anonymization," In *Proc. of PerCom*, Kassel, Germany (virtual), March 2021.
[20] S. De Capitani di Vimercati, S. Foresti, S. Jajodia, S. Paraboschi, and P. Samarati, "Authorization Enforcement in Distributed Query Evaluation," In *Journal of Computer Security*, vol. 19, no. 4, pp. 751–794, 2011.
[21] K.Y. Oktay, M. Kantarcioglu, and S. Mehrotra. "Secure and Efficient Query Processing over Hybrid Clouds," In *Proc. of ICDE*, San Diego, CA, USA, April 2017.
[22] G. Salvaneschi, M. Köhler, D. Sokolowski, P. Haller, S. Erdweg, and M. Mezini. "Language-Integrated Privacy-Aware Distributed Queries," In *Proc. of ACM on Programming Language*, vol. 3, pp. 1–30, 2019.
[23] Q. Zeng, M. Zhao, P. Liu, P. Yadav, S. Calo, and J. Lobo. "Enforcement of Autonomous Authorizations in Collaborative Distributed Query Evaluation," In *IEEE Transactions on Knowledge and Data Engineering*, vol. 27, no. 4, pp. 979–992, 2015.
[24] S. De Capitani di Vimercati, S. Foresti, S. Jajodia, G. Livraga, S. Paraboschi, and P. Samarati. "An Authorization Model for Multi-Provider Queries," In *Proc. of the VLDB Endowment*, vol. 11, no. 3, pp. 256–268, 2017.
[25] S. De Capitani di Vimercati, S. Foresti, G. Livraga, and P. Samarati, "Digital Infrastructure Policies for Data Security and Privacy in Smart Cities," In *Smart Cities Policies and Financing*. J. Vacca (ed.), Elsevier, 2022.
[26] S. De Capitani di Vimercati, S. Foresti, G. Livraga, and P. Samarati. "Practical Techniques Building on Encryption for Protecting and Managing Data in the Cloud," In *The New Codebreakers: Essays Dedicated to David Kahn on the Occasion of His 85th Birthday*, P. Ryan, D. Naccache, J.-J. Quisquater (eds.), Springer, 2016.

[27] S. De Capitani di Vimercati, S. Foresti, G. Livraga, and P. Samarati, "Supporting Users in Data Outsourcing and Protection in the Cloud," In *Proc. of CLOSER*, Rome, Italy, April 2016.

[28] H. Singh Lallie, L. A. Shepherd, J. R.C. Nurse, A. Erola, G. Epiphaniou, C. Maple, and X. Bellekens. "Cyber security in the age of COVID-19: A timeline and analysis of cyber-crime and cyber-attacks during the pandemic," In *Computers & Security*, vol. 105, pp. 102248, 2021.

[29] CYNET, "What Are Endpoint Detection and Response (EDR) Tools? Definition, Features & Top 6 Tools", 2022. Accessed: April 6, 2023, [Online], Available: https://www.cynet.com/endpoint-protection-and-edr/top-6-edr-tools-compared/.

[30] S. Rommer, P. Hedman, M. Olsson, L. Frid, S. Sultana, and C. Mulligan, "Chapter 11 – Network slicing", in *5G Core Networks,* S. Rommer, P. Hedman, M. Olsson, L. Frid, S. Sultana, and C. Mulligan Eds, Academic Press, pp. 247–264, ISBN 9780081030097, 2020.

[31] T. Hewa, G. Gür, A. Kalla, M. Ylianttila, A. Bracken, and M. Liyanage, "The Role of Blockchain in 6G: Challenges, Opportunities and Research Directions," In *2020 2nd 6G Wireless Summit (6G SUMMIT)*, 17–20 March 2020.

[32] H. Xu, P. V. Klaine, O. Onireti, B. Cao, M. Imran, and L. Zhang. "Blockchain-enabled resource management and sharing for 6G communications," In *Digital Communications and Networks*, vol. 6, no. 3, pp. 261–269, 2020.

[33] J. Wang, X. Ling, Y. Le, Y. Huang, and X. You, "Blockchain-enabled wireless communications: a new paradigm towards 6G," In *National Science Review*, vol. 8, no. 9, Sept. 2021.

[34] W3C, "Verifiable Credentials Data Model v1.1," 2023. Accessed: April 6, 2023. [Online]. Available: https://www.w3.org/TR/vc-data-model/.

[35] Sovrin Foundation, Accessed April 6, 2023. [Online], Available: https://sovrin.org/.

[36] IDunion, Accessed April 6, 2023. [Online], Available: https://idunion.org/?lang=en.

[37] Hyperledger Foundation, Accessed April 6, 2023. [Online], Available: https://www.hyperledger.org/.

[38] G. Fedrecheski, J. M. Rabaey, L. C. P. Costa, P. C. Calcina Ccori, W. T. Pereira, and M. K. Zuffo, "Self-Sovereign Identity for IoT environments: A Perspective," 2020, Accessed: April 6, 2023. [Online], Available: https://arxiv.org/abs/2003.05106.

[39] S.K. Gebresilassie, J. Rafferty, P. Morrow, L. Chen, M. Abu-Tair, and Z. Cui, "Distributed, Secure, Self-Sovereign Identity for IoT Devices," In *2020 IEEE 6th World Forum on Internet of Things (WF-IoT)*, June 2–16, 2020.

[40] M. Sabt, M. Achemlal, and A. Bouabdallah. "Trusted Execution Environment: What It is, and What It is Not," In *2015 IEEE Trustcom/BigDataSE/ISPA*, Helsinki, Finland, 2015. DOI: 10.1109/Trustcom.2015.357.

[41] NIST "Hardware-Enabled Security for Server Platforms: Enabling a Layered Approach to Platform Security for Cloud and Edge Computing Use Cases," White Paper, 2020. Accessed: April 6, 2023. [Online], Available: https://doi.org/10.6028/NIST.CSWP.04282020-draft.

[42] Nokia, "Security and trust in the 6G era," White paper, 2021. Accessed April 6, 2023. [Online], Available: https://d1p0gxnqcu0lvz.cloudfront.net/documents/Nokia_Security_and_trust_in_the_6G_era_White_Paper_EN.pdf.

[43] Common Vulnerabilities and Exposures (CVE) Programme, CVE ID "CVE-2019-5736". Accessed April 6, 2023. [Online], Available: https://cve.mitre.org/cgi-bin/cvename.cgi?name=CVE-2019-5736.

[44] D. Williams, R. Koller, M. Lucina, and N. Prakash, "Unikernels as Processes". In *Symposium on Cloud Computing (SoCC 2018)*, NY, USA, 2018.

[45] L. Urciuoli, T. Männistö, J. Hintsa, and T. Khan, "Supply Chain Cyber Security–Potential Threats," In *Information & Security: An International Journal*, vol. 29, no. 1, pp. 51–68, 2013. https://en.wikipedia.org/wiki/Supply_chain_attack.

[46] Solo5, "The Solo5 Unikernel". Accessed: April 6, 2023. [Online], Available: https://github.com/solo5/solo5.

[47] A. Madhavapeddy, R. Mortier, C. Rotsos, D. Scott, B. Singh, T. Gazagnaire, S. Smith, S. Hand, and J. Crowcroft, "Unikernels: Library operating systems for the cloud," In *ACM SIGARCH Computer Architecture News*, vol. 41, no. 1 pp. 461–472, March 2013, DOI: 10.1145/2490301.2451167.

[48] A. Kantee and J. Cormack. "Rump Kernels: No OS? No Problem!" In *Login Usenix Mag*, 2014.

[49] D. Williams and R. Koller, "Unikernel monitors: extending minimalism outside of the box," In *8th USENIX Conference on Hot Topics in Cloud Computing*, pp. 71–76, June 2016.

[50] A. Agache, M. Brooker, A. Florescu, A. Iordache, A. Liguori, R. Neugebauer, P. Piwonka, and D.-M. Popa, "Firecracker: Lightweight Virtualization for Serverless Applications," In *17th USENIX Symposium on Networked Systems Design and Implementation*, pp. 419–434, Feb. 2020.

[51] E. Jonas, J. Schleier-Smith, V. Sreekanti, C.-C. Tsai, A. Khandelwal, Q. Pu, V. Shankar, J. Carreira, K. Krauth, N. Yadwadkar, J. Gonzalez, R. Popa, I. Stoica, and D. Patterson. "Cloud Programming Simplified: A Berkeley View on Serverless Computing," University of California at Berkely, February 2019.

[52] Redhat and IBM. "Kata containers", 2023. Accessed: April 6, 2023. [Online], Available: https://katacontainers.io.

[53] S. Kuenzer, V.-A. Badoiu, H. Lefeuvre, S. Santhanam, A. Jung, G. Gain, C. Soldani, C. Lupu, Ş. Teodorescu, C. Răducanu, C. Banu, L. Mathy, R. Deaconescu, C. Raiciu, and F. Huici. "Unikraft: fast, specialized unikernels the easy way," In *16th European Conference on Computer Systems EuroSys '21*, New York, NY, USA, 2021.

[54] L. Smith. "Architectures for secure computing systems," MITRE CORP BEDFORD MASS, 1975.

[55] J. Won and E. Bertino. "Robust Sensor Localization against Known Sensor Position Attacks," In *IEEE Transactions on Mobile Computing*, vol. 18, no. 12, pp. 2954–2967, Dec. 2019. DOI: 10.1109/TMC.2018.2883578.

[56] A. Hern. "Berlin artist uses 99 phones to trick Google into traffic jam alert". The Guardian. Accessed: April 6, 2023. [Online], Available: https://www.theguardian.com/technology/2020/feb/03/berlin-artist-uses-99-phones-trick-google-maps-traffic-jam-alert.

[57] S. Fu, Z. Jiang, S. Zhang, S. Xu, B. Han, and H. D. Schotten, "Data-Injection-Proof Predictive Vehicle Platooning: Performance Analysis with Cellular-V2X Sidelink Communications," In *IEEE Internet of Things Journal*, 2021. DOI: 10.1109/JIOT.2021.3122125.

[58] O. Holmes and G. Oladipo. "iPhones calling 911 from owners' pockets on rollercoasters," The Guardian, October 2022. Accessed: April 6, 2023. [Online], Available: https://www.theguardian.com/technology/2022/oct/11/iphones-calling-911-from-owners-pockets-on-rollercoasters.

[59] B. Han, D. Krummmacher, Q. Zhou, and H. D. Schotten, "Trust-Awareness to Secure Swarm Intelligence from Data Injection Attack". In *IEEE International Conference on Communications (ICC)*, Rome, Italy, June 2023.

[60] H. Yu, Z. Shen, C. Leung, C. Miao, and V. R. Lesser. "A Survey of Multi-Agent Trust Management Systems". In *IEEE Access*, vol. 1, pp. 35–50, 2013. DOI: 10.1109/ACCESS.2013.2259892.

[61] T. Bhattasali, R. Chaki, and N. Chaki. "Secure and Trusted Cloud of Things". In *Annual IEEE India Conference (INDICON)*, pp. 1–6, 2013. DOI: 10.1109/INDCON.2013.6725878.

[62] T. H. Noor, Q. Z. Sheng, L. Yao, S. Dustdar, and A. H. H. Ngu. "CloudArmor: Supporting Reputation-Based Trust Management for Cloud Services". In *IEEE Transactions on Parallel and Distributed Systems*, vol. 27, no. 2, pp. 367–380, February 2016. DOI: 10.1109/TPDS.2015.2408613.

[63] I. R. Chen, J. Guo, D. C. Wang, J. J. P. Tsai, H. Al-Hamadi, and I. You. "Trust as a Service for IoT Service Management in Smart Cities," In *IEEE 20th International Conference on High Performance Computing and Communications;*

IEEE 16th International Conference on Smart City; IEEE 4th International Conference on Data Science and Systems (HPCC/SmartCity/DSS), pp. 1358–1365, June 28–30, 2018. DOI: 10.1109/HPCC/SmartCity/DSS.2018.00225.

[64] L. Liu and M. Loper. "Trust as a Service: Building and Managing Trust in the Internet of Things," In *IEEE International Symposium on Technologies for Homeland Security (HST)*, pp. 1–6, 2018. DOI: 10.1109/THS.2018.857 4169.

[65] F. Miltiadis, L. Vasiliki, M. Jafar, M. Mattia, E. U. Soykan, B. Tamas, R. Nandana, R. Nuwanthika, L. Le Magoarou, and P. Pietro, "Pervasive Artificial Intelligence in Next Generation Wireless: The Hexa-X Project Perspective," In *Proc. 1st Int. Workshop Artif. Intell. Beyond 5G 6G Wireless Netw.(AIG) Co-Located With IEEE World Congr. Comput. Intell.(WCCI)*, 2022.

[66] European Commission High-Level Expert Group on AI. "Ethics guidelines for trustworthy AI," Accessed: April 6, 2023. [Online], Available: https://digi tal-strategy.ec.europa.eu/en/library/ethics-guidelines-trustworthy-ai.

[67] E. Ustundag Soykan, L. Karaçay, F. Karakoc, and E. Tomur, "A Survey and Guideline on Privacy Enhancing Technologies for Collaborative Machine Learning," In *IEEE Access*, vol. 10, pp. 97495–97519, 2022. DOI: 10.1109/ACCESS.2022.3204037.

[68] ISO/IEC 23894:2023, "Information technology – Artificial intelligence – Guidance on risk management", February 2023.

[69] NIST AI 100-1, "NIST AI Risk Management Framework," January 2023. DOI: https://doi.org/10.6028/NIST.AI.100-1.

[70] ETSI Industry Specification Group on Securing Artificial Intelligence (ISG SAI). Accessed: April 6, 2023. [Online], Available: https://www.etsi.org/com mittee/sai.

DOI: 10.1561/9781638282396.ch9

Chapter 9

6G Outlook and Timeline

By Mauro Boldi, Mikko Uusitalo, Patrik Rugeland, et al.[1]

9.1 Introduction

Research on 6th Generation (6G) has been going on for several years. A major milestone after concurrent and focused research efforts was the establishment of the first pan-European 6G Flagship EU-funded collaborative research project Hexa-X [1], which started in January 2021 and collected industry and academia together to create momentum and direction for 6G. The next phase of such a broad initiative started in January 2023, when the second pan-European 6G Flagship project Hexa-X-II started, and is planned to last until June 2025 [2]. Simultaneously with Hexa-X-II, many other European Union (EU)-funded projects within the Smart Networks and Services Joint Undertaking (SNS JU) program [3] kicked off in January 2023, addressing several aspects of the forthcoming 6G system. It is expected that at the end of Hexa-X-II, the momentum on 6G actions will shift gradually from research to pre-productization and the creation of technical specifications in standardization bodies.

This final chapter will present the current status of 6G with reference to the standardization and regulation framework, aiming for the adoption of the agreed specifications worldwide around 2030.

1. The full list of chapter authors is provided in the Contributing Authors section of the book.

9.2 The Foreseen 6G Standardization Process

The value proposition of the communication service provider (CSP) ecosystem depends heavily on economies of scale. For networking, this is captured by the well-known Metcalfe's law [4] that states that the value of the telecommunications network is proportional to the square of the number of connected end users. Standardization is a key means to maximize this value proposition by ensuring the interoperability of various subsystems through the creation of global technical specifications, where all ecosystem stakeholders are represented and recognize a common added value. Global standards ensure service and vendor interoperability by defining a set of well-defined interfaces that the system providers comply with, and interoperation is verified by a set of tests that the vendors need to pass. In this way, standardization creates mass-market economies of scale, without vendor lock-in, while still enabling innovation within the system components.

De facto standards are established when a critical mass of users and providers favour one particular solution that starts dominating the market. Quite often evolution and innovation of the de facto standards technology are constrained by a single or very few stakeholders. Formal standards are developed by Standard Development Organizations (SDOs) through official fair procedures and membership bylaws defining the scope, decision process, and stakeholders of the SDO. De facto standards can become formal standards if they are adopted and approved by a formal SDO. In addition to de facto and formal standards, there are also "de jure" standards. A standard is a "de jure" standard if a legally binding regulation (e.g., EU regulation) makes an explicit reference to a specific standard.

In this section, we present the foreseen process of 6G standardization. Such a process considers the allocation and use of radio spectrum for 6G, network architectures, and systems functionality as well as interfaces, services, and terminal capabilities in key SDOs that have been playing a central role in defining the previous cellular network generations and are expected to continue doing so for 6G as well.

The International Organization for Standardization (ISO) [5] and the International Electrotechnical Commission (IEC) [6] define global standards for electronic equipment and products covering a wide range of application sectors. For example, ISO has developed standards for life cycle assessment guidance and circular economy as part of a set of actions focusing on increasing the sustainability of future systems. IEC also defines electromagnetic compatibility (EMC) standards that provide guidelines for communication equipment as well. The membership of these global SDOs is for national bodies only.

9.2.1 ITU, 3GPP, and ETSI

The International Telecommunication Union (ITU), an agency of the United Nations, oversees telecommunication-specific standards. The members of the ITU are national bodies as well as sector members, such as vendors, operators, and other ICT companies. It is divided into the following sectors: the Telecommunication Standardization Sector, (ITU-T) [7] for information and communication technologies, ITU-R [8] for radiocommunications, and ITU-D that is mandated to create policies for developing countries [9]. ITU regularly organizes the World Radiocommunication Conferences (WRC) that review and update the Radio Regulations covering the use of the radio-frequency spectrum as well as geostationary and non-geostationary satellite orbits. 6G spectrum regulation is expected to be decided at the WRC-27 in 2027. ITU-R is also responsible to specify the requirements that define each new generation of mobile networks, e.g., IMT-Advanced for 4G [10], and IMT-2020 for 5G [11]. Currently, a number of studies and recommendations need to be prepared by ITU-R to justify spectrum assignments as well as to lay out the foundation for the creation of the International Mobile Telecommunications-2030 (IMT-2030) Standard, expected around 2030 (see Figure 9.1) that will define 6G networks, devices, and services, as well as the requirements they must fulfil.

The drafting process of IMT-2030 has several milestones. First, the IMT Vision of IMT towards 2030 [12] and beyond is expected to be completed in 2023. Such a document explores and documents foreseen 6G use cases, applications, capabilities, and technology enablers and is complemented with feasibility studies of spectrum characteristics of additional frequency bands suggested for 6G. Then the vision document and related feasibility studies will be followed by the identification and collection of technical requirements that will trigger other relevant SDOs to define and develop fitting functional technologies, architecture, and solutions for

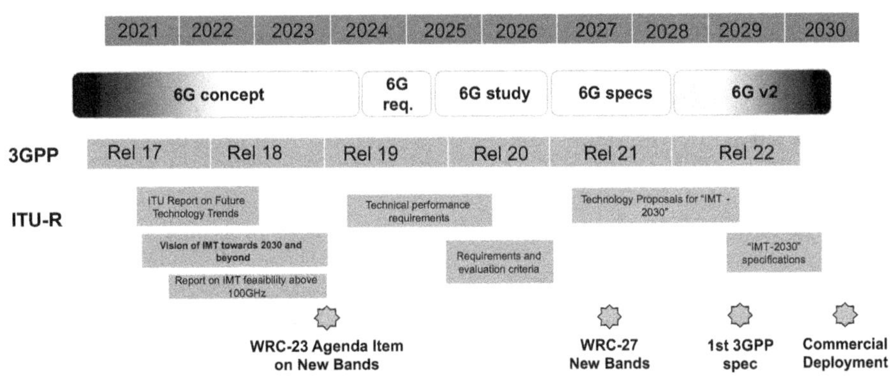

Figure 9.1. Expected 6G standardization timeline – note that the timeline may vary based on the progress in SDOs.

the next-generation networks. The outcome of these efforts will be integrated and documented in the IMT-2030 recommendation.

The 3rd Generation Partnership Project (3GPP) has been the main SDO for 3G, 4G, and 5G specifications and is expected to play the same role for 6G, covering radio access, core network, UE, and service aspects. The term "partnership" in the name of 3GPP refers to the fact that multiple regional or technology-specific SDOs, with more than 750 companies, participate in the 3GPP work. 3GPP specifications are grouped into a series of releases that define the features and capabilities of the cellular system. A new release is usually specified in 15–24 months. The first release for 6G has not yet been defined, but from ITU timelines, one can deduce that 6G studies will start in 3GPP around 2025 under Release 20 (see Figure 9.1).

For each release, 3GPP follows a staged process that starts with overall service descriptions from the user's standpoint leading to requirements. This is the responsibility of Services and Systems Aspects (SA) Working Group (WG) 1 – Services. Based on the results of SA WG1, a description of service requirements of the network functions and mapping to network capabilities is defined by SA WG2 – System Architecture and Services, which is then used to define switching and signalling capabilities and protocols in Core Network and Terminals (CT) WGs and radio access network (RAN) WG.

The European Telecommunications Standards Institute (ETSI) is one of the European Standards Organizations officially mandated to support EU regulation and policies [13]. ETSI defines ICT standards for systems, applications, and services across multiple industry sectors. The work in ETSI is organized into Technical Committees (TC), sector-specific ETSI projects, and Industry Specification Groups (ISGs). It is working on multiple enabler technologies essential for 6G systems. ETSI has established an ISG for terahertz communication (THZ ISG [14]), with the target of laying out the technical foundation for THz communications (0.1–10 THz) including RF characteristics and channel models. Other 6G relevant ETSI groups are the Experiential Networked Intelligence ISG that defines a cognitive network management architecture, the Zero-touch network and Service Management (ZSM) group that automates network operations, and the Network Functions Virtualization (NFV) that enables cloud orchestration of native network functions.

9.2.2 Other Standardization Efforts

In addition to the three previously mentioned SDOs, there are a number of 6G relevant industry alliances that define and develop a specific technology or systems that will play a part in some or all 6G network deployments. O-RAN ALLIANCE [15] is one of them and is expected to have an impact on the definitions of how the radio access of the 6G networks will unfold. O-RAN Alliance is an industry specification group focusing on the next-generation RAN infrastructure. It is a community of around 300 mobile operators, vendors, and research institutions with a mission to develop RAN towards more intelligent, virtualized cloud-based network elements with standardized open interfaces. For 5G, O-RAN has defined among others fronthaul solutions and AI-enabled RAN Intelligent Controller solutions (i.e., non-real-time and near-real-time RAN intelligent controllers). O-RAN has started pre-standardization research activities to investigate the foreseen 6G use cases and technologies and their impact on the RAN architecture in the Next Generation Research Group (nGRG). This research group is expected to publish white papers and research reports based on their studies. The group is not chartered for any normative specification work which is left for later phases in alignment with ITU and 3GPP schedules in O-RAN working groups that are in charge of the actual specification work.

Next Generation Mobile Networks (NGMN) Alliance [16] is an open forum founded by world-leading mobile network operators. NGMN has different membership categories with different roles, rights, and obligations. Operators are in the member category, vendors and software companies are in the contributor category, and research institutes are in the advisor category. NGMN creates deliverables that provide guidelines to equipment developers and SDOs, proposals, and requirements to the industry and other SDOs [17, 19]. It is also acting as a venue for information sharing on critical concerns and sharing experiences and lessons learnt. NGMN has by so far started 6G considerations by publishing white papers on 6G drivers and vision [18] and 6G use cases with their technical challenges and implications to 6G requirements [19]. Based on their previous work on 5G, we expect that NGMN will continue to be an active player in shaping the 6G end-to-end architecture and the ecosystem around it. Particularly, NGMN is likely to work on 6G migration matters, security and operational aspects, overall ecosystem aspects, identification of new requirements for 6G features, and testing the network capabilities.

GSM Association (GSMA) is a mobile operator-driven organization with the goal of achieving scale and interoperability for new mobile technologies. It is perhaps most known by the annual Mobile World Congress trade show that takes place in Barcelona usually at the end of February or early March. GSMA develops and publishes technical documents that have a high impact and are accepted as standards across the communication industry. These technical documents include among others "Definitions of generic slice templates" [20], "5GS Roaming Guidelines" [21], and "Requirements for Multi-SIM Devices" [22], and so on. In addition to the technical documents, GSMA provides its members with tools, statistics, and forecasts.

9.3 Regulatory Trends Towards 2030 and Beyond

While the telecommunications sector has been liberalized and privatized in the 1990s, sector regulation continues to be important in conjunction with efficient spectrum access rules, aspects of EMF, and assurance of level playing field with platform and cloud operators beyond the telco context. Towards 2030, this trend will continue. For example, spectrum management is at the heart of future networks and any wireless technology development, and governments and regulators will have new opportunities due to a wide variety of spectrum bands with highly distinct deployment characteristics and spectrum access models with different levels and needs of spectrum sharing. Another rising issue is electromagnetic field (EMF) exposure. The deployment of 5G technology has started in different areas of the world, and in some regions, (including Europe) concerns over EMF exposure fuel the opposition of the public to its rollout [23, 24]. The exposure to EMF is and will be regulated, based on the guidelines from the International Commission on Non-Ionizing Radiation Protection (ICNIRP) [25]. Since the beginning of telephony, regulations have played an important role in shaping innovation and the operation of the telecommunications industry, for example, setting the industry to be monopolies in the 1960s, liberalizing the sector with privatization in the 1990s, and setting up new regulations for 5G local and private networks. Future networks will likely combine a range of RAN technologies from macro cells to small cells with very high-capacity short-range links. This calls for refining regulations to resolve inconsistent local approval processes and frequency band assignments to enable dense small-cell deployments.

9.4 European 6G Research and Innovation Activities

The work on 6G networks and services has started, and multiple activities are taking place around the world. In Europe, the 6G Smart Networks and Services Industry Association (6G-IA) [26], in collaboration with the supporting Associations, namely, Networld Europe [27], Alliance for Internet of Things Innovation (AIOTI) [28], CISPE.cloud [29], and NESSI [30], prepared a proposal for a European Partnership for Smart Networks and Services to be implemented in the context of Horizon Europe Framework (HEU).

In November 2021, the SNS Joint Undertaking (JU) [31] was established as a legal and funding entity to drive the 6G digital transition. The SNS JU is a public–private partnership jointly run by the European Commission and the 6G-IA. The initiative enables the pooling of EU's industrial and academic/research resources in the area of smart networks and services. In addition, it fosters alignment with EU

Figure 9.2. SNS JU strategic objectives.

Member States for 6G Research and Innovation and deployment of advanced 5G networks.

The SNS JU sets out an ambitious mission with an EU funding budget of € 900 million for the period 2021–2031 (Figure 9.2):

- Foster Europe's technology sovereignty for 6G smart networks and services by designing and providing advanced applications and services together with competitive 6G connectivity solutions. The research and innovation actions are designed to consider the full value chain from end devices to service providers and from softwarized and virtualized solutions down to innovative hardware implementations.
- Improve the digital industry's operation by identifying the needs of verticals and provide them with well-needed solutions that will efficiently address their pain points and improve their productivity. One of the main goals of the SNS JU is to act as the catalyst for the creation of a fertile multi-domain business ecosystem.
- Address, beyond the business and technological needs, societal needs like key UN's Sustainable Development Goals (SDGs) [32], the European Green Deal [33], as well as safeguard safety and security European policies.

The SNS JU aims at mobilizing the European Stakeholders within ICT as well as the vertical sectors and making Europe the centre of gravity for global collaboration and consensus. The partnership is based on the 6G-IA Vision for 6G Networks and Services [34] and follows a solid research and innovation (R&I) roadmap [35] that is being implemented through a dedicated call for projects.

As shown in Figure 9.3, the SNS JU is covering multiple research topics across all network domains. These topics range from typical networking research areas in the radio access, transport, and core networks, up to areas that are applicable on an

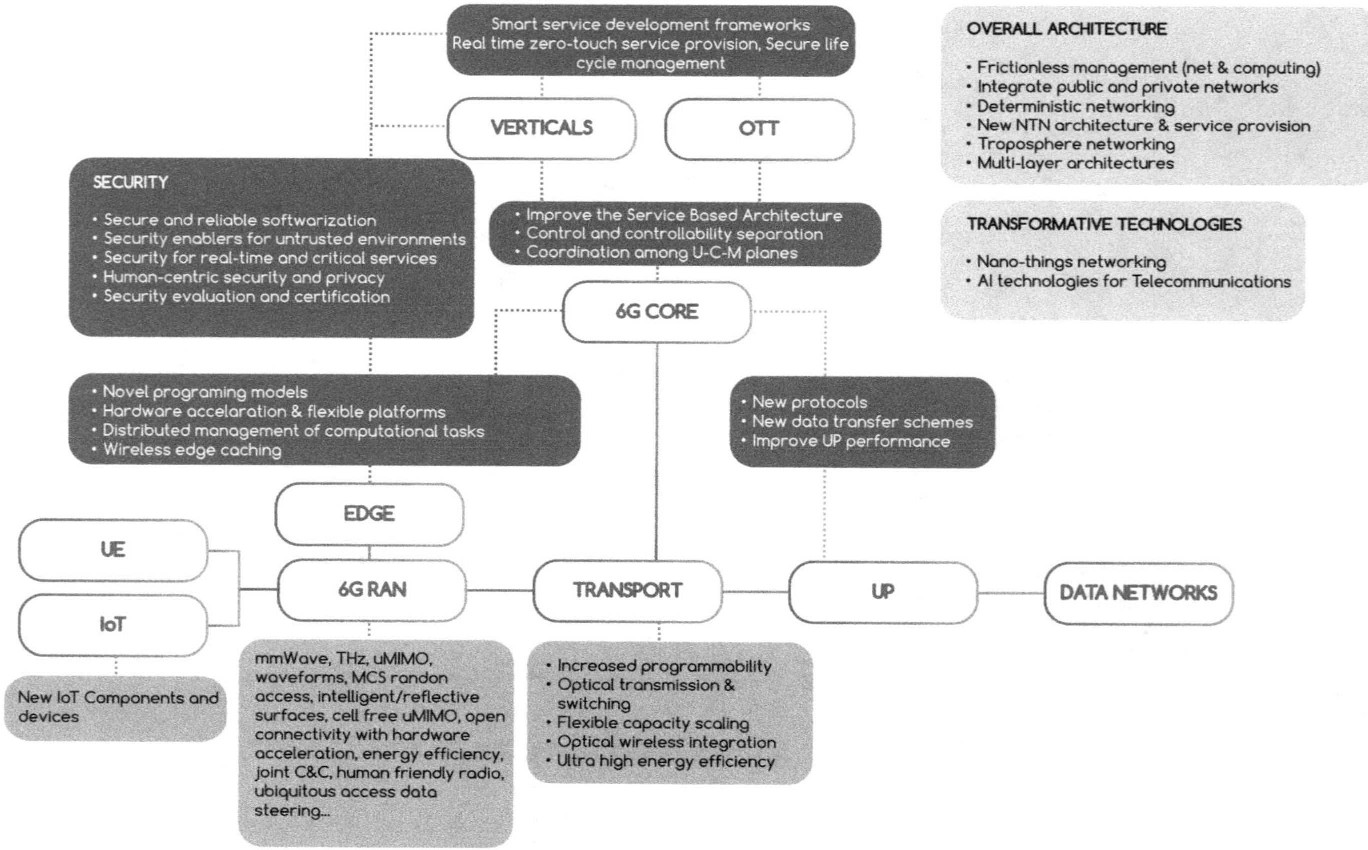

Figure 9.3 SNS JU indicative R&I topics.

end-to-end basis (e.g., security, agile deployment and management of services, new northbound interfaces, architectural topics, and transformative technologies). One key point for 6G networks is the current expectation that the further softwarization of the network may blur the current boundaries between the current network domains, bringing thus significant advancements in the overall architecture.

9.5 Summary and Outlook

This chapter concludes and completes this book, presenting the current status of 6G in terms of standardization and regulation initiatives, towards issuing specifications accepted worldwide as "the new 6G system." In doing so, this chapter has outlined the situation as it is known at the time of writing, and as the road towards standardization is still long, changes are of course possible.

As said, this chapter completes the book, which, in its overall layout, has reported on the current paths towards the new 6G system, deemed to be a sustainable and trustworthy system, with human needs at its core. The book has covered several aspects, all being investigated in Europe and elsewhere, in order to take strategical and technical decisions on the features of 6G, within the framework of the mentioned needs of sustainability and trustworthiness. After the prefaces from European Commission and 6G IA, a chapter has been dedicated to the overall use cases, deemed relevant for 6G, and the visioned E2E architecture evolution towards 6G. The following chapters have been dedicated to specific features, novel enabling technologies, and peculiarities that this system shall present in order to correspond to the desired performance and value indicators.

References

[1] Hexa-X, "A flagship for B5G/6G vision and intelligent fabric of technology enablers connecting human, physical, and digital worlds," 2023. Accessed: March 31, 2023. [Online]. Available: https://cordis.europa.eu/project/id/101015956.

[2] Hexa-X-II, "A holistic flagship towards the 6G network platform and system, to inspire digital transformation, for the world to act together in meeting needs in society and ecosystems with novel 6G services," Cordis, 2023. Accessed: March 31, 2023. [Online]. Available: https://cordis.europa.eu/project/id/101095759.

[3] 6G Smart Networks and Services – Joint Undertaking (SNS-JU), 2023. [Online]. Available: https://smart-networks.europa.eu/.

[4] A. Madureira, F. den Hartog, H. Bouwman and N. Baken, “Empirical validation of Metcalfe’s law: How Internet usage patterns have changed over time,” *Information Economics and Policy*, vol. 25, no. 4, pp. 246, 256, 2013.
[5] International Organization for Standardization (ISO), 2023. Accessed: March 31, 2023. [Online]. Available: https://www.iso.org/home.html.
[6] International Electrotechnical Commission (IEC), 2023. Accessed: March 31, 2023. [Online]. Available: https://iec.ch/homepage.
[7] ITU Telecommunication Standardization Sector (ITU-T), “ITU-T in breif,” 2023. Accessed: March 31, 2023. [Online]. Available: https://www.itu.int/en/ITU-T/about/Pages/default.aspx.
[8] ITU Radiocommunication Sector (ITU-R), “ITU-R Information,” 2023. Accessed: March 31, 2023. [Online]. Available: https://www.itu.int/en/ITU-R/information/Pages/default.aspx.
[9] ITU Telecommunication Development Sector (ITU-D), “About ITU-D and the BDT,” 2023. Accessed: March 31, 2023. [Online]. Available: https://www.itu.int/en/ITU-D/Pages/About.aspx.
[10] *Requirements related to technical performance for IMT-Advanced radio interface(s)*, M.2134-0, ITU Radiocommunications sector (ITU-R), 2008.
[11] *Detailed specifications of the terrestrial radio interfaces of International Mobile Telecommunications-2020 (IMT-2020)*, M.2150 ITU Radiocommunication sector (ITU-R), 2022.
[12] A. Weissberger, “Summary of ITU-R Workshop on “IMT for 2030 and beyond” (aka “6G”),” In *IEEE ComSoc*, June 20, 2022. Accessed: March 31, 2023. [Online]. Available: https://techblog.comsoc.org/2022/06/20/summary-of-itu-r-workshop-on-imt-for-2030-and-beyond-aka-6g/.
[13] European Telecommunications Standards Institute (ETSI), “About ETSI,” 2023. Accessed: March 31, 2023. [Online]. Available: https://www.etsi.org/about.
[14] European Telecommunications Standards Institute (ETSI), “Industry Specification Group (ISG) Terahertz (THZ),” 2023. Accessed: March 31, 2023. [Online]. Available: https://www.etsi.org/committee/2124-thz.
[15] O-RAN, “O-RAN Alliance,” 2023. [Online]. Accessed: March 31, 2023. Available: https://www.o-ran.org/.
[16] Next Generation Mobile Networks (NGMN) Alliance, 2023. Accessed: March 31, 2023. [Online]. Available: https://www.ngmn.org/.
[17] Next Generation Mobile Networks (NGMN) Alliance, “Publications,” 2023. Accessed: march 31, 2023. [Online]. Available: https://www.ngmn.org/publications.html.
[18] Next Generation Mobile Networks (NGMN) Alliance, “NGMN 6G Drivers and vision,” 1 April 2023. Accessed: march 31, 2023. [Online]. Available: https://www.ngmn.org/work-programme/ngmn-6g-drivers-and-vision.html.

[19] Next Generation Mobile Networks (NGMN) Alliance, "6G Use cases and analysis," 22 February 2023. Accessed: March 31, 2023. [Online]. Available: https://www.ngmn.org/publications/6g-use-cases-and-analysis.html.

[20] GSMA Association, "Generic Network Slice Template, Version 7.0," 17 June 2022. Accessed: March 31, 2023. [Online]. Available: https://www.gsma.com/newsroom/wp-content/uploads/NG.116-v7.0-1.pdf.

[21] GSM Association, "5GS Roaming Guidelines, version 6.0," 16 May 2022. Accessed: March 31, 2023 [Online]. Available: https://www.gsma.com/newsroom/wp-content/uploads/NG.113-v6.0.pdf.

[22] GSM Association, "Requirements for Multi SIM Devices, Version 9.0," 21 February 2022. Accessed: March 31, 2023. [Online]. Available: https://www.gsma.com/newsroom/wp-content/uploads/TS.37-v9.0.pdf.

[23] Arcep, "Networks and the Environment," 9 September 2020. Accessed: March 31, 2023. [Online]. Available: https://en.arcep.fr/news/press-releases/view/n/networks-and-the-environment.html.

[24] The Connexion, "Lille issues 'moratorium' on 5G technology," 11 October 2020. Accessed: March 31, 2023. [Online]. Available: https://www.connexionfrance.com/article/French-news/Lille-issues-moratorium-on-controversial-5G-technology-pending-2021-Anses-report.

[25] International Commission on Non-Ionizing Radiation Protection (ICNIRP), 2023. Accessed: March 31, 2023. [Online]. Available: https://www.icnirp.org/.

[26] 6G Smart Networks and Services Industry Association (6G-IA), Accessed: Accessed: March 31, 2023. [Online]. Available: https://6g-ia.eu/.

[27] Networld Europe European Technology Platform, Accessed: Accessed: March 31, 2023. [Online]. Available: https://www.networldeurope.eu/.

[28] Alliance for IoT and Edge Computing Innovation (AIOTI), Accessed: March 31, 2023. [Online]. Available: https://aioti.eu/.

[29] Cloud Infrastructures Services Provides in Europe (CISPE.cloud), Accessed: March 31, 2023. [Online]. Available: https://cispe.cloud/.

[30] NESSI, Accessed: March 31, 2023. [Online]. Available: https://nessi.eu/.

[31] Smart Networks and Services Joint Undertaking, Accessed: March 31, 2023. [Online]. Available: https://smart-networks.europa.eu/.

[32] United Nations, Sustainable Development Goals, Accessed: March 31, 2023. [Online]. Available: https://sdgs.un.org/goals.

[33] EC, The European green deal, Accessed: March 31, 2023. [Online]. Available: https://eur-lex.europa.eu/resource.html?uri=cellar:b828d165-1c22-11ea-8c1f-01aa75ed71a1.0002.02/DOC_1&format=PDF.

[34] 6G-IA, European Vision for the 6G Network Ecosystem, Version 1.0, DOI: 10.5281/zenodo.5007671, June 2021, available at: https://5g-ppp.eu/wp-content/uploads/2021/06/WhitePaper-6G-Europe.pdf.

[35] SNS JU, Strategic Research and Innovation Agenda, Accessed: March 31, 2023. [Online]. Available: https://smart-networks.europa.eu/wp-content/uploads/2022/10/122021_sns_gb_decision_sria_including_annexdocx_89dnouztkolqi0m6dij7feh9da_82079_compressed-1.pdf.

Index

Contributing Authors[1]

Chapter 1

Mauro Boldi
TIM S.p.A., Italy

Ömer Bulakçı
Nokia, Germany

John Cosmas
Brunel University London, UK

Mårten Ericson
Ericsson AB, Sweden

Gines Garcia-Aviles
I2CAT Foundation, Spain

Anastasius Gavras
Eurescom GmbH, Germany

Mir Ghoraishi
Gigasys Solutions, UK

Marco Gramaglia
Universidad Carlos III de Madrid, Spain

Alexandros Kostopoulos
Hellenic Telecommunications Organization S.A. (OTE), Greece

Xi Li
NEC Laboratories Europe, Germany

Bahare Masood Khorsandi
Nokia, Germany

Agapi Mesodiakaki
Aristotle University of Thessaloniki, Greece

Md Arifur Rahman
IS-Wireless, Poland

Patrik Rugeland
Ericsson AB, Sweden

Dimitris Tsolkas
Fogus Innovations & Services P.C.; National & Kapodistrian University of Athens, Greece

Mikko Uusitalo
Nokia Bell Labs, Finland

1. Authors are listed in alphabetical order per chapter.

Chapter 2

Mehdi Abad
Ericsson GmbH,
Germany

Mohammad Asif Habibi
Rheinland-Pfälzische Technische Universität Kaiserslautern-Landau, Germany

Riccardo Bassoli
Technische Universität Dresden, Germany

Giacomo Bernini
Nextworks, Italy

John Cosmas
Brunel University London, UK

Xavier Costa
NEC Laboratories Europe/Fundació i2CAT, Spain

Mårten Ericson
Ericsson, Sweden

Miltos Filippou
Intel, Italy

Hannu Flinck
Nokia, Finland

Pål Frenger
Ericsson, Sweden

Adrian Gallego
Atos, Spain

Mir Ghoraishi
Gigasys Solutions, UK

Marco Gramaglia
Universidad Carlos III de Madrid, Spain

Omer Haliloglu
Ericsson, Turkey

Marie-Helene Hamon
Orange, France

Alexandros Kostopoulos
OTE – Hellenic Telecommunications Organization, Greece

Giada Landi
Nextworks, Italy

Xi Li
NEC Laboratories Europe, Germany

Christofer Lindheimer
Ericsson, Sweden

Bahare Masood Khorsandi
Nokia, Germany

Agapi Mesodiakaki
Aristotle University of Thessaloniki, Greece

Cédric Morin
B-COM, France

Giovanni Nardini
Università di Pisa, Italy

José Antonio Ordoñez Lucena
Telefónica Investigación y Desarrollo, Spain

Ignacio Labrador Pavón
Atos, Spain

Petteri Pöyhönen
Nokia, Finland

Md Arifur Rahman
IS-Wireless, Poland

Bjoern Richerzhagen
Siemens, Germany

Patrik Rugeland
Ericsson, Sweden

Lucas Scheuvens
Technische Universität Dresden, Germany

Hans D. Schotten
Rheinland-Pfälzische Technische Universität Kaiserslautern-Landau, Germany

Merve Seimler
Ericsson, Turkey

Tommy Svensson
Chalmers University of Technology, Sweden

Dimitris Tsolkas
Fogus Innovations & Services P.C.; National & Kapodistrian University of Athens, Greece

Janne Tuononen
Nokia, Finland

Mikko Uusitalo
Nokia Bell Labs, Finland

Panagiotis Vlacheas
WINGS ICT Solutions, Greece

Stefan Wänstedt
Ericsson, Sweden

Andreas Wolfgang
Qamcom Research and Technology AB, Sweden

Chapter 3

Angeliki Alexiou
University of Piraeus, Greece

Tezcan Cogalan
Samsung Research, UK

Jean-Marc Conrat
Orange, France

Mar Francis De Guzman
Aalto University, Finland

Francesco Devoti
NEC Laboratories Europe, Germany

Geoffrey Eappen
Brunel University London, UK

Chao Fang
Chalmers University of Technology, Sweden

Pål Frenger
Ericsson, Sweden

Mir Ghoraishi
Gigasys Solutions, UK

Adam Girycki
IS-Wireless, Poland

Hao Guo
Chalmers University of Technology, Sweden

Hardy Halbauer
Nokia Solutions and Networks, Germany

Omer Haliloglu
Ericsson, Turkey

Katsuyuki Haneda
Aalto University, Finland

Israel Koffman
RunEL, Israel

Pekka Kyösti
University of Oulu, Finland

Marko Leinonen
University of Oulu, Finland

Yinggang Li
Ericsson, Sweden

Charitha Madapatha
Chalmers University of Technology, Sweden

Behrooz Makki
Ericsson, Sweden

Jorge Navarro-Ortiz
University of Granada, Spain

Le Hang Nguyen
Nokia Solutions and Networks, Germany

Ahmad Nimr
Technische Universität Dresden, Germany

Aarno Pärssinen
University of Oulu, Finland

Sofie Pollin
Katholieke Universiteit Leuven, Belgium

Simon Pryor
Acceleran, Germany

Rafael Puerta
Ericsson, Sweden

Md Arifur Rahman
IS-Wireless, Poland

Juan J. Ramos-Munoz
University of Granada, Spain

Vida Ranjbar
Katholieke Universiteit Leuven, Belgium

Kilian Roth
Intel Deutschland GmbH, Germany

Muris Sarajlic
Ericsson, Sweden

Vincenzo Sciancalepore
NEC Italy, Italy

Tommy Svensson
Chalmers University of Technology, Sweden

Nuutti Tervo
University of Oulu, Finland

Andreas Wolfgang
Qamcom Research and Technology AB, Sweden

Chapter 4

Kareem Ali
Brunel University
London, UK

Marco Araújo
Capgemini, Portugal

Bastien Béchadergue
Oledcomm, France

Hui Chen
Chalmers University of Technology, Sweden

John Cosmas
Brunel University London, UK

Diego Andres Dupleich
Fraunhofer Institute for Integrated Circuits, Germany

Geoffrey Eappen
Brunel University London, UK

Musa Furkan Keskin
Chalmers University of Technology, Sweden

Meysam Goodarzi
IHP – Leibniz Institute for High Performance Microelectronics, Germany

Hao Guo
Chalmers University of Technology, Sweden

Israel Koffman
RunEL, Israel

Simon Lindberg
Qamcom Research and Technology AB, Sweden

Ali Mahbas
Brunel University London, UK

Bruno Mendes
Capgemini, Portugal

Ben Meunier
Brunel University London, UK

Ehsan Moeen Taghavi
University of Oulu, Finland

Alejandro Ramirez
Siemens, Germany

Vladica Sark
IHP – Leibniz Institute for High Performance Microelectronics, Germany

Kim Schindhelm
Siemens, Germany

Tommy Svensson
Chalmers University of Technology, Sweden

Liesbet Van der Perre
Katholieke Universiteit Leuven, Belgium

Thomas Wilding
Technische Universität Graz, Austria

Henk Wymeersch
Chalmers University of Technology, Sweden

Vijaya Yajnanarayana
Ericsson, India

Xun Zhang
Institut supérieur d'électronique de Paris, France

Hongxiu Zhao
Institut supérieur d'électronique de Paris, France

Chapter 5

Jose M. Alcaraz-Calero
University of the West of Scotland, UK

Anttonen Antti
VTT Technical Research Centre of Finland, Finland

Marco Araújo
Capgemini, Portugal

Raul Barbosa
Capgemini, Portugal

Sokratis Barmpounakis
Wings ICT Solutions, Greece

Sergio Barrachina-Muñoz
Centre Tecnològic de Telecomunicacions de Catalunya, Spain

Riccardo Bassoli
Technische Universität Dresden, Germany

Giacomo Bernini
Nextworks, Italy

Luis Blanco
Centre Tecnològic de Telecomunicacions de Catalunya, Spain

Anne-Marie Bosneag
Ericsson, Ireland

Ashima Chawla
Ericsson, Ireland

Loizos Christofi
eBOS Technologies, Cyprus

Dilin Dampahalage
University of Oulu, Finland

Panagiotis Demestichas
Wings ICT Solutions, Greece

Mårten Ericson
Ericsson AB, Sweden

Hamed Farhadi
Ericsson AB, Sweden

Marco Fiore
IMDEA Networks Institute, Spain

Frank H.P. Fitzek
Technische Universität Dresden, Germany

Hannu Flinck
Nokia, Finland

João Fonseca
Capgemini, Portugal

Martti Forsell
VTT Technical Research Centre of Finland, Finland

Victor Gabillon
Thales, France

Ginés García-Avilés
i2CAT Foundation, Spain

Andres Garcia-Saavedra
NEC Laboratories Europe, Germany

Marco Gramaglia
Universidad Carlos III de Madrid, Spain

Bin Han
Rheinland-Pfälzische Technische

Universität Kaiserslautern-Landau, Germany

Mikko Honkala
Nokia Bell Labs, Finland

Alexandre Kazmierowski
Thales, France

Charalambos Klitis
eBOS Technologies, Cyprus

Dani Korpi
Nokia Bell Labs, Finland

Slawomir Kuklinski
Orange, Poland

Ignacio Labrador Pavón
Atos, Spain

Vasiliki Lamprousi
Wings ICT Solutions, Greece

Giada Landi
Nextworks, Italy

Haeyoung Lee
University of Hertfordshire, UK

Xi Li
NEC Laboratories Europe, Germany

Josep Mangues-Bafalluy
Centre Tecnològic de Telecomunicacions de Catalunya, Spain

Bahare Masood Khorsandi
Nokia, Germany

Mattia Merluzzi
Université Grenoble Alpes/CEA-Leti, France

Jafar Mohammadi
Nokia Bell Labs, Germany

Cédric Morin
B-COM, France

José Antonio Ordoñez Lucena
Telefónica Investigación y Desarrollo, Spain

Petteri Pöyhönen
Nokia, Finland

Nuwanthika Rajapaksha
University of Oulu, Finland

Premanandana Rajatheva
University of Oulu, Finland

Farhad Rezazadeh
Centre Tecnològic de Telecomunicacions de Catalunya, Spain

Roberto Riggio
Università Politecnica delle Marche, Italy

Lucas Scheuvens
Technische Universität Dresden, Germany

Merve Seimler
Ericsson, Turkey

Janne Tuononen
Nokia, Finland

Ricard Vilalta
Centre Tecnològic de Telecomunicacions de Catalunya, Spain

Qi Wang
University of the West of Scotland, UK

Stefan Wunderer
Nokia Networks, Germany

Lanfranco Zanzi
NEC Laboratories Europe, Germany

Engin Zeydan
Centre Tecnològic de Telecomunicacions de Catalunya, Spain

Xun Zhang
Institut supérieur d'électronique de Paris, France

Chapter 6

Sergio Barrachina
Centre Tecnològic de Telecomunicacions de Catalunya, Spain

Benjamin Deutschmann
Technische Universität Graz, Austria

Navideh Ghafouri
Iquadrat Informatica, Spain

Roberto Gonzalez
NEC Laboratories Europe, Spain

Panagiotis Kokkinos
Institute of Communication & Computer Systems, Greece

Mattia Merluzzi
Université Grenoble Alpes/CEA-Leti, France

Agapi Mesodiakaki
Aristotle University of Thessaloniki, Greece

Miquel Payaró
Centre Tecnològic de Telecomunicacions de Catalunya, Spain

Soumplis Polyzois
Institute of Communication & Computer Systems, Greece

Md Arifur Rahman
IS-Wireless, Poland

Kostas Ramantas
Iquadrat Informatica, Spain

Lucas Scheuvens
Technische Universität Dresden, Germany

Esteban Selva
Orange, France

Giuseppe Siracusano
NEC Laboratories Europe, Germany

Karthik Upadhya
Nokia Bell Labs, Finland

Liesbet Van der Perre
Katholieke Universiteit Leuven, Belgium

John Vardakas
Iquadrat Informatica, Spain

Emmanouel Varvarigos
Institute of Communication & Computer Systems, Greece

Christos Verikoukis
University of Patras, Greece

Thomas Wilding
Technische Universität Graz, Austria

Klaus Witrisal
Technische Universität Graz, Austria

Stefan Wunderer
Nokia Networks, Germany

Chapter 7

Mehdi Abad
Ericsson GmbH, Germany

Jose M. Alcaraz-Calero
University of the West of Scotland, UK

Marco Araújo
Capgemini, Portugal

Raul Barbosa
Capgemini, Portugal

Loizos Christofi
eBOS Technologies, Cyprus

Marius-Iulian Corici
Fraunhofer Institute for Open Communication Systems, Germany

Paulo Duarte
Capgemini, Portugal

Mårten Ericson
Ericsson AB, Sweden

João Fonseca
Capgemini, Portugal

Dimitrios Fragkos
National Centre of Scientific Research Demokritos, Greece

Andreas Gavrielides
eBOS Technologies, Cyprus

Hasanin Harkous
Nokia, Germany

Ta Dang Khoa LE
EURECOM, France

Harilaos Koumaras
National Centre of Scientific Research Demokritos, Greece

Xi Li
NEC Laboratories Europe, Germany

Håkon Lønsethagen
Telenor, Norway

Thomas Luetzenkirchen
Intel, Germany

Bruno Mendes
Capgemini, Portugal

Renxi Qiu
University of Bedfordshire, UK

Marios Sophocleous
eBOS Technologies, Cyprus

Christos Tranoris
University of Patras, Greece

Dimitris Tsolkas
Fogus Innovations & Services P.C.; National & Kapodistrian University of Athens, Greece

Qi Wang
University of the West of Scotland, UK

Lanfranco Zanzi
NEC Laboratories Europe, Germany

Chapter 8

Loizos Christofis
eBOS, Cyprus

Ioannis Chochliouros
Hellenic Telecommunications Organization S.A. (OTE), Greece

John Cosmas
Brunel University London, UK

Sabrina De Capitani di Vimercati
Universita degli Studi di Milano, Italy

Ioannis Giannoulis
National Technical University of Athens, Greece

Roberto Gonzalez
NEC, Germany

Bin Han
Rheinland-Pfälzische Technische Universität Kaiserslautern-Landau, Germany

Charalambos Klitis
eBOS, Cyprus

Alexandros Kostopoulos
Hellenic Telecommunications Organization S.A. (OTE), Greece

Anastassios Nanos
Nubificus, UK

Jawad Nawar
Brunel University London, UK

Samia Oukemeni
Institut supérieur d'électronique de Paris, France

Sebastian Robitzsch
InterDigital, UK

Pierangela Samarati
Universita degli Studi di Milano, Italy

Giuseppe Siracusano
NEC, Germany

Elif Ustundag Soykan
Ericsson AB, Sweden

Matthias Weh
Deutsche Telekom, Germany

Ming Yin
Deutsche Telekom, Germany

Chapter 9

Mauro Boldi
TIM S.p.A., Italy

Hannu Flinck
Nokia, Finland

Valerio Frascolla
Intel, Germany

Alexandros Kaloxylos
6G Infrastructure Association, Greece

Bahare Masood Khorsandi
Nokia, Germany

Patrik Rugeland
Ericsson AB, Sweden

Kostas Trichias
6G Infrastructure Association, Greece

Mikko Uusitalo
Nokia Bell Labs, Finland

Colin Willcock
6G Infrastructure Association, Germany

Editor Short Bios

Dr. Ömer Bulakçı is the 6G Program Manager at Nokia, Germany and the Chair of the 5G PPP Architecture Working Group.

Dr. Xi Li is a Senior Researcher at NEC Laboratories Europe in Germany and the Vice-Chair of the 5G PPP Architecture Working Group.

Dr. Marco Gramaglia is a visiting professor at University Carlos III of Madrid.

Anastasius Gavras is a programme manager at Eurescom GmbH in Germany.

Dr. Mikko Uusitalo is Head of Research Department Radio Systems Research Finland at Nokia Bell Labs and overall lead in the European level 6G Flagship projects Hexa-X and Hexa-X-II.

Dr. Patrik Rugeland is a Master Researcher at Ericsson Research, Sweden and is the Technical Manager in the European 6G flagship projects Hexa-X and Hexa-X-II.

Mauro Boldi is a Project Manager in the Innovation department at TIM, Italy, and is currently leading dissemination activities in the European 6G flagship projects Hexa-X and Hexa-X-II.

www.ingramcontent.com/pod-product-compliance
Lightning Source LLC
LaVergne TN
LVHW010720210926
848485LV00002B/83

9781638282389